Surface Phenomena in Fusion Welding Processes

Surface Phenomena in
Fusion Welding
Processes

German Deyev
Dmitriy Deyev

CRC Press
Taylor & Francis Group
Boca Raton London New York

CRC Press is an imprint of the
Taylor & Francis Group, an **informa** business

A TAYLOR & FRANCIS BOOK

Published in 2006 by
CRC Press
Taylor & Francis Group
6000 Broken Sound Parkway NW, Suite 300
Boca Raton, FL 33487-2742

ISBN-13: 978-0-367-45393-0 (pbk)
ISBN-13: 978-0-8493-9883-4 (hbk)

Library of Congress Card Number 2005052851

Library of Congress Cataloging-in-Publication Data

Deyev, German F.
  Surface phenomena in fusion welding processes / by German F. Deyev and Dmitriy G. Deyev
      p. cm.
  Includes bibliographical refrences and index.
  ISBN 0-8493-9883-5 (alk. paper)
  1. Welding. 2. Surfaces (Technology) I. Deyev, Dmitriy G. II. Title.

TS227.D49 2005
671.5'2--dc22   200505285

Visit the Taylor & Francis Web site at
http://www.taylorandfrancis.com

and the CRC Press Web site at
http://www.crcpress.com

Preface

The industry is currently applying highly diverse welding processes. However, the main technological process applied in manufacture of weldments is fusion welding. Fusion welding is used to join practically all metals and alloys of similar and dissimilar materials of any thickness. In the application of fusion welding, the parts are joined either by melting of the parts' edges or as a result of simultaneous melting of the base and additional (electrode) metal. A weld pool forms from the metal melt, its subsequent solidification, proceeding without application of external pressure, leads to the formation of the weld.

Depending on the kind of the applied heat source, a distinction is made between the following fusion welding processes: electric arc, electron beam, electroslag, laser, and gas welding. Irrespective of the actual welding process applied, the main aim of any process is producing a sound welded joint. This is only possible if a sufficiently comprehensive allowance is made for the main physicochemical processes proceeding in the welding zone.

Heating, melting, evaporation, solidification, and subsequent cooling of solid metal, the transition of a substance from one phase into another, initiation of new phases, substance distribution inside a phase, interaction of the contacting phases — all these and many other processes are characteristic of fusion welding. Since all or part of the above processes proceed simultaneously in different regions of the weld pool and welded joint, and their occurrence and progress is determined by a large number of factors, studying the process as a whole is a real challenge. On the other hand, there is no doubt that many of the above processes are related, to varying degrees, to the surface phenomena and surface properties of the applied materials.

For instance, formation of various defects (pores, non-metallic inclusions, lack-of-penetration, solidification cracks, etc.), hydrodynamic processes in the weld pool, formation of deposited metal and weld root, electrode-metal transfer, interaction of molten metal with the gas and slag, and a number of other processes are, to a certain extent, related to surface phenomena. Surface phenomena largely determine the strength of the bonds joining the dissimilar metals. The role of surface phenomena is particularly high in welding in space, when the influence of gravity forces becomes minor.

Therefore, in the development of welding technologies, as well as welding consumables, it is necessary to know and take into account both the surface properties (surface and interphase tension) of the applied materials, and surface phenomena (wettability, electrocapillary phenomena, adsorption, etc.), that affect the boundaries of contacting phases.

Authors

German Deyev earned a B.S. degree in civil engineering from Cheljabinsk State Polytechnical Institute (Cheljabinsk, Russia, 1963), an M.S. degree in engineering of welding technology and production from the Cheljabinsk State Polytechnical Institute (Cheljabinsk, Russia, 1965), and a Ph.D. degree in civil engineering with specialization in 'the role of surface properties and phenomena in the formation of welds and deposited metal' from Cheljabinsk State Polytechnical Institute (Cheljabinsk, Russia, 1971).

Professor Deyev served as the chairman of the Department of Welding at the University of Lipetsk from 1976 until 1999. In December 1993, he was appointed dean of the Metallurgical Faculty of the University of the Lipetsk. During his academic career, Professor Deyev has received 24 awards. He is a member of the American Welding Society. Professor Deyev has to his credit 6 books and over 120 publications and articles in professional journals.

Dmitriy Deyev earned a B.S. degree in civil engineering in 1994, an M.S. degree in engineering of welding technology and production from the Lipetsk State Polytechnical Institute (Lipetsk, Russia, 1995), and a Ph.D. degree in Civil Engineering from Voronezh State University (Voronezh, Russia, 1998). He served as the leading specialist at the Department of Transportation of the Lipetsk State Government from 1995 until 1999. Dmitriy Deyev has written many publications and articles in professional journals.

Contents

Chapter 1   Characterization of the Fusion Welding Process   1

1.1   Some Features of Fusion Welding   1
   1.1.1   The Basics of Weld Formation   1
   1.1.2   The Influence of Gases   2
   1.1.3   The Influence of Slag   3
1.2   Welding Heat Sources   4
   1.2.1   The D.C. vs. A.C. Welding   4
   1.2.2   Electron Beam Welding   7
   1.2.3   Laser Welding   8
1.3   Fusion Welding Processes   9
   1.3.1   Electric Arc Welding   9
      1.3.1.1   Gas Shielding   10
   1.3.2   Electroslag Welding   11
   1.3.3   Electron Beam and Laser Welding   12
1.4   Physicochemical Processes of Welded Joint Formation   13
   1.4.1   Direct Interaction of The Metal and The Gas   14
      1.4.1.1   Theoretical Models   14
      1.4.1.2   The Drawback of the Theoretical Models   16
   1.4.2   Influence of Surfactants   17
   1.4.3   The Gas-Adsorption Ability   18
   1.4.4   The Influence of Vacuum   19
   1.4.5   The Influence of Fluxes   20
   1.4.6   The Role of the Slag   21
   1.4.7   The Physiochemical Processes at the Solid–Molten
      Metal Boundary   23
      1.4.7.1   The Nucleation Process   23
      1.4.7.2   Melt Overcooling   24
      1.4.7.3   Presence of Refractory Impurities   24
      1.4.7.4   Shape of the Interface   25
      1.4.7.5   Effect of Metal Stirring   25
   1.4.8   Calculation of σ_{sol-1} Values   26
   1.4.9   The Crystallization Phenomena   28
      1.4.9.1   Electromagnetic Stirring and Ultrasonic
           Oscillations   29
      1.4.9.2   Segregation Processes   30

Chapter 2 Surface Properties and Phenomena 35

2.1 Surface Properties and Phenomena on the Boundaries of
  Contacting Phases 35
  2.1.1 The Concept of Surface Energy 35
  2.1.2 Surface Energy vs. Surface Tension 37
  2.1.3 Interphase Tension 37
  2.1.4 The Adsorption Process 40
  2.1.5 The Electrochemical Phenomena 42
2.2 Determination of Surface Properties of Melts 44
  2.2.1 Pendant Drop Method 44
  2.2.2 Drop Weight Method 44
  2.2.3 Sessile Drop Method 45
  2.2.4 Method of Maximum Pressure in the Bubble 46
  2.2.5 Large Drop Method 48
  2.2.6 Method of Metal Melting in the Pendant State 49
  2.2.7 Method to Determine the Surface Tension of a Metal in
      the Zone of the Arc Active Spot 49
2.3 Methods of Measurement of the Surface Tension of Solids Metals ... 51
  2.3.1 Method of Multiphase Equilibrium 51
  2.3.2 Method of "Zero Creep" 52
  2.3.3 Method of Groove Smoothing 53
  2.3.4 Method of Autoelectronic Emission 54
  2.3.5 Method of Thermal Etching of Grain Boundaries 54
2.4 Methods to Determine the Surface Energy of the Interphases 56
  2.4.1 The Sessile Drop Method 56
  2.4.2 The Drop Weighing Method 57
  2.4.3 Turned Capillary Method 57
  2.4.4 Drops Contact Method 58
  2.4.5 X-ray Photography of a Sessile Drop 58
2.5 Wettability, Spreading, and Electrocapillary Phenomena 59
  2.5.1 General Considerations 59
  2.5.2 A Modern Unit for Studying the Welding Phenomena 59
  2.5.3 Studying the Electrocapillary Phenomena 61
  2.5.4 A Comparison of the Different Methods 64

Chapter 3 Results of Studying the Surface Properties
      and Phenomena 65

3.1 Surface Tension of Molten and Solid Metals 65
  3.1.1 Influence of Alloy Composition 65
  3.1.2 The Influence of Gases 67
    3.1.2.1 Hydrogen, Nitrogen, and Oxygen 67
    3.1.2.2 The Behavior of CO 69
    3.1.2.3 The Behavior of CO_2/Air 70
    3.1.2.4 The Behavior of Binary Gas Mixtures 76

		3.1.2.4.1	Armco-Iron and Steels	76
		3.1.2.4.2	Aluminum and its Alloys	76
		3.1.2.4.3	Copper and its Alloys	77
		3.1.2.4.4	Titanium and its Alloys	79
		3.1.2.4.5	Nickel and its Alloys	80
		3.1.2.4.6	The Role of Metal Temperature	81

3.2 Interphase Surface Energies in Metal–Slag and
Solid Metal–Melt Systems 82
  3.2.1 Composition of the System 83
     3.2.1.1 Carbon 84
     3.2.1.2 Sulfur 84
     3.2.1.3 Phosphorus 84
     3.2.1.4 Oxygen 84
     3.2.1.5 Cr, Mo, W, and Si 85
     3.2.1.6 Individual Slag Components 85
  3.2.2 Temperature of the System 86
  3.2.3 Experimental Determination of Interphase Tension 86
     3.2.3.1 The Available Information on Welding
          Processes 87
     3.2.3.2 Solidification Processes 88

3.3 Wettability and Spreading of Metal Melts over the
Surface of Solids 88
  3.3.1 Influence of Material Composition 90
  3.3.2 Influence of Surface Roughness 92
  3.3.3 Influence of External Fields 92
  3.3.4 Isothermal Conditions 95
     3.3.4.1 Steel/Copper and Copper-Based alloys 95
     3.3.4.2 Steel Aluminum 96
     3.3.4.3 Copper/Aluminum 98
     3.3.4.4 Titanium and Titanium Alloys/Aluminum 99
  3.3.5 Behavior under Non-Isothermal Process 100
     3.3.5.1 Influence of Arc Parameters 101
     3.3.5.2 Gas-Shielded Welding 102
     3.3.5.3 The Fluoc Effect 102
     3.3.5.4 The Role of Preheading 103

3.4 Influence of Welding Heat Sources on Wettability
and Spreading 105
  3.4.1 Influence of Arc Discharge 105
     3.4.1.1 Direct Arc 107
     3.4.1.2 Indirect Arc 107
     3.4.1.3 The Role of Substrate Composition 109
     3.4.1.4 Composite Materials 110
  3.4.2 General Mechanistic Considerations 113
     3.4.2.1 Experimental Determination of Δ_m 117
     3.4.2.2 The Nature of the Gas Flame 118

3.5   Wettability of the Solid Metal by Nonmetal Melts 119
      3.5.1   The Influence of Sulfur/Sulfides 119
      3.5.2   Influence of Chlorides/Fluorides of Fluxes 122
              3.5.2.1   The Experimental Procedure 123
3.6   Electrocapillary Phenomena in a Metal–Slag System 123
      3.6.1   Electrochemical Nature of Interactions 124
      3.6.2   The Approach 124
      3.6.3   Experimental Methods 125
      3.6.4   The Influence of Cell Dimensions 128
      3.6.5   Influence of an External Electric Field 131

Chapter 4   Electrode-Metal Transfer and Surface Phenomena 135

4.1   Forces Influencing Electrode-Metal Transfer 135
      4.1.1   The Nature of Electrode-Metal Transfer 135
      4.1.2   The Process of Drop Formation 136
      4.1.3   The Electromagnetic Force 137
      4.1.4   The Reactive Forces 138
      4.1.5   Other Forces 138
4.2   Transfer and Spatter of Electrode Metal in Welding in a
      Gaseous Medium 138
      4.2.1   Stages in Drop Transfer 138
      4.2.2   Drop Detachment 140
      4.2.3   Formation of Metal Spatter 141
      4.2.4   The Role of Part Preheating 142
      4.2.5   Wettability and Welding Current 145
      4.2.6   The Role of Oxidizing Gases 146
      4.2.7   Influence of Electrode Wire Composition 148
      4.2.8   Flux-Cored Wires and Filler Metal 149
4.3   Transfer of the Electrode Metal Contacting the Slag 150
      4.3.1   The Underlying Factors 150
      4.3.2   The Impact of Interphase Tension 150
              4.3.2.1   Basic-Type Coating 150
              4.3.2.2   Acidic Coating 152
      4.3.3   The Influence of Welding Polarity 153
              4.3.3.1   Polarity and Single-Component Coatings 153
              4.3.3.2   Polarity Effect with Flux 155
                        4.3.3.2.1   Submerged-Arc Welding with a Flux 155
                        4.3.3.2.2   Electroslag Welding 155

Chapter 5   Formation of Weld and Deposited Metal 157

5.1   Weld and Serviceability of a Welded Structure 157
      5.1.1   Causes of Strength Deterioration 157
              5.1.1.1   Cracks and Crack-Like Defects 160
              5.1.1.2   Undercuts 161

       5.1.1.3 Craters 162

       5.1.1.4 Rolls 163

5.2 Formation of the Penetration Zone 164

    5.2.1 Shape and Dimensions of the Penetration Zone 164

       5.2.1.1 Penetration Depth 165

    5.2.2 The Influence of Base-Metal Surface Properties 166

    5.2.3 The Action of Surfactants 168

    5.2.4 Thermocapillary Phenomena 169

       5.2.4.1 Evidence from Laser Welding and Arc Welding 171

       5.2.4.2 Consumable Electrode Welding 172

    5.2.5 Effect of Current Polarity 173

    5.2.6 Experimental Findings 177

       5.2.6.1 Influence of Welding Current 178

       5.2.6.2 Duration of Arc Discharge Action 178

       5.2.6.3 Dimensions and Spacing of the Steel Plates 179

       5.2.6.4 Calculations 181

5.3 Formation of the Weld Bead and Root 182

    5.3.1 Issues in the Formation of Butt Welds 182

    5.3.2 The Graphical Method 182

    5.3.3 The Role of Temperature Fields and Inclination Angles 184

       5.3.3.1 Angle of Inclination 185

    5.3.4 Surface Tension vs. Gravitational Force 187

       5.3.4.1 Angle of Transition 188

5.4 Formation of the Deposited Metal 189

    5.4.1 Wetting Angle 189

    5.4.2 Experiments with Electron Beam Surfacing 191

       5.4.2.1 The Critical Rate of Deposition 191

       5.4.2.2 Electron Beam Oscillations 194

    5.4.3 Temperature Gradient and the Speed of Displacement 194

5.5 Defects of Weld Shape and Surface Phenomena 198

    5.5.1 Groove Filling and Unfilling 198

       5.5.1.1 Extending the Time of Weld-Pool Existence 199

       5.5.1.2 Surface Roughness 199

    5.5.2 Multilayer Welding 200

       5.5.2.1 Formation of Burns-Through 201

       5.5.2.2 Formation of Craters 201

       5.5.2.3 Formation of Rolls 202

Chapter 6 Non-Metallic Inclusions 205

6.1 Non-Metallic Inclusions in the Welds 205

    6.1.1 Types of Non-Metallic Inclusions 205

    6.1.2 The Influence of Non-Metallic Inclusions 208

       6.1.2.1 Thermal−Structural Factors 208

       6.1.2.2 Strength of Welded Joints 208

6.2 Nucleation of Non-Metallic Inclusions in the Weld Pool 211
    6.2.1 The Kinetics of Nucleation 211
    6.2.2 Oversaturation and its Effects 212
        6.2.2.1 Precipitation Phenomena 213
    6.2.3 Electrochemical Considerations 214
    6.2.4 Thermodynamic Considerations 216
6.3 Coarsening of Non-Metallic Inclusions in the Weld Pool 218
    6.3.1 Fundamentals of Nuclei Growth 218
    6.3.2 Coalescence and Coagulation 219
    6.3.3 The Effect of Oxides 220
        6.3.3.1 The Role of the Gaseous Environment 222
    6.3.4 The Effect of Stirring 222
    6.3.5 Disjoining Pressure 223
    6.3.6 Determination of Adhesion Forces 225
6.4 Removal of Non-Metallic Inclusions from the Weld Pool 227
    6.4.1 The Basics of Removal 227
    6.4.2 Theoretical Considerations 227
    6.4.3 The Combined Effect of a Gas Bubble and the Inclusion 229
    6.4.4 Deformation and Disruption 231
    6.4.5 Some Important Calculations 233

Chapter 7 Porosity in Welds 237

7.1 Pores and Performance of a Welded Joint 237
7.2 Formation of Gas Bubble Nuclei 239
    7.2.1 Surface Tension and Radius of Curvature 241
    7.2.2 Existence of Different Gases in the Pores 242
    7.2.3 Effect of Oxygen on the Nucleation of Pores 246
7.3 Formation of Gas Bubble Nuclei in a Heterogeneous Environment ... 248
    7.3.1 Intermolecular Adhesion Forces 248
    7.3.2 The Influence of Electric Field 250
        7.3.2.1 Energy of Formation of the Gas Nucleus 252
    7.3.3 The Influence of Non-Metallic Inclusions 254
        7.3.3.1 Important Calculations 257
    7.3.4 Other Salient Features 260
        7.3.4.1 Forces of Physical Interactions 261
7.4 Growth of Gas Bubbles in the Weld Pool 262
    7.4.1 Thermodynamic Considerations 263
    7.4.2 Effect of Oscillations and Stirring 264
    7.4.3 Diffusion Phenomena 265
    7.4.4 Surface vs. Volume Concentration of Components 268
7.5 Escape of Gas Bubbles from the Molten Metal 269
    7.5.1 The Need for Advanced Mathematical Treatment 269
    7.5.2 Providing Allowance for Many Bubbles 271
    7.5.3 Destruction of the Film 273

| | 7.5.3.1 | Experimental Findings | 274 |
| | 7.5.3.2 | The Influence of FeS and FeO | 275 |

Chapter 8   Solidification Cracking 279

8.1   Solidification Cracks: Factors Affecting their Formation
        and Test Methods 280
        8.1.1   The Influence of Carbon Content 280
        8.1.2   The Role of the Metal Structure 281
        8.1.3   Determination of Susceptibility to
                  Solidification Cracking 282
8.2   Mechanism of the Effect of Metal Melts on the Fracture of
        Solid Metals 284
        8.2.1   Interaction of the Metal Atoms at the Crack Apex 285
                  8.2.1.1   The Complex Nature of the Mechanism 287
        8.2.2   The Role of the Melt 288
                  8.2.2.1   Investigation of Mechanical Properties 288
                  8.2.2.2   The Corrosion Effect 291
8.3   Effect of Low-Melting Non-Metallic Melts on the
        Mechanical Properties of Solid Metals 292
        8.3.1   Experimental Findings on Mechanical Properties 292
                  8.3.1.1   The Effect of Comfort Area 294
                  8.3.1.2   The Effect of Ductility 296
        8.3.2   Experimental Study of the Corrosion 298
                  8.3.2.1   Carbon-Rich Steels 300
                  8.3.2.2   Austenitic Steels 301
        8.3.3   Important Observations in Metallography 301
        8.3.4   The Mechanism of the Corrosion Effect 304
        8.3.5   Other Parameters 305
8.4   Effect of Low-Melting Melts on Solidification Cracking 306
        8.4.1   Crack Formation 306
        8.4.2   The Role of the Surface Phenomena 309
        8.4.3   Combined Effect of Oxygen and Sulfur 311

Chapter 9   Development of Welding and Surfacing Technologies
                  with Allowance for Surface Phenomena 313

9.1   Welding of Composite Fiber-Reinforced Materials 313
        9.1.1   Problems in the Fusion Welding of Composite Materials 314
        9.1.2   The Heterogeneous Nature of the Surface 315
                  9.1.2.1   The Capillary Forces 316
        9.1.3   Experimental Observations 320
9.2   Electron Beam Surfacing 323
        9.2.1   Surfacing — Experimental Findings 324
                  9.2.1.1   Effect of Composition 325
                  9.2.1.2   Effect of Gases 325

	9.2.1.3	Effect of Deposited Metal	326
	9.2.1.4	Effect of Beam Power	327
	9.2.1.5	Important Conclusions	329
9.3	Welding and Surfacing of Copper–Steel Parts		330
	9.3.1	Copper–Steel Pipes	331
	9.3.1.1	Crack Resistance	332
	9.3.2	Wetting and Spreading	333
	9.3.2.1	Important Parameters	334
	9.3.2.2	Deflection of Electron Beam	336
9.4	Manufacture of Metal Cutting Tools with a Deposited Cutting Edge		337
	9.4.1	Hardness	337
	9.4.2	Cracks	339
	9.4.3	Ductility	340
9.5	The Role of Surface Phenomena in the Submerged-Arc Welding and Surfacing Processes		341
	9.5.1	Method of Increasing the Productivity	341
	9.5.2	Slag Adhesion	344
9.6	Surface Phenomena Occurring in Welding and Surfacing in Space		347
	9.6.1	Major Comparisons with Earth Conditions	347
	9.6.2	The Force of Acceleration	347
	9.6.3	Microgravity	349
	9.6.4	The Question of Convection	352
	9.6.5	Growth of Gas Bubbles and Non-Metallic Inclusions	353

Appendix 355

References 361

Index 387

1 Characterization of the Fusion Welding Process

A feature of fusion welding is melting of the base metal or of the base and filler metals and producing a welded joint without application of external pressure. This involves heating to high temperatures a small region of the metal, which is surrounded by the cold metal whose volume is much larger than that of the molten metal. This requires use of powerful heat sources for welding, the presence of which essentially influences the physicochemical processes, proceeding in the welding zone.

Fusion welding also includes surfacing, which is a technology, involving application of a layer of molten metal on the surface, of an item effecting either restoration of the initial dimensions of the worn out part, or imparting specially required properties to the surface of a new item.

Since the welding or surfacing process is responsible for the nature of the physicochemical processes proceeding in the weld zone, it is necessary to consider the features of various welding processes.

1.1 SOME FEATURES OF FUSION WELDING

In fusion welding, melting of just the base metal and of the base and additional (filler or electrode) metals may take place simultaneously. In the first case, weld formation is only due to the base metal (Figure 1.1a), and in the second it is due to the base and additional metals (Figure 1b). In both the cases, a molten pool is formed from the molten metal, where the boundaries are the partially melted regions of the base metal and the prior produced weld.

1.1.1 THE BASICS OF WELD FORMATION

Weld-pool volume in different fusion welding processes may vary from 0.1 to 10 cm^3, depending on the kind and mode of the welding process; in electroslag welding, it may be up to 80 cm^3 and more. Time of weld-pool metal staying in the molten condition will be different for different pool zones. Now the average time of weld-pool existence may be approximately determined from the following expression:[1]

$$t_{\mathrm{w}} = L/v_{\mathrm{w}}, \qquad (1.1)$$

where t_{w} is the average time of weld-pool existence; L is the weld-pool length; v_{w} is the speed of heat-source displacement.

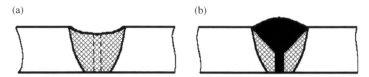

FIGURE 1.1 Weld formed due to melting: (a) base metal; (b) base and electrode (filler) metals.

From Equation (1.1) it follows that the time of weld-pool existence may be increased by lowering the speed of heat-source displacement, or by using certain techniques, that allow the length of the weld pool to be increased, for instance, by preheating of the parts being welded.

Since fusion welding process is a high-temperature process, it is accompanied by intensive interaction of the contacting phases. Such interacting phases in fusion welding include: (i) molten and solid metals (ii) metal melt–gas or slag, depending on the applied welding process. The latter forms as a result of melting of the slag-forming components, included into the electrode coating or flux, as well as in the interaction of the metal and gas.

It should be noted that the processes of interaction of the contacting phases first proceed under the conditions of temperature rise, and then under the conditions of temperature lowering. In the first case, the endothermal processes and in the second case, the exothermal processes are more intensive. Eventually, due to a short time of existence of the weld pool, there is usually not enough time in fusion welding for the equilibrium between the reacting substances to set in.

The most intensive interaction between the contacting phases is observed in the cases of electric arc, electroslag, and gas welding. This is because in these cases the time of weld-pool existence is long enough.

1.1.2 THE INFLUENCE OF GASES

In fusion welding, the liquid metal dissolves a certain amount of gases from the air and the gaseous products of decomposition of the components of the electrode coating or flux. The main gases, influencing the metal properties and which are most often present in the metal, are oxygen, hydrogen, and nitrogen. In this case, hydrogen is physically dissolved in the molten metal, and oxygen and nitrogen enter into a chemical interaction with the most of the metals. Therefore while a considerable amount of hydrogen, is present in the metal in the dissolved condition, oxygen is present in the form of compounds. Nitrogen forms nitrides under appropriate conditions.

Intensity of interaction of the gases with metals depends on many factors: chemical affinity of the contacting phases, temperature, pressure, size of contact surface, and so on.

In welding, conditions are in place, which enhance the interaction of gases with metals. So, with electric arc processes, penetration of gases into the metal melt is promoted by a high temperature of the molten metal, quite a large metal–gas contact surface at a comparatively small volume of the molten metal, intensive stirring of the melt, and presence of electric and magnetic fields.

Most of these factors, although to varying degrees are also characteristic of other fusion welding processes except electron beam welding, if it is performed in vacuum. Therefore, gas content in the weld metal is often markedly higher than that in the base metal, and sometimes it is close to or even higher than the limit of gas solubility in the metal. This is further promoted by the high rate of weld-metal cooling.

During cooling, evolution of gases proceeds due to lowering of their solubility in the metal. Gases may be removed from the metal through desorption from the weld-pool surface or as a result of formation of gas bubbles in the volume of the molten metal. If gas bubbles are removed from the weld pool, this will simply promote metal degassing. However, if they form during completion of solidification of the weld-pool metal, such bubbles will remain in the metal in the form of pores, the presence of which impairs the mechanical properties of the welded joint. The risk of pore formation is further increased due to an abrupt reduction of N_2 and H_2 solubility in the metal during its solidification.

Since in practice one has to deal with multi-component alloys, the deleterious influence of oxygen and nitrogen, present in the metal, is enhanced by the possibility of their interaction with most of the alloying components. This may result in formation of both gaseous products, for instance, CO, which will influence the process of pore formation, and of non-metallic inclusions, namely nitrides and oxides. The main cause for formation of these inclusions is lowering of the element's solubility in the metal as temperature drops. In addition, the de-oxidizing ability of the elements is enhanced with temperature lowering, this further leading to higher intensity of oxide formation in the weld metal.

1.1.3 THE INFLUENCE OF SLAG

In fusion welding processes, when a slag is used to protect the welding zone, the duration of interaction of slag with the molten metal is usually short and varies from several seconds in arc welding processes up to several minutes for electroslag welding. Nonetheless, the importance of the process of interaction of the molten metal and slag is quite high. This is due to the high temperature of their interaction, as well as a considerable area of contact at a comparatively small volume of the metal pool and great mass of the slag.

Slag interaction with the molten metal in welding is manifested primarily in exchange redox reactions, allowing the elements to go into the metal and back. In this case, the composition of the slag and of the metal is important. For instance, lowering of the content of active oxides (FeO, MnO, SiO_2) in the slag and increase of the concentration of strong oxides (Al_2O_3, MgO) lower the oxidizing ability of the slag. Basic slags containing considerable amounts of CaO and MgO promote removal of S from the metal, thus lowering sulfide content in the metal. Presence of slag on the weld-pool surface also promotes removal of oxide inclusions from the metal melt.

Processes of interaction of the solid and liquid metals have a significant role. So, weld-metal dendrites, as indicated by the results of metallographic examination, are a continuation of the base-metal grains, located along the fusion line.

Therefore, the section of columnar crystallites of the weld metal is largely determined by the size of base-metal grains. More over, during weld-metal solidification, different solubility levels of an element in the solid and the molten metals results in a considerable chemical inhomogeneity in the fusion zone, as well as in the boundary regions of the solidification zones, which is attributable to an intermittent motion of the solidification front.

The nature and degree of microscopic chemical inhomogeneity has a significant influence on weld-metal resistance to cracking and on its mechanical properties. Chemical inhomogeneity, formed during welding, cannot be eliminated, as the diffusion process, which promotes equalizing of the composition does not have enough time for completion before weld-metal solidification is over.

In addition to chemical inhomogeneity produced by fusion welding, physical inhomogeneity is also observed in the metal of welds, that is related to the formation of secondary boundaries, running through the regions, where imperfections of crystalline lattices are accumulated.

It should be noted that in fusion welding, the heat evolved by the heat source is consumed not only for metal melting, but also for heating of base-metal regions, adjacent to the weld. Heating and cooling of the base-metal regions change their structure and may lead to deterioration of the mechanical properties of the metal. The change in the heat-affected zone (HAZ) structure is dependent on the temperature of heating (as felt at a particular region) and its cooling rate, as well as on the thermo-physical properties of the material.

Thus, various physical and chemical processes proceed in fusion welding, which largely determine the quality of the weld and welded joint as a whole. The nature of these processes depends essentially on the kind of the applied heat sources.

1.2 WELDING HEAT SOURCES

At present, the most widely used heat source in fusion welding processes is the welding arc, which is an electric discharge in gases and vapors of comparatively high pressure. It is characterized by a high density of current in the electrically conducting gas channel and low inter-electrode voltage. The electrically conducting channel between the electrodes has the shape of a truncated cone or cylinder. Its properties change with the increase of the distance from the electrodes. Thin layers of gas, adjacent to the electrodes have a comparatively low temperature.

1.2.1 THE D.C. VS. A.C. WELDING

In d.c. welding, two variants of the process may be used, namely straight-polarity welding, when the electrode is connected to the negative pole and the base metal to the positive pole of the power source, and reverse-polarity welding. In the latter case, the electrode is connected to the positive, and the base metal to the negative pole of the power source. Depending on electrode polarity, a distinction is made between the anode and cathode regions of the arc (Figure 1.2). Having a short length ($l_c = 10^{-4}-10^{-5}$ cm; $l_a = 10^{-3}-10^{-4}$ cm), these regions are

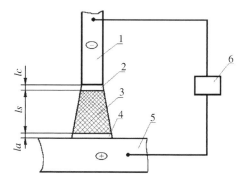

FIGURE 1.2 Schematic of arc discharge: (1) electrode (cathode); (2) cathode region; (3) arc column; (4) anode region; (5) part (anode); (6) current source.

characterized by a high intensity of the electric field. For instance, in the cathode region of 10^{-5} cm length with voltage drop of 10–20 V, field intensity reaches the value of $2 \cdot 10^6$ V/cm. In the anode region, the electric field intensity rises up to 10^4 V/cm.[1] Therefore, the processes proceeding in near-electrode regions are of paramount importance in the conversion of electric energy into thermal energy and its transfer to the electrodes.

The region located between the near-electrode regions is called the arc column. All the parameters of the arc column, namely, temperature, field intensity, and current density depend primarily on the magnitude of the ionization potential (U_i) of the gas of the arc. For instance, in the case of an arc running in potassium vapors ($U_i = 4.33$ eV) at a current of 200 A, the temperature of the arc column, $T = 3460$ K; electric field intensity, $E = 8$ V/cm; current density, $i = 300$ A/cm^2. At the same current, but for an arc running in iron vapors ($U_i = 7.83$ eV), the arc column parameters will be as follows: $T = 6320$ K, $E = 28$ V/cm, and $i = 1800$ A/cm^2.

Welding arcs usually run in mixtures of vapors and gases, which may be regarded as some homogeneous gaseous phase with an effective ionization potential, the magnitude of which depends on the ionization potentials of the components of the mixture and on their relative concentration in the mixture. The component with the smallest value of ionization potential has the strongest influence on the magnitude of effective ionization potential. So, in the case of an arc running in a mixture of potassium and iron vapors at the same concentration of them in the arc column, the magnitude of the effective potential of ionization is equal to 4.61 eV, which is much closer to the value of ionization potential of potassium. Therefore, addition to the arc column gas of just 5–8% of components with low ionization potential leads to a noticeable lowering of the magnitude of effective potential of ionization of the mixture, and significantly changes the arc-column parameters.[1]

Heating, melting, and evaporation of the electrode material depend primarily, on the processes, proceeding in the anode and cathode regions. Therefore, let us consider these processes in greater detail.

Gas temperature in the anode region varies by several thousand degrees from the arc-column temperature T_{col}, to the temperature of the active spot surface, that is, that part of the anode surface, through which the welding current flows. As a rule, the anode material evaporates in the active spot zone. Therefore, the temperature gradient in the anode zone is given by the following expression:

$$T = T_{col} - T_{boil},$$

where T_{boil} is the boiling temperature of anode material.

At the anode region extent of 10^{-4} cm and $T = 3 \cdot 10^4$ K, the temperature gradient ΔT will be $3 \cdot 10^7$ K/cm. Presence of a temperature gradient of such a magnitude will lead to appearance of a powerful heat flow Q_a in the direction of the anode, the flow being equal to

$$Q_a = -\eta \Delta T F_a / l_a,$$

where η is the coefficient of heat conductivity of gas; F_a is the area of the anode active spot.

Stationary condition of the anode region is possible, if this amount of heat is compensated by the applied power, which is equal to $U_a \cdot I_a$, where U_a is the voltage drop in the anode region, I_a is the current.

It should be noted that U_a value is markedly influenced by the magnitude of the coefficient of heat conductivity of the gas — the higher the heat conductivity of the gas, the greater the value of voltage drop in the anode region. Under the conditions of welding, however, the value of voltage drop in the anode region changes comparatively little, and is usually equal to 4–6 V, even at considerable changes of the magnitude of effective ionization potential of the gas.

The cathode region, similar to the anode region, is characterized by high temperature gradients, which may reach 10^8 K/cm. In this case, the magnitude of the heat flow, directed to the cathode is equal to

$$Q_c = -\eta \Delta T F_c / l_c$$

where F_c is the area of the cathode active spot.

Similar to the anode region, the stationary condition of gas at the cathode surface, is possible, if a balance of its energies is observed:

$$U_c I_c = -\eta \Delta T F_c / l_c$$

In the cathode region, unlike the anode one, increase of the magnitude of effective ionization potential of the gas leads to a marked rise of cathode voltage. Similar to the anode voltage, the value of cathode voltage depends on the heat conductivity of the gas. In arcs running in helium atmosphere, having a high heat conductivity, the cathode and anode voltage are higher than those in argon, the heat conductivity of which is lower than that of helium.

Experimental studies showed that at currents above 50 A, powerful flows of ionized gas arise in the welding arcs. These flows are directed mainly along the arc axis. Velocities of these flows are up to $75-150$ m/s, when steel electrodes are used. At current above 300 A, the heat power, transferred by the gas flow to the item being welded is equal to about 40% of the power, coming through the active spot. In addition, these flows also have a force impact on the weld-pool metal.

If a.c. is used for powering the arc, each of the electrodes alternatively is either the anode, or the cathode, and the frequency of such alternation depends on that of the applied current. A feature of a.c. arcs is the fact that during each period, the current reaches zero value two times and the arc goes out at this moment each. Although the period of time during which the arc does not run, is very short, it still leads to cooling of the arc column and deionization of the arc gap. Therefore, each re-ignition of the arc requires a short-time increase of the electrode voltage. Under practical conditions, such features of an a.c. arc do not change the thermal characteristics.

Heat power of the arc may be increased by arc constriction in the radial direction. In this case, the temperature of the arc column reaches 33,000 K, which is much higher than the temperature of a regular welding arc. A constricted or plasma arc is characterized by a smaller area of the anode spot, leading to energy concentration on the anode. In this case, the specific power of the arc may be higher than 5.0 kW/mm^2. In addition, the power consumed in anode heating and melting is made up of the power transferred to the anode by the arc, as well as the power, transferred to the anode by a plasma jet.

Thus, the arc discharge, running between the electrodes, has both a thermal and a mechanical impact on them due to the gas and plasma flows. It should be noted that the impact of the arc discharge also results in the removal of contamination and oxide films from the part surface. This is particularly important in welding of active metals, for instance, aluminum, magnesium, and alloys on their base.

The welding processes may also use the slag pool as the heat source, in which the electrical energy is converted into thermal energy. In this case, the heat, evolved in the slag pool due to current flow is consumed in metal heating and melting. The base and the electrode metals thus get molten and flow to the bottom of the slag pool to form the weld pool. The weld pool solidifies with the increase of the distance from the heat source. As the main purpose of the slags in this case is conversion of electrical energy into thermal energy, their major characteristic is electrical conductivity and its dependence on temperature. In addition, the slags in this case should have a high temperature of boiling, should form no gases when heated to high temperatures, and should provide reliable protection of the molten metal from interaction with the environment.

1.2.2 ELECTRON BEAM WELDING

In fusion welding processes, the electron beam may also be a heat source, which is a focused flow of electrons. The electron flow is accelerated due to the difference of potentials between the anode and the cathode, and is then focused into a spot, the dimensions of which may vary from centimeters to several millimeters.

At the braking of accelerated electrons near the surface of the metal part, their kinetic energy is converted into thermal energy. The higher the power density in the point of electron braking, the more is a body heated.

Depending on the value of the accelerating voltage and the properties of the metal being welded, the electrons may penetrate into the metal to a depth of several tens of micrometers. Such an electron runs into multiple collisions and loses energy, changes the velocity and direction of its motion, thus spending the main part of its energy in the final portion of the path. The thickness of a layer of substance, in which the electron completely loses its velocity, is called the electron path. According to B. Shenland's data,[2] the length of the electron path, δ is equal to:

$$\delta = 2.1 \cdot 10^{-12} U_{acc}^2 / \rho,$$

where ρ is the density of the substance; U_{acc} is the accelerating voltage.

Thus, metal heating proceeds in the metal proper, unlike the electric arc, wherein the item's surface is heated. At the impact of an electron flow on a part, most of the heat evolves at the depth of electron path. This heat is sufficient to melt the edges of the parts being welded. Energy concentration with this method of heating is very high, leading to a narrow and deep weld being produced, as well as a narrow HAZ. A feature of electron beam welding is a very small pressure of the electron beam on the weld-pool metal. In this case, the main impact on the pool is produced by reactive forces, arising from metal evaporation during welding. The intensity of metal evaporation is so high, that according to the results of research by V.I. Nasarov, N.V. Shiganov, and E.D. Raimond,[2] the mechanical impact on the weld-pool metal at the same power of the heat source is by an order of magnitude higher for electron beam welding, than for consumable electrode argon-arc welding. The magnitude of mechanical impact in electron beam welding is inversely dependent on the product of the coefficient of heat conductivity and the temperature of melting of the metal being welded.

1.2.3 LASER WELDING

Recently, laser welding has become accepted in welding fabrication, where the heat source is a powerful light beam, generated by a quantum generator. This heat source is contactless, and, therefore, welding may be performed in media, transparent for this radiation or in closed cylinders. Owing to a low divergence of radiation, a light beam with the spot diameter of up to several hundreds of a millimeter may be produced after focusing. As a result, the light beam features a high concentration of heating.

It should be noted that when the light beam hits a metal surface, only a part of the light flux is absorbed by the metal and is consumed in its preheating. The rest of the light flux is reflected and lost. Light absorption by the metal proceeds in a thin surface layer of the thickness of several wave lengths. The absorbed share of the light flux is determined by the absorpance, A of the metal and its magnitude depends on the kind of the metal and the conditions of its surface. So, for polished plates, the magnitude of coefficient A varies from 0.05 for silver to 0.3−0.5 for most of the metals.[1] Thus, just a part of the light flux power is used for heating the metal item. Effectiveness of using the heat power in laser welding may be enhanced by multiple return of reflected radiation.

Gas flame has been used quite often earlier and is now used to a smaller degree now as a heat source in the fusion welding processes. In this case, the heat used is generated by the combustion of various combustible gases, most often in combination with oxygen. Acetylene, propane–butane mixtures, hydrogen, methane, natural gas, and so on, are applied as combustible mixtures. Flame temperature (T) depends on the kind of the gas used; for an acetylene–oxygen mixture, $T = 3373$–3473 K, for a propane–butane–oxygen mixture, $T = 2673$–2973 K, and so on. This temperature is sufficient to melt the metal. However, the gas flame is characterized by the lowest concentration of heating among the all considered heat sources, used in fusion welding processes. For this heat source, the maximal power density in the heated spot is $5 \cdot 10^4$ W/cm^2, while for a free running electric arc, this value is $1 \cdot 10^5$ W/cm^2, for the electron beam, $5 \cdot 10^8$ W/cm^2, and for the laser beam, $1 \cdot 10^9$ W/cm^2.

1.3 FUSION WELDING PROCESSES

Depending on the applied heat source, a distinction is made between the following kinds of fusion welding: electric arc welding, electroslag welding, electron beam, or laser welding, as well as gas welding. These welding processes are also used for surfacing the most diverse parts and products.

1.3.1 ELECTRIC ARC WELDING

In electric arc welding, a direct arc is most often used, which runs between the parts being welded and the rod — electrode, further on called the electrode (Figure 1.3a). In this case, the heat, evolved in the arc column and on the electrodes, is used to melt the metal. An indirect arc (Figure 1.3b), running between two electrodes, is not much frequently used for welding. Recently, however, it has become quite often used for surfacing, as this greatly enhances the process efficiency with significant lowering of power consumption. In the case of an indirect arc, the heat required for base metal melting comes from the arc column, which contacts the surface of the part being welded or surfaced, physically as well as by the heat, transferred to the

(a) (b) (c)

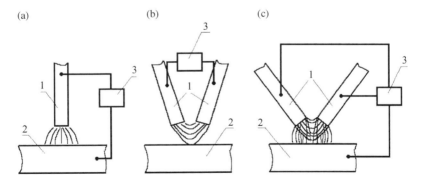

FIGURE 1.3 Kinds of welding arcs: (a) direct arc; (b) indirect arc; (c) combined arc. [1] electrode; [2] part; [3] current source.

electrode (metallic drops). The temperature of the part surface heating is controlled by varying the distance between the electrodes and the part, as well as the process of electrode-metal transfer. Therefore, the depth of penetration is also controlled this being particularly important in surfacing of dissimilar materials. Sometimes, particularly in the welding of thick parts, or in the deposition of a large quantity of the metal, a combined or three-phase arc is used (Figure 1.3c). In this case, heating of the parts is due to the heat of the arcs, running between the electrodes, as well as between the electrodes and the part. A feature of three-phase arcs is only a.c. is used for powering the arc, while both d.c. and a.c. are used for powering the direct and indirect arcs.

In electric arc processes of welding and surfacing, both consumable and nonconsumable electrodes are used. In nonconsumable electrode welding, the weld forms through melting of just the base metal or of the base and the filler metals. In consumable electrode welding, the weld forms due to melting of the base and electrode metals.

1.3.1.1 Gas Shielding

Arc welding processes involve an intensive interaction of the weld-pool metal and molten electrode metal with oxygen and nitrogen of the air. This leads to deterioration of weld performance. Therefore, it is necessary to provide shielding of the welding zone from contact with the ambient air. For manual stick-electrode arc welding (Figure 1.4), a metal rod is covered with a special coating for this purpose. The coating melts in welding, thus providing gas or gas–slag shielding, depending on the composition of the applied coating. In automatic welding processes, the welding zone is often protected by fluxes, which create a slag coverage in melting (Figure 1.5). This method of shielding is applied also in semi-automatic welding, but much more seldom.

In both semi-automatic and automatic welding processes, gas shielding is also frequently used (Figure 1.6). The role of the gas is mostly restricted to physical

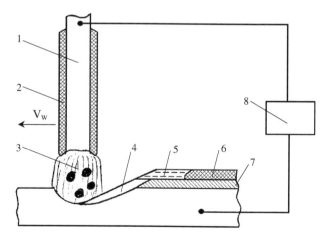

FIGURE 1.4 Covered-electrode welding: (1) metal rod; (2) coating; (3) arc; (4) molten metal of weld pool; (5) slag; (6) slag crust; (7) weld; (8) current source.

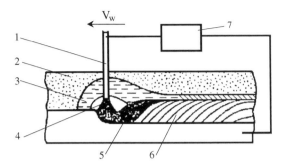

FIGURE 1.5 Submerged-arc welding: (1) electrode; (2) flux; (3) slag; (4) arc; (5) molten metal of weld pool; (6) weld; (7) current source.

isolation of the molten metal from the ambient air. Inert and active gases, as well as gas mixtures are used as shielding gases. In this case, the electrodes may be either consumable or nonconsumable.

Welding can be also conducted without any additional protection, if the used solid welding wire or flux-cored wire contains a sufficient amount of deoxidizer-elements or elements capable of binding nitrogen into stable nitrides.

Constricted- or plasma-arc welding also belongs to the category of gas-shielded arc welding. In this case, two welding schematics are possible, namely, by a direct and by an indirect arc. In either case, an additional shielding gas is fed for shielding the molten metal from the interaction with the environment around the periphery of the constricted arc.

1.3.2 ELECTROSLAG WELDING

Electroslag welding and surfacing, wherein a slag pool is used as the heat source, are usually conducted by the method, shown in Figure 1.7. Slag pool (4) forms as a

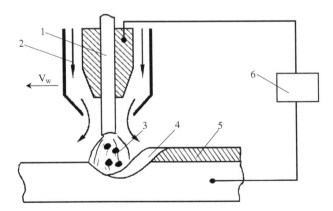

FIGURE 1.6 Gas-shielded welding: (1) electrode; (2) shielding gas; (3) arc; (4) molten metal of weld pool; (5) weld; (6) current source.

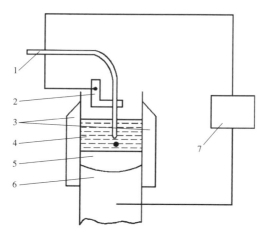

FIGURE 1.7 Electroslag welding: (1) electrode; (2) nozzle; (3) shoes; (4) slag pool; (5) metel pool; (6) weld; (7) source.

result of the flux melting by an arc, struck in the initial period of welding between the electrode and the base metal.

After a sufficient amount of flux has been melted, the welding current, which is fed from the source (7) to the electrode (1), using a nozzle (2), is shunted by the slag pool, and the arc is extinguished. The slag pool (4), induced at the start of welding, moves from the beginning of the part to its end, as weld (6) is formed. Slag and weld pools are contained by the shoes (3). The amount of heat, evolving in the slag pool is sufficient to melt the edges of the base metal and the electrode, which is fed into the slag pool at a speed equal to the rate of electrode melting. Wires, plates, rods, pipes, or consumable nozzles can be used as electrodes. Molten slag, which is above the metal pool (5), also provides reliable protection to the molten metal preventing its interaction with air. This welding process is characterized by a high efficiency and allows welding the parts of practically unlimited thickness.

1.3.3 ELECTRON BEAM AND LASER WELDING

Electron beam energy is used mainly for welding and sometimes for surfacing. This process is most often conducted in a vacuum chamber by a focused electron beam. Presence of vacuum allows avoiding contact of the metal being welded with the air environment, and creates conditions for producing a sound welded joint, but makes the technological process more complicated. Powerful accelerators of electrons, which eliminate the application of vacuum, are presently being used, mainly for surfacing. In this case, however, it is necessary to solve the problem of protection of the working personnel from x-rays generated from electron braking on the part.

Development of optical quantum generators allowed using the energy of the laser beam in the processes of welding and surfacing. A block diagram of a laser welding system is shown in Figure 1.8. Laser (1) generates radiation (2), which is formed by the laser system (3) into a beam with certain spatial characteristics and

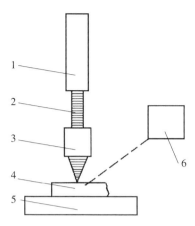

FIGURE 1.8 Block diagram of a laser-welding system: (1) laser; (2) radiation; (3) laser system; (4) the part; (5) a device that ensures fastening; (6) a device that improves the heating.

is aimed at the part being welded (4). The device (5) ensures fastening of the component being welded and its movement during welding, and the device (6) is used to improve the effectiveness of using the heat power of the light beam (ultrasound generator, the unit for returning reflected radiation, etc.).

At present, laser welding can be conditionally divided into three groups by its purpose: microwelding (penetration depth below 100 μm); miniwelding (penetration depth of 0.1–1 mm); macrowelding (penetration depth of more than 1 mm).

Gas welding, in which a gas flame is used as the heat source, is now rather seldom used in practical work. The gas flame is characterized by a low concentration of heating and comparatively low temperatures. Maximal temperature is achieved in the central part of the flame and depends on the applied combustible gas. Temperature values for different combustible gases, burnt in a mixture with oxygen were given earlier (Section 1.2). Distinction is made between reducing (carbonizing) and oxidizing gas flames, depending on the ratio of oxygen and combustible gas in the gas mixture used.

1.4 PHYSICOCHEMICAL PROCESSES OF WELDED JOINT FORMATION

The final aim of all the applied fusion welding processes is producing a sound welded joint or deposited metal, characterized by a high performance under specified conditions. Achieving this condition is largely dependent on the physicochemical processes proceeding during welding. One of them is the interaction of gases with the molten metal. In general, gases penetrating into a metal may noticeably change its properties. So, with increase in the concentration of nitrogen and, particularly, oxygen, in the metal, its mechanical properties deteriorate considerably (Figure 1.9). Numerous studies[3–5] indicate that hydrogen present in the metal also impairs its mechanical properties and, first of all, ductility.

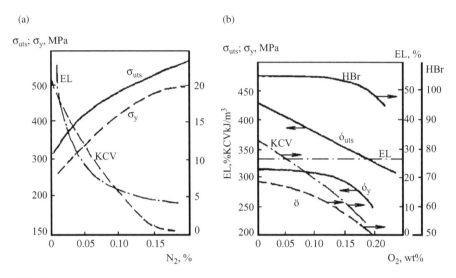

FIGURE 1.9 Influence of nitrogen and oxygen on the mechanical properties of low-carbon steel. [From: Seferian, D., *Welding Metallurgy*, Mashgiz, Moscow, 1963, p. 347.]

Adverse influence of hydrogen is enhanced with increase of carbon content in the metal. Deleterious influence of hydrogen in steels is manifested already at its content of $1-2 \text{ cm}^3/100$ g. Hydrogen present in the metal also lowers its cold-cracking resistance.

It is known that interaction of gases with the metal is in fact a set of physical and chemical processes. This includes material transfer in the contacting phases, processes of adsorption and desorption, and chemical transformations on transfer surfaces, and so on.

As was noted earlier, fusion welding processes, depending on the applied method of molten metal protection, are characterized by two types of systems, namely, (i) metal–gas, when gas directly contacts the metal and (ii) metal–slag–gas, when the gas, before coming to the metal, should pass through a layer of molten slag. Let us first consider the process of direct interaction of the metal with the gas.

1.4.1 DIRECT INTERACTION OF THE METAL AND THE GAS

1.4.1.1 Theoretical Models

In this case, interaction of gases with the metal consists of the following successive stages: mass transfer in the gas phase (external-diffusion stage); phase transformation, consisting of the processes of dissociation of the molecules or molization of atoms, chemisorption of atoms, or their desorption, chemical transformations on the melt surface, transition of atoms or molecules into boundary layers of contacting phases, and so on.

With respect to internal mass transfer in the melt, processes of gases' interaction with the liquid are currently described using three models of mass transfer, namely,

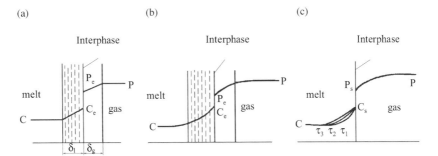

FIGURE 1.10 Schematic of distribution of the absorbed component on melt–gas interphase: (a) two-film theory; (b) theory of boundary diffusion layer; (c) theory of surface restitution.

two-film theory, theory of the boundary diffusion layer, and theory of surface restitution. Schematics of distribution of the absorbed component on the interphase for each of the given models, are presented in Figure 1.10.

According to the two-film theory, in the substance transfer from one phase to another, the mass transfer rates are equal to:

$$V_g = \beta_g^m (P - P_e) = D_g(P - P_e)S/\delta;$$

$$V_l = \beta_l^m (C_e - C) = D_l(C_e - C)S/\delta,$$

where S is the surface area of the phases; P, P_e are the actual and equilibrium concentration of substance in the gas phase, respectively; C, C_e are the actual concentration and equilibrium concentration of substance in the liquid phase, respectively; β^m is the coefficient of mass transfer; D is the diffusion coefficient of the absorbed component; δ is the thickness of boundary film; g, l are the indices, denoting the gas and liquid phase, respectively.

This theory has the disadvantage of neglecting the influence of movement of the contacting phases on film thickness and substance transfer in these films. On the other hand, according to modern concepts, substance transfer in the volume of a phase undergoing intensive stirring proceeds mainly due to turbulent diffusion, which leads to instantaneous equalizing of substance concentration in the volume. The closer is the surface to the interface, where the convective flows are weaker, the smaller is the role of turbulent diffusion, and the greater is the role of molecular diffusion. The layer in which the molecular mechanism of diffusion predominates over the turbulent mechanism, is called the boundary diffusion layer. Unlike the two-film theory, where film thickness is considered to be a constant value, in the theory of boundary diffusion layer, the thickness of this layer is a variable, dependent on flow velocities, diffusion coefficient of the substance, kind of contacting phases, and so on.

Recently, the theory of permeation and restitution of the surface,[6] proposed by Higbi,[6] has become applied. According to this theory, at short-term contact of phases, the diffusion process will be non-steady, and hence the presence of films on the phase surface is eliminated. In this theory, the rate of mass transfer is determined by the rate of surface restitution. As under the conditions of welding, there is a continuous restitution of the surface of the reaction zone, this theory should better characterize the process of mass transfer between the gas phase and molten metal.

1.4.1.2 The Drawback of the Theoretical Models

A common drawback of all the theories is the presence in the expressions of quantities, which often cannot be determined either experimentally or by calculation; for instance, boundary layer thickness, diffusion layer thickness, rate of surface restitution, and so on. At present, the theory of boundary diffusion layer is used more frequently which allows for the influence of various factors on the diffusion layer thickness, and rather accurately describes the process of gases interaction with metal melts.

According to Fick's first law, the number of moles dn_m, diffusing through the surface normal to it, is equal to

$$dn_m = \pm DSd\tau(dc/dx), \tag{1.2}$$

where dc/dx is the concentration gradient in the direction of axis x, normal to the contact surface; $d\tau$ is the process duration. In this case, dn_m, depending on the change of concentration along axis x, may be either positive or negative, that is, saturation or removal of substance will take place.

The concentration gradient is formed only on the interphase region of the boundary layer; denoting the instantaneous concentration of the impurity in the melt on one side of the layer as C and that on the other side as C_s, we obtain $dc/dx = (C - C_s)/\delta_1$, where δ_1 is the thickness of diffusion layer. Then, Equation (1.2) will be written as follows:

$$dn_m/d\tau = DS(C - C_s)/\delta_1. \tag{1.3}$$

If the mole number in Equation (1.3) is replaced by the number of moles in a unit of volume dn_m/V, that is, concentration and ratio D/δ_1 is replaced by the magnitude of the coefficient of mass transfer, β^m, we obtain an approximate formula for determination of the rate of transfer of the dissolved gas from the melt surface into the volume at adsorption and from the volume onto the surface at desorption

$$dc/d\tau = \beta^m(C - C_s)S/V. \tag{1.4}$$

Validity of Equation (1.4) for description of the interaction, molten metal−gas is confirmed by numerous studies. Although such data, demonstrate the processes of nitrogen desorption, according to Shimmyo and Takami[7] adsorption and desorption are described by equations of second order in terms of the concentration of the gas dissolved in the metal.

So far, from the experimental data on the processes of metal interaction with gases in welding, it has not been possible to establish the true sequence of the reactions. In line with most of the researchers, let us assume that this process is described by first-order equations, and, therefore, the diffusion processes have the main role particularly in the metal melt. Then from Equation (1.4) it follows that the rates of metal saturation with the gas and its removal depend on β^m values, and are also related to the size of the interface and gas concentration in the surface layer of the metal, which, in turn, depend on the presence of surfactants in the metals.

1.4.2 INFLUENCE OF SURFACTANTS

Influence of surfactants on gas content in the metal is noted in a number of works.[8-14] In this case, the observed phenomena cannot be attributed to the change of viscosity of the metal melt or to the magnitude of the diffusion coefficient and the coefficient of mass transfer. So, with increase of oxygen content in iron from 0.002 to 0.04 wt.%, the viscosity of the metal melt increases approximately 1.2 times,[15] which should lower the mass transfer constant for nitrogen by approximately 1.15 times.[16,17] Actually, the value of the coefficient of mass transfer in this case was reduced 2.2 times.[12] Such a lowering of β^m value cannot be attributed to the influence of oxygen on the value of diffusion coefficient of nitrogen in molten iron.[6]

Therefore, the influence of surfactants on saturation of the metal melt with gases is, apparently, related to the processes of their adsorption on the interface and resulting change of the area of contact of the metal with gases.

In arc welding processes, metal interaction with gases has certain features, which are due to the following reasons: presence of arc column is confined only to a part of the interface, limited by the active spot, having a higher temperature and the plasma, containing dissociated and partially ionized hydrogen, nitrogen, and oxygen. Allowing for the existence of two regions of metal–gas interface (low-temperature and high-temperature regions) Equation (1.4) becomes:

$$\mathrm{d}c/\mathrm{d}\tau = \beta_1^m(C_s^a - C)S_1/V + \beta_2^m(C_s^p - C)S_2/V. \tag{1.5}$$

Here β_1^m and β_2^m are the coefficients of mass transfer in the high- and low-temperature parts of the weld pool, respectively; S_1, S_2 the area of weld-pool surface, limited by the active spot and the other surface, respectively; C_s^a, C_s^p are the gas concentration on the pool surface in the active spot and on the low-temperature surface of the pool, respectively.

In arc welding processes, it is possible for the gas concentration in the metal-pool volume to become higher than that on the surface of the low-temperature part of the weld pool, that is, $[C] > [C_s^p]$. This is attributable to the fact that in the active spot, $[C_s^a]$ may reach high values due to the high temperature. Thus, according to calculations[18,19] at the temperature of iron surface of 2500 K, $[C_s^a] \approx$ 10% of nitrogen. Therefore, it is quite possible that, since $[C] > [C_s^p]$, the second summand in Equation (1.5) will become a negative value. Then, the rate of metal saturation with the gas will be determined by the difference of the two flows, namely the flow, bringing the gas into the metal through the surface (limited by the active spot), and the flow removing the gas through a relatively low-temperature surface of the tail part of weld pool.

Such a mechanism of metal–gas interaction in arc welding processes is confirmed by the experimental data,[20] according to which at unchanged electric parameters of the arc, increase of the total surface of weld pool is equivalent to the enlargement of the low-temperature part of weld pool and lowers the rate of nitrogen absorption by the metal melt.

Therefore, presence of surfactants in the metal should influence the process of melt–gas interaction in arc welding processes. However, the influence of surfactants on gas absorption by the metal in the high-temperature zone of the weld pool will be manifested to a smaller degree, than in gases evolution from the metal. This is mostly due to two reasons: (i) lowering of surface activity of the components with temperature increase which was for instance, noted by S.I. Popel',[21] for oxygen and (ii) intensive stirring of the metal melt in the high-temperature zone, leading to surfactants drifting from the melt surface. The presence of surfactants should have a more prominent role in the evolution of gases from the metal, in a less mobile region of comparatively low temperature. This is confirmed in particular, by the data of I. Pokhodnya[22] who noted an increase of hydrogen content in weld metal in the presence of surface-active silicon in the metal.

1.4.3 THE GAS-ADSORPTION ABILITY

Rates of saturation and evolution of gases are to a certain extent dependent also on the gas-adsorption ability because in each of the processes involved, one of the intermediate stages is gas-adsorption in the surface layer of the metal. In this case, the higher the gas-adsorption ability, (in other words, the higher its surface activity), the easier should be both the absorption and evolution of the gas by the metal. Calculation of nitrogen and hydrogen adsorption in iron-based alloys[10] showed that with the increase of the content of surfactants (C, Si, Cr, Mn) in the metal, there is a lowering of surface concentration of both nitrogen and hydrogen. This is usually attributed to a decrease of the number of vacancies in the surface layer, which could be taken up by nitrogen and hydrogen.

The process of transfer of gases from the atmosphere into the metal is also influenced by the composition of the atmosphere and rate of formation of gaseous products in the reactions proceeding on the melt surface. According to the data of Whiteway, in the presence of complex gases in the atmosphere, containing bound oxygen, for instance, CO_2, CO, and SO_2, reagent transfer in the gas phase will be difficult. This could be the reason for the observed lowering of gas content in weld metal in gas welding, for instance, in CO_2 welding, the shielding atmosphere consists mainly of CO_2 and CO.[23]

In addition, in welding in oxidizing gas media, CO may form at the metal–gas interphase, as a result of the following reaction:

$$[C] + \{O\} = \{CO\}. \tag{1.6}$$

In addition, the surface curvature of the electrode metallic drops and the weld pool will result in a much higher intensity of CO formation, than on a flat surface.

This follows from the equation:

$$\ln P^r_{CO} - \ln P^s_{CO} = 2\sigma_{m-g}M_{CO}/r\ RT\rho_{CO},$$

where P^r_{CO} and P^s_{CO} are the equilibrium pressures of CO above the convex and flat surfaces, respectively; σ_{m-g} is the surface tension of the metal melt; M_{CO} and ρ_{CO} are the molecular mass and density of CO; R is the gas constant; T is the metal temperature; r is the curvature radius of the reaction surface.

As is seen, the smaller the radius of curvature of the metal surface, the higher is the partial pressure of the formed carbon monoxide, and hence, the lower the probability of transition of the atmospheric gases into the metal. It is known[1,24] that in consumable electrode arc welding processes, the reactions between the metal melt and the gas atmosphere or slag proceed largely at the drop stage. Therefore, with the reduction in the size of electrode metallic drops, the content of nitrogen, hydrogen, and oxygen in the weld metal should be lower.

1.4.4 THE INFLUENCE OF VACUUM

When electron beam welding and surfacing proceed in vacuum, interaction of the metal melt with the gas atmosphere is of a different nature. First of all, in this case, no transition of gases into the metal occurs, but gas evolution from the metal will take place. The process of gas evolution will be noticeably influenced by the presence of surfactants in the melt. Second, degassing in welding of steels results in an increase of the de-oxidizing ability of carbon.[25] This leads to shifting of the equilibrium of reaction (1.6) toward CO formation; further, at oxygen contents below the equilibrium level, with a given element deoxidizer, dissolution of the respective non-metallic oxide inclusions may start. With respect to dissociation of all kinds of oxides, the amount of oxygen, forming per unit time may be found from the following equation,[26] provided the size of oxides does not exceed 140 μm:

$$d[O]/d\tau = \beta^m S_{in}([O]_e - [O]_{ac})/V, \tag{1.7}$$

where $d[O]/d\tau$ is the amount of oxygen, formed at oxide dissociation per unit time; S_{in} is the surface area of the dissolving inclusion; V is the volume of the metal pool; β^m is the coefficient of mass transfer; $[O]_e$ is the oxygen content in the metal in equilibrium with the given oxide; $[O]_{ac}$ is the actual oxygen content in the metal.

From Equation (1.7), it follows that the amount of oxygen, formed as a result of the oxide dissociation, increases with the reduction both of the actual content of oxygen in the metal and the weld-pool volume and increase of the surface area of non-metallic inclusions.

Calculations showed (Figure 1.11 and Figure 1.12) that SiO_2 and Cr_2O_3 inclusions of up to 10 μm radius may completely dissolve in carbon–iron in 3–5 sec. With the increase of Si and Cr content in the melt, the dissolution rate noticeably decreases, which is related to reduction of $[O]_e - [O]_{ac}$ difference.

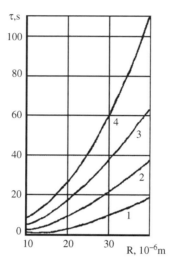

FIGURE 1.11 Dependence of the time of complete dissolution of SiO_2 inclusions on their size with different content of Si in the metal: (1) Si (0.01); (2) Si (0.03); (3) Si (0.06); (4) Si (0.1) (wt%).

1.4.5 THE INFLUENCE OF FLUXES

In covered-electrode arc welding, in submerged-arc electroslag welding and sometimes also in gas welding, various fluxes are used for metal protection. Therefore, let us consider the features of gas transfer from the atmosphere into the metal and back for gas–slag–metal system. In this system, the general case of metal–gas

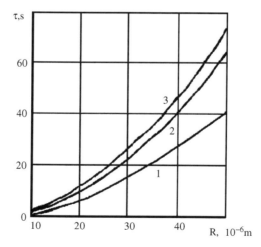

FIGURE 1.12 Dependence of the time of complete dissolution of Cr_2O_3 inclusions on their size with different content of Cr (wt%) in the metal: (1) Cr (1); (2) Cr (3); (3) Cr (5).

interaction consists of the following stages: mass transfer in the gas phase; phase transfer through gas–slag interphase; mass transfer in the slag; phase transfer through slag–metal interphase; mass transfer in the metal. The described schematic is more characteristic for nitrogen. Hydrogen comes into the weld-pool metal mostly from the flux.[27] This is related to the fact that a certain amount of moisture is always present in the fluxes, which evaporates in heating, and penetrates into the arcing zone and decomposes with hydrogen evolution. Nitrogen, probably, mainly penetrates into the metal from the atmosphere, passing through a layer of slag. This is confirmed, to a certain extent, by studies[28] of the influence of flux-layer thickness on N_2 and H_2 content in the weld metal (Table 1.1).

Oxygen comes to the weld metal mostly due to exchange reactions between the metal and slag, for instance, as a result of the following reactions:

$$(SiO_2) + 2[Fe] \longrightarrow 2(FeO) + [Si],$$
$$(MnO) + [Fe] \longrightarrow (FeO) + [Mn].$$

Appearance of ferrous oxide in the slag, which dissolves easily in liquid iron, leads to an increase of oxygen content in the metal (Figure 1.13). As is seen, oxygen content in the metal depends on the composition of the slag, formed in welding.

1.4.6 THE ROLE OF THE SLAG

Let us now consider the factors influencing the interaction of the metal melt with gases in the presence of slag and what the nature of this influence is. It is found[29] that basic slags have better hydrogen permeability, than acidic ones. So, at the same slag basicity, addition of such components as FeO, MnO, and MgO leads to lowering of hydrogen permeability of the slag. Addition to slags of both basic and acidic components, increases the slag viscosity and lowers their hydrogen permeability.

Intensive removal of hydroxyl hydrogen from the slag is promoted by an addition of fluorides to the flux. This leads to a marked lowering of hydrogen content in the weld metal. Hydrogen may be removed from the slag, and, hence, its content in the weld metal may be reduced due to electrolysis of the liquid slag.

TABLE 1.1
Dependence of H_2 and N_2 Content in the Weld Metal on the Thickness of Flux Layer

Thickness of flux layer (mm)	ΣH_2 (cm^3/100 g)	N_2 (cm^3/100 g)
45	7.3	7.8
30	7.14	13.82
20	5.1	35.0
10	3.4	97.5

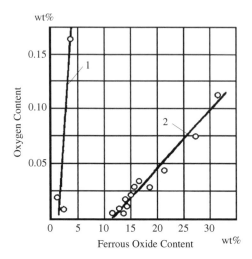

FIGURE 1.13 Dependence of oxygen content in the weld metal on ferrous oxide content in the slag in covered-electrode welding: (1) basic coating; (2) acid coating.

In electroslag remelting, conducted in a.c., it was possible to reduce the hydrogen content in the metal by $4 \, \text{cm}^3/100 \, \text{g}$, due to passage of electricity directly through a layer of the slag.[30]

Since nitrogen penetrates into the metal from the atmosphere, having first passed through the slag, the process of metal interaction with nitrogen will depend on nitrogen concentration in the slag and on its diffusion mobility in the slag melt. According to Morozov,[4] the amount of nitrogen dissolved in the slag, increases with decrease in its oxidizing potential. Nitrogen content in the slag also increases upon passage of d.c. through the slag.[29] Therefore, in d.c. welding, nitrogen content in the slag should be greater, than in a.c. welding.

Investigation of the influence of slag composition on the diffusion mobility of nitrogen showed[29] that at comparatively small changes of slag composition, the value of the diffusion coefficient of nitrogen may vary considerably. For instance, for $Al_2O_3-SiO_2-CaO$ system even at small changes of the ratio of slag components, the diffusion coefficient of nitrogen varies from $0.5 \cdot 10^{-2}$ to $5.2 \cdot 10^{-2} \, \text{m}^2/\text{sec}$ (i.e., by an order of magnitude). Such a noticeable dependence of D_N on slag composition leads to the conclusion that addition of small amounts of oxides or other components to the flux permits achieving a significant improvement of the protective properties of the flux, without deterioration of its process properties.

Weld-metal saturation with nitrogen is also influenced by the structure of flux particles. From the data of Lubavsky,[31] in welding with glass-type flux OSTs-45, nitrogen content in weld metal in some experiments was almost three times lower, compared to the welding using a pumice-like flux. Addition of fluoride compounds to the flux also lowers nitrogen content in the weld metal,[32] which is, probably, caused by driving the air away from the molten slag due to evolution of fluorine compounds from it.

Thus, interactions with gases in metal–gas and metal–slag–gas systems are multi-factorial processes, which should be taken into account in the development of technologies and welding consumables, to reduce the deleterious influence of gases on the performance of the welded joint.

1.4.7 THE PHYSIOCHEMICAL PROCESSES AT THE SOLID–MOLTEN METAL BOUNDARY

A significant influence on the properties of the welded joint is produced also by the physicochemical processes in the boundary between the solid and molten metals. For instance, they are responsible for grain size, which in its turn, influences the metal susceptibility to cracking and its mechanical properties. Grain size is known to be largely determined by the progress of initial solidification, which depends on many factors.

1.4.7.1 The Nucleation Process

In the formation of any new phase, the initial stage of the appearance of crystals in the melt is nuclei formation. According to Chalmers,[33] under equilibrium conditions, atomic complexes with a structure corresponding to that of the solid phase are present in the liquid. These complexes may grow into a crystal only if at a given temperature, their size is greater than the critical size of the nuclei. Radius of a stable nuclei of a spherical shape is given by Chalmers[33]:

$$r_{cr} = 2\sigma_{sol-1}T_E/L_m\Delta T, \tag{1.8}$$

where σ_{sol-1} is the interphase tension on the boundary of the crystal nuclei with the melt; T_E is the melting temperature; L_m is the heat of melting; ΔT is the degree of overcooling.

The work of such a nuclei formation on metal-solidification process, according to Efimov[34] is equal to:

$$W = 16\pi\sigma_{sol-1}^3 T_E^2/L_m^2(\Delta T)^2. \tag{1.9}$$

The number of stable nuclei, formed in unit time in unit volume of the liquid is equal to[32]:

$$I = (nkT/h)e^{-U/kT}e^{-W/kT},$$

where n is the number of atoms in unit liquid volume; k is the Boltzmann constant; h is the Plank's constant; U is the energy of activation of atoms transition from the liquid to the crystal; T is the absolute temperature.

As the rate of nuclei formation depends on the energy of activation of the atom transition from the liquid to the crystal, an intensification of the diffusion processes in the melt (for instance due to electromagnetic stirring), should lower the U value. In this case, the intensity of nuclei formation will tend to the following value:

$$I = (nkT/h)e^{-W/kT}. \tag{1.10}$$

1.4.7.2 Melt Overcooling

From Equation (1.8) through Equation (1.10), it follows that the critical sizes of the nuclei of the crystal and intensity of their formation essentially depend on the degree of the melt overcooling and on the magnitude of interphase tension on the boundary of the crystal nucleus and the melt. Therefore, presence of elements in the melt, capable of changing σ_{sol-l} value should influence the rate of formation of crystal nuclei.

In addition, presence of surfactants in the metal may also lower the degree of melt overcooling.[35,36] By the data of Braun and Stock,[35] addition of 0.1% of cerium or lanthanum to iron, characterized by surface activity in the iron melt, lowers the degree of its overcooling from 593 to 313–323 K.

1.4.7.3 Presence of Refractory Impurities

The process of crystal nucleation in the metal may be significantly influenced by the presence of ready interfaces. Under the conditions of welding, crystals may form on the boundaries with base-metal grains, located on the fusion line, as well as with crystals of the lower-lying layer of the metal in multi-pass welding or surfacing. In addition, various refractory particles, present in the weld-pool metal may play the role of solidification centers. In this case, the greater the number of such particles in the melt, the higher the rate of solidification nuclei formation, which follows from equation[34]: $I = N \cdot e^{-w/(\Delta T)^2}$, where N is the number of particles of the refractory impurity.

The nucleation process will be influenced in the melt of particles of different composition, primarily by those of the particles with a smaller magnitude of ΔT. The degree of overcooling, required for heterogeneous nucleation of solidification centers, depends on particle size. This is understood if Equation (1.8) is re-arranged as follows:

$$\Delta T = 2\sigma_{sol-l}T_E/L_m r_{cr}.$$

Deep overcooling of the metal is required for all the particles present in the melt, to become the solidification centers.

Crystal formation on a ready surface is largely dependent on wettability of the solid surface by the metal melt, and it will be more intensive, when the particles are better wetted by the melt. In addition, the work of formation of solidification centers decreases, if the particles have a crystalline lattice, having a structure similar to that of the solidifying metal. Such particles, for instance, will be the oxides of the solidifying metal.[37,38]

Influence of iron oxides on iron structure was studied in Ovsienko and Kostyuchenko,[38] who established that the presence of iron oxides in the melt lowers the degree of iron overcooling 3–3.5 times. On the other hand, the influence of soluble additives (Ni, Al, Ti) practically does not change the degree of overcooling.

However, as the melting temperature of iron oxides is close to that of iron melting, the influence of the presence of oxides in the melt on the rate of the solidification process is manifested only at a certain degree of overcooling, when the oxides pass into the solid state.

1.4.7.4 Shape of the Interface

The rate of formation of solidification centers is further influenced by the shape of the interface. If the shape of the surface on which the crystal nuclei form is concave, the nuclei appear at a lower degree of overcooling, than on a flat or convex substrate. This is attributable to the fact that the following dependence exists between the work of nucleation on a flat surface of a solid and the volume of the nucleus, V^{39}:

$$W = \sigma_{sol-1}V/R, \tag{1.11}$$

where R is the radius of nucleus curvature.

Simultaneous consideration of Figure 1.14 and Equation (1.11) (which in the case of crystal nucleation on a curved surface will be approximate), confirms the correctness of the earlier conclusion.

A high probability of nucleus formation in a heterogeneous medium is usually associated with reduction of the work required for its formation. However, in the formation of the crystal nucleus on a solid substrate, its appearance is also facilitated by the fact that in this case, the exchange of atoms between the nucleus and molten metal will be more intensive than in the formation of the same nucleus in a homogeneous medium. This leads to larger factor, $e^{-U/kT}$ and therefore, also to a higher probability of appearance of stable crystal nuclei.

1.4.7.5 Effect of Metal Stirring

As noted earlier, the value of U decreases upon metal stirring, which further promotes formation of crystal nuclei. However, intensive stirring of the molten metal due to electromagnetic forces or ultrasonic oscillations, may under the impact of the viscous friction force, change the state of the transition layer and the value of interphase tension on the melt–crystal interphase.[40] Calculations

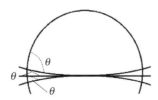

FIGURE 1.14 Schematic of crystal nuclei formation on the surface of a solid.

show that for a steel with a crystal of radius equal to $0.1 \cdot 10^{-2}$ m, the value of interphase tension on melt–crystal interphase decreases by $\Delta\sigma_{sol-1} = 80$ mJ/m^2 at ultrasonic oscillations of intensity, $1 \cdot 10^4$ W/m^2 and frequency, $2 \cdot 10^4$ Hz. Therefore, for the case of weld-pool metal stirring, the value of actual interphase tension σ_{ac}, which may be found from the expression $\sigma_{ac} = \sigma_{sol-1} - \Delta\sigma_{sol-1}$, should be substituted into Equation (1.8), Equation (1.10), and Equation (1.11), where σ_{sol-1} is the interphase tension on metal–crystal interphase in a quiet melt. It turned out,[40] that the value σ_{sol-1} depends only slightly on the nature of the metal. Therefore, the smaller is σ_{sol-1} value, the more will the solidification process be affected by ultrasonic oscillations or electromagnetic stirring of the melt.

1.4.8 CALCULATION OF σ_{sol-1} Values

Since the solidification process and critical dimensions of the nucleus are largely dependent on the value of interphase tension at the melt–crystal interphase, it is necessary to know σ_{sol-1} values. Experimental values of σ_{sol-1} have so far been derived, mostly by overcooling methods and by different variations of the method of lowering the melting temperature of fine particles, compared to massive samples.[41–43]

In addition, using various approaches, different authors suggested calculation formulas to determine the σ_{sol-1} value. So, L.M. Shcherbakov,[464] assuming that capillary effects of the second kind cannot be neglected, in view of the small dimensions of crystal nuclei, derived the following expression:

$$\sigma_{sol-1} = \sigma_{sol-st}Q_{mel}/L, \tag{1.12}$$

where L is the heat of sublimation; Q_{mel} is the molar heat of melting.

However, the presence in this formula of value σ_{sol-st} as the surface energy of the solid phase on the boundary with saturated vapor, which is rather difficult to be determined, lowers the accuracy of finding the value of σ_{sol-1}.

Formula, derived by J. Taylor,[44] on the assumption that surface energy increases by 30% at transition from the liquid to the solid phase, is simple:

$$\sigma_{sol-1} = 4\sigma_{1-st}/27, \tag{1.13}$$

but it is valid only for a small number of metals (copper, silver, gold).

Assuming that the mutual potential energy of the atoms is only dependent on their location and is independent (in the first approximation) of the distance between them, B.Ya. Peanes[44] obtained the following expression for the determination of interphase tension on crystal–melt boundary

$$\sigma_{sol-1} = \frac{1}{4}\left[\sigma_{1-st}(Q_{met}/L_{evap})^2\right] + \left[4\sigma_{1-st}P^2/9\right], \tag{1.14}$$

where P is the relative change of volume at melting.

Regarding the melt as a solid with a high concentration of dislocation centers and taking these dislocations to be in contact at the interphase, Kotze and Kalman[44] proposed finding value σ_{sol-l} from the following formula:

$$\sigma_{sol-l} = 0.0258\ \tau b, \tag{1.15}$$

where τ is the modulus of elasticity (shear); b is the Burgers vector.

Assuming that the number and configuration of the nearest neighbors in the liquid phase are the same as in the solid phase, and that the angle of wetting of a solid metal by its own melt is zero, A. Skapski[44] obtained equations for the calculation of σ_{sol-l} value of the boundary of the face centered cubic (fcc) lattice plane (1 1 1)−melt

$$\sigma_{sol-l} = (Q_{mel}/4\omega) + (2P\sigma_{l-st}/3) + 0.3 \cdot 10^7\ T_{mel}/\omega, \tag{1.16}$$

and of the body centered cubic (bcc) lattice plane (1 1 0)−melt

$$\sigma_{sol-l} = (Q_{mel}/4\omega) + (2P\sigma_{l-st}/3) + 0.5 \cdot 10^7\ T_{mel}/\omega, \tag{1.17}$$

where T_{mel} is the melting temperature of the material in K; ω is the area taken up by a mole of the substance on the surface.

Average σ_{sol-l} values of polycrystal−melt boundary can be found from expressions, obtained by S.N. Zadumkin:[44]

- For metal with a fcc lattice:

$$\sigma_{sol-l} = 1.82\sigma_{l-st}[(Q_{mel}/L) + (2P/3)]; \tag{1.18}$$

- For metal with a bcc lattice

$$\sigma_{sol-l} = 1.89\sigma_{l-st}[(Q_{mel}/L) + (2P/3)]; \tag{1.19}$$

- For metals with a hexagonal close-packed (hcc) structure

$$\sigma_{sol-l} = \sigma_{l-st}[(Q_{mel}/L) + 2P/3] \tag{1.20}$$

Construction of an electronic-statistical theory of the interface energy on the solid metal−melt interface enabled S.N. Zadumkin to derive a more general equation for determination of value σ_{sol-l}, allowing for all kinds of energy of the metallic bond:

$$\sigma_{sol-l} = 5.43[1 - \chi(0)]\sigma_{l-st}, \tag{1.21}$$

where

$$\chi(0) = 3(1 + P)[1 - 1/(1 + P)^{5/3}]/5P.$$

An equation, which in the opinion of the authors, T.M. Taova and M.Kh. Khokonov,[45] is valid for all the substances, is given subsequently:

$$\sigma_{\text{sol}-1} = L_{\text{mel}}\rho_{\text{sol}}\sigma_{1-\text{g}}T_{\text{boil}}/L\rho_1 T_{\text{mel}}. \tag{1.22}$$

Here, the values of the temperatures of boiling T_{boil} and melting T_{mel} are substituted in K.

Results of calculation of $\sigma_{\text{sol}-1}$ values by the above formulas for some metals are given in Table 1.2.

As follows from the earlier discussion, the intensity of formation of the crystal nuclei and their critical size depend on the surface properties of the melt and melt–crystal boundary, as well as on the degree of metal overcooling, its physical properties, presence of refractory particles in the metal melt, their structure, and intensity of melt stirring.

1.4.9 THE CRYSTALLIZATION PHENOMENA

Final size of the crystal grain, d largely depends on the ratio of the rates of formation of solidification centers, I and linear speed of growth of these centers, v[46]: $d = h_{\text{g}}$ $\sqrt[4]{v/I}$, where h_{g} is a dimensionless value, dependent on grain shape; the values for a cubic grain, and a spherical grain are respectively 1.093 and 1.29.

In addition, the rate of crystal growth and their shape are greatly influenced by the presence of surfactants in the melt. This is related to the fact that surfactants, adsorbing on the crystal surface, change their surface properties noticeably, even if this layer has a thickness of the order of a monoatomic layer.[47] Connection between the growth of individual faces of the crystal and the magnitude of interphase tension on the respective faces, according to Wuelff's rule, is established by the following relationship:

$$\sigma_{\text{i}}/l_{\text{i}} = \text{const}, \tag{1.23}$$

TABLE 1.2
Results of the Calculation of Values $\sigma_{\text{sol}-1}$ (mJ/m^2) by Formulas from Various Authors Presented in the Different Equations in this Chapter

Metal	\multicolumn{7}{c}{Formula Number}						
	(1.12)	(1.14)	(1.15)	(1.16)–(1.17)	(1.18)–(1.20)	(1.21)	(1.22)
Cu	62.1	1.55	141.0	128.0	170.0	132.0	134.0
Al	33.6	1.65	113.0	97.1	95.0	97.1	115.0
Ag	46.0	1.25	107.0	101.0	119.0	85.0	—
Ni	110.9	2.92	270.0	207.1	280.0	250.0	—
Fe	84.9	1.33	255.0	152.0	203.1	184.8	136.0

where l_i is the length of a perpendicular, lowered from the solidification center onto the ith face or its continuation; σ_i is the interphase tension on the boundary of the melt with ith face of the crystal.

Dependence shown by Equation (1.23) is indicative of a considerable influence of surfactants on the process of crystal growth. The action of surfactants is usually manifested at their quite low concentrations in the melt, which points to a significant role of adsorption phenomena in the considered process.

From Equation (1.23), it follows further that the crystalline faces, characterized by a greater value of the interphase tension, grow faster, than those with a low value of σ_{sol-l}. Therefore, in the presence of surfactants in the weld pool, which are adsorbed on the melt–crystal interphase, the linear velocity of crystal growth decreases, leading to the grain refinement.

Connection between the crystal dimensions and the presence of surfactants in the melt is confirmed by numerous experiments. According to the data, given by Kunin,[48] a relationship has been established between the influence of Al, Mn, Zn, Na, and some other elements on the surface tension and cast structure of tin.

Influence of calcium on surface tension and initial solidification of steel Kh15N25M6 was noted by R.I. Zaletaeva, N.S. Kreshchanovski and L.L. Kunin.[49] Dependence of structure of the deposited metal on the presence of surfactants in the electrode coating was considered by V.N. Skvortsov.[50]

From the practical work, it is known that the grain size is also influenced by the rate of cooling of the weld-pool metal and the intensity of melt stirring. Grain size becomes smaller with the increase of these parameters. Once the influence of the cooling rate (the increase of which leads to shorter time of grain growth) is understandable, the action of stirring would demand a more detailed consideration.

1.4.9.1 Electromagnetic Stirring and Ultrasonic Oscillations

According to Nernst's diffusion theory, the crystal surface is coated with a thin layer of slow-moving melt. This layer of thickness, δ_l is what offers the main resistance to diffusion transition of a substance to the surface of a growing crystal. It is obvious that crystal growth may be enhanced if δ_l is decreased. This may be achieved by applying electromagnetic stirring of the metal melt, or ultrasonic oscillations to the weld pool.

In the study of Gasyuk and Savchuk,[51] the influence of ultrasonic oscillation amplitude A_o on δ_l value is considered for the following three cases: (1) $A_o \gg \delta_l$; (2) $A_o = \delta_l$; (3) $A_o \ll \delta_l$. For each of these cases, the thickness of diffusion layer at ultrasonic oscillations will be equal to, respectively: (1) $\delta_{us} \approx 0.1\delta_l\ (\delta_l/A_o)^3$; (2) $\delta_{us} \approx \delta_l\ (1 - 2/\pi)$; (3) $\delta_{us} = \delta_l - 2\ A_o/\pi$.

As is seen, the greatest reduction of the diffusion layer thickness, and hence, the greatest increase of the crystal growth rate will be observed in the first case. Crystal growth becomes more intensive, if metal stirring is applied in combination with ultrasonic oscillations, as it leads to additional reduction of δ_l value.

Thus, the use of weld-pool metal stirring due to electromagnetic forces and ultrasonic oscillations, should lead to an increase of the rate of grain growth and of its dimensions. In practice, however, application of such techniques results in grain refinement. This is attributable to two causes. First, as was shown, the number of crystal nuclei increases at metal-melt stirring. Second, melt stirring may lead to dispersion of the grown crystals, as a result of collapse of cavitation bubbles. In addition, shattering of the grown crystals may be due to the forces of viscous friction, generated by the motion of solid particles relative to the melt.

Strength of impact of the oncoming body (crystal, weld-pool walls, refractory particles) on the formed crystal[34] is determined from the expression: $P = \rho_m v^2 / 2\,g$, where ρ_m is the density of crystal material; g is the free fall acceleration; v is the rate of melt stirring.

For the case of steel welding, it may be assumed that $\rho_m = 7.38 \cdot 10^3 \ \mathrm{kg/m^3}$. Then for $v = 0.1 \ \mathrm{m/sec}$, $P = 35.3$ Pa; for $v = 0.3 \ \mathrm{m/sec}$, $P = 44.1$ Pa; for $v = 1.0 \ \mathrm{m/sec}$, $P = 353$ Pa.

In view of a very low strength of the barely solidified crystals, the above forces, which become stronger with the increase of metal stirring rate, can turn out to be sufficient for crystal fragmentation, which will lead to a greater number of solidification centers and, eventually, to a fine-grained structure. In addition, at stirring of the weld-pool metal, structure refinement may result from the change of the temperature condition of the weld pool.[52] This is because structure formation in steels depends[34,53,54] on the relative position of the isochrones of liquid metal liquidus and the temperature field near the solidification front.

1.4.9.2 Segregation Processes

Properties of a welded joint, in particular, the formation of solidification cracks in the weld, are largely dependent on the segregation processes. It is known,[33] that in the solidification of molten metals, the first portions of the solidified metal contains the smallest amount of impurities. Impurities, remaining in the liquid metal, are driven away from the surface of the growing crystals. As the probability of impurity entrapment by the crystals will become lower, concentration of the solute ahead of the growing crystals will become higher. As the grains draw closer to each other, a certain amount of impurities precipitates on the grain surface, depending on the cooling conditions. The greatest accumulation of impurities will be found on the surface of grains, which are the last to solidify.

As the process of motion of impurity atoms is a diffusion process, that is, it is related to time, segregation processes will be markedly influenced by the rates of metal cooling and solidification. The lower these rates, the higher the degree of segregational inhomogeneity.

In welding processes, characterized by high rates of solidification and cooling, segregation processes proceed to a smaller degree, than in large ingots. However, segregational intergranular inhomogeneity[55] is also observed under the conditions of welding; the degree of segregation depends on metal composition, content of the segregating component, welding modes, and so on.

Let us consider the features of segregation process in welds for the case of sulfur, as this component is characterized by a strong segregation ability, and sulfur segregation may greatly influence the process of solidification cracking in welds.

By the data of,[56,57] dendritic and layer segregation are rare phenomena. Usually in the upper part of welds of a large volume, zone segregation is found.

According to the data of V. Pfank, B. Chalmers, and M. Fleming,[33] sulfur concentration in the molten part of the metal during its solidification can be found from the following formula:

$$C_L = C_o(1 - g + \alpha_s k_{eq})^{k_{eq}-1}. \tag{1.24}$$

Here C_o is the initial concentration of sulfur in the melt; g is the fraction of solidified metal; k_{eq} is the equilibrium coefficient of sulfur distribution between the molten and solid metals; $\alpha_s = 4 D_s \tau_a / L^2$, where D_s is the diffusion coefficient of sulfur in the solidified metal; τ_a is the time of solidification of a given interdendritic region of the metal; L is the distance between dendrite boughs.

Equation (1.24) allows determination of the average value of sulfur concentration for a given volume of molten metal. Change of sulfur concentration in the volume proper can be found from the following expression[33]:

$$C_L^x = C_L'[1 + \{(1 + k_{eg})/k_{eg}\}e^{-vx/D}], \tag{1.25}$$

where C_L' is the impurity concentration in a solidifying volume of the metal at a given moment of time; v is the linear speed of metal solidification; x_L is the distance from the interface to the point in the melt, where the impurity concentration is equal to C_L^x.

From the latter expression it follows that the highest concentration of sulfur will be observed at the melt–crystal interphase region, and its distribution in the melt depends on the rate of melt solidification, diffusion coefficient of sulfur in the melt, D_s, and distribution coefficient of sulfur between the solid and molten metals.

Table 1.3 gives the change of relative segregation of the impurity, $C_s(x)/C_o$,[58] depending on the rate of melt solidification v, with the magnitude of the coefficient

---

TABLE 1.3

Influence of the Rate of Metal Solidification on the Magnitude of Relative Segregation of the Impurity on the Melt–Crystal Interphase Region

v (m/s)	$k_p = 0.5$	$k_p = 0.1$	$k_p = 0.01$	$k_p = 0.001$
16.6	50/16	1000/400	1000/1000	1000/1000
100	18.5/6	1000/32	1000/45	1000/55
200	14/4	300/17.5	800/23	1000/25
400	10/3	150/9	200/12	300/13
800	6.5/2.5	38/4.52	66/6	72/6.5

Note: $C_s(x)/C_o$ values at $x = 0.1$ μm are before the slash symbol and at $x = 1.0$ μm after the slash symbol.

of element distribution k_p at the distance from the crystal surface, equal to 0.1 and 1.0 μm. In this case, it was assumed that the diffusion coefficient is a constant and is equal to $5 \cdot 10^{-1}$ m^2/sec.

As is seen, the value of relative segregation of the impurity increases with the decrease of the solidification rate and decrease of value of the coefficient of impurity distribution. This effect is manifested to a greater extent near the crystal surface. Therefore, let us consider the thermodynamic features of grain–melt system.

If the metal contains impurities ($i = 1, 2, \ldots$) in the amount of N_i and their chemical potentials are different from the chemical potential of the matrix μ_0, then the system energy is equal to:[59,60]

$$G = TS_e + \Sigma\mu_i N_i - PV + \sigma_{m-g}A_g, \qquad (1.26)$$

where T is the absolute temperature; S_e is the entropy; P is the pressure; V is the volume; σ_{m-g} is the surface tension; A_g is the area of grain boundary surface.

For the sake of simplicity let us a binary solution ($i = 0; 1$) and comparing the energy of a system, having interfaces G with a homogeneous system G_o. Under the condition that they contain the same amount of atoms of matrix N_0 and impurity N_1, and $\mu_0 = \mu_1$, we obtain:

$$G_o = TS_e^o + \mu_o N_o + \mu_1^o N_1^o - P_o V, \qquad (1.27)$$
$$G = TS_e + \mu_o N_o + \mu_1 N_1 - PV + \sigma_{m-g}A_g. \qquad (1.28)$$

Then, the excess energy of a system, having interfaces will be:

$$\Delta G = T\Delta S_e + \mu_1(N_1 - N_1^o) - P\Delta V + \sigma_{m-g}A_g, \qquad (1.29)$$

and for unit area

$$\Delta G' = T\Delta S_e' + \mu_1 \Delta N' - P\Delta V' + \sigma_{m-g}. \qquad (1.30)$$

From Equation (1.30) we have

$$\Delta S_e' dT + \Delta N' d\mu_1 - \Delta V' dP + d\sigma = 0,$$

whence,

$$\Delta S_e' = -(\partial\sigma/\partial T)_{P,\mu,N};$$
$$\Delta N' = -(\partial\sigma/\partial\mu_1)_{T,P,V}; \qquad (1.31)$$
$$\Delta V' = -(\partial\sigma/\partial P)_{T,\mu,N}.$$

Equation (1.31) is the Gibbs's adsorption isotherm. Therefore, the change of energy in a system with interfaces is due to adsorption of the components at the intergranular boundary. Similarity of segregation and adsorption processes[61] allows using of the formulas, applied to describe the processes of adsorption, for element segregation. This is particularly important to establish a mutual influence of components on their segregation ability.

Thus, interaction of gases, slag, and solid metal with metal melts, impurity segregation in the weld, and primary solidification of metal – all these processes, responsible for the properties of the welded joint, are in many aspects related to the surface properties of the metal and slag, as well as the surface phenomena, proceeding at the boundaries of contacting phases. Let us, therefore, consider the surface properties and phenomena: their nature, methods for their determination, and some results of studying the surface properties and phenomena as applied to welding processes. These are dealt with in the following chapters.

2 Surface Properties and Phenomena

Surface properties (surface and interphase tension) and surface phenomena (wettability, adsorption electrocapillary phenomena) have a significant influence on many aspects of welding and surfacing processes. Weldability of dissimilar metals and composite materials, formation of the weld and the metal deposit, and of various defects (pores, non-metallic inclusions, solidification cracks, etc.), the hydrodynamic processes in the weld pool, metal structure, and many other properties are to varying degrees related to the surface properties and phenomena. Therefore, it is necessary to know them and take them into account in the development of welding and surfacing technologies, as well as welding consumables.

This chapter sets forth the modern concepts of the nature of surface properties and phenomena and describes the methods of their investigation.

2.1 SURFACE PROPERTIES AND PHENOMENA ON THE BOUNDARIES OF CONTACTING PHASES

One of the main properties of liquid and solid bodies is the availability of free surface energy, which is manifested in a well-known tendency of the liquid to take up a shape with minimum surface. This tendency of the liquid surface to reduction is due to the forces of interatomic interaction and has the following explanation. Molecules, present inside of the liquid volume are surrounded on all sides by other molecules, and on average they experience the same attraction in all the directions. Molecules, located on the liquid surface or near it at a distance smaller than the radius of molecular force action, are under a different condition. In this case, each molecule experiences attraction, directed inside and to the sides, but does not experience the balancing attraction from the side of the gaseous phase. This is because the number of molecules in the vapor phase is much smaller than in the liquid. As a result, transition of molecules from the liquid surface into its bulk prevails over and above the transition of molecules from the bulk to the surface. This leads to a smaller number of molecules on the surface, and, hence, also reduction of the surface proper, which goes on, until the surface area becomes minimum.

2.1.1 THE CONCEPT OF SURFACE ENERGY

In solid bodies, a surface atom may have from three up to nine nearest neighbors, while an atom, being inside the lattice, has twelve neighbors in a face-centered cubic (fcc) and hexagonal close-packed (hcp) lattices and eight in a body-centered cubic (bcc) lattice. For this reason, the energy of the surface atom increases by a value proportional to the number of absent links, leading to an excess of energy on the crystal surface, depending on the structure of this surface. Total surface energy of a crystal face is determined by arrangement of atoms, forming this face. Therefore, each face will have its own magnitude of surface energy. Total surface energy of the crystal is given by the sum of surface energies of all its faces.

Spontaneous reduction of the liquid surface without energy consumption, is indicative of the existence of free surface energy. It means that some work should be done to increase the liquid surface. Value, numerically equal to the work of reversible isothermal formation of a unit of the surface of a solid or liquid body, is called the surface tension coefficient, or specific free surface energy. The term "surface tension" is usually used in publications, just as a mathematical concept, equivalent to the concept of free surface energy of the liquid on a boundary with gas. However, when this term is used, one should bear in mind that no actual tension exists over the liquid surface.

As was noted above, solids also have the free surface energy, the magnitude of this energy being related to the surface structure. Features of the structure of a free surface of a solid can be more conveniently considered in the case of an outer face of a single-crystal, where the basic element is a plane, uniformly and densely populated by atoms. Distinction is made between three types of crystal surfaces, namely singular, vicinal, and non-singular. Singular surfaces are absolutely smooth on the atomic scale and do not have any steps. These surfaces have the lowest free surface energy. Vicinal surfaces are not ideally flat and steps (terraces) are imposed on them. Specific surface energy of vicinal plane, A_v only slightly differs from that of singular plane, A_s. According to Geguzin,[62] the relationship between A_v and A_s is described by the following expression:

$$A_v = A_s + mR, \qquad (2.1)$$

where m is the number of steps per unit of length of the main face; R is the energy, related to the presence of one step. As

$$m = (1/l)tg\alpha, \qquad (2.2)$$

where l is the height of the step; then substituting the value of m into Equation (2.1), we have:

$$A_v = A_s + (R/l)tg\alpha. \qquad (2.3)$$

It is known that for vicinal faces the magnitude of the angle α is small. Therefore, as follows from Equation (2.3) that, A_v only slightly differs from A_s.

Non-singular surfaces are characterized by a high density of steps and will have a higher value of free surface energy.

2.1.2 Surface Energy vs. Surface Tension

The actual metallic surface differs from the ideal surface of the crystal by the availability of a large number of crystallographic defects and impurity atoms in it, which influence the condition of the surface and its energy.

Speaking of the surface properties of solids, it is necessary to address the difference between the surface energy and surface tension of a solid. This is important because for crystalline bodies, unlike liquids, the value of surface tension and surface energy may differ considerably, and they have a different physical sense.

So, the work, required to create a unit of the surface, is equal to

$$dW_1 = \sigma_1 \, dS, \tag{2.4}$$

where σ_1 is the free specific surface energy, which is most often called simply the surface energy. Its magnitude depends on the surface area, S and does not depend on the initial phase volumes.

The work, required to increase the surface per unit area by stretching the surface layer, is equal to

$$dW_2 = \sigma_2 \, dS, \tag{2.5}$$

where σ_2 is the surface tension.

In this process, the number of atoms forming the surface should be unchanged and the average distance between them on the surface increases with stretching.

Values σ_1 and σ_2 are bound by the following relationship:

$$\sigma_2 = \sigma_1 + S(\partial \sigma_1 / \partial S).$$

It should be noted that the conditions, corresponding to the process described by Equation (2.5) are seldom satisfied, because the arrival of new atoms from the liquid bulk always occurs simultaneously with the stretching of the surface layer. The driving force for the transition of atoms or molecules from the bulk to the surface is the gradient of chemical potential. If atoms' arrival to the surface layer proceeds at a speed sufficient for their surface density to remain constant, the numerical values of σ_1 and σ_2 become equal, as in this case, $\partial \sigma_1 / \partial S = 0$. Such a behavior is characteristic of liquids and in a number of cases of solids at high temperatures and long time of measurements. Equality of values σ_1 and σ_2 for liquids is the reason for identification of the notions of surface tension and surface energy.

2.1.3 Interphase Tension

Interfaces of any two contacting bodies, which may be both liquids and solids, also have free surface energy. In this case, the excess of free energy is most often called the interphase free energy or interphase tension. The value of the interphase tension is determined by a predominant attraction of particles of the boundary layer by one of the contacting phases. The value of the interphase tension will be the higher, the stronger is the asymmetry of force fields of the interacting phases.

Formation of molecular bonds on the interphase of the contacting phases requires work to be done for their separation. This work, called the work of adhesion, in the case of contact of two immiscible liquids is equal to the sum of surface tensions of the contacting liquids minus the interphase tension on the interphase and is given by Dupre equation:

$$W_A = \sigma_A + \sigma_B - \sigma_{A-B}, \qquad (2.6)$$

where σ_A and σ_B are the surface tension of phases A and B, respectively; σ_{A-B} is the interphase tension between phases A and B.

Work of adhesion cannot be determined directly, it can just be calculated by Equation (2.6), having first found values σ_A, σ_B, and σ_{A-B}. In the case of breaking up of a homogeneous liquid, the work of new surface formation, called the work of cohesion, is calculated by the following formula:

$$W_K = 2\sigma_A. \qquad (2.7)$$

Therefore, the work of adhesion is a measure of interaction of the particles on the boundary of the two different phases or bodies, and the work of cohesion characterizes the forces of the bond inside the body.

The notions of adhesion and cohesion are applicable both to liquid and solid bodies. A knowledge of these quantities is often highly important, as they are a measure of the ratio of the forces of attraction between the molecules of an individual phase and between the molecules of different phases. In particular, wetting of a solid with the melt, largely responsible for the quality of the welded and brazed joints, is connected with the ratio of the melt cohesion and melt adhesion to the solid.

When a liquid penetrates onto the surface of a solid, the liquid may spread or may not spread over the solid, depending on the surface tension of the phases and the interphase tension on the contact boundary. At the spreading of the liquid, the total change of the surface energy per unit surface area is equal to

$$dF = \sigma_{l-g} + \sigma_{sol-l} - \sigma_{sol-g}, \qquad (2.8)$$

where σ_{l-g} is the surface tension of the liquid; σ_{sol-l} is the interphase tension on the solid–liquid boundary; σ_{sol-g} is the surface tension of a solid. It is seen that the free energy of the system decreases only if the following inequality is fulfilled:

$$\sigma_{sol-g} > \sigma_{l-g} + \sigma_{sol-l}. \qquad (2.9)$$

Therefore, the spreading process proper is only possible if the inequality represented by Equation (2.9) is satisfied. Incorporating the Dupre equation,

Equation (2.9) may be written as follows:

$$W_A > 2\sigma_{1-g}. \tag{2.10}$$

From Equation (2.10), it follows that the intensity of the liquid spreading over the solid surface increases with the increase of the liquid adhesion to a given body and with lowering of cohesion of the liquid.

The so-called coefficient of spreading, $k_p = W_A - W_K$, is very often used to evaluate the intensity of the liquid spreading, and with the greater the value of this coefficient, the better the spreadings.

In the contact of a liquid with a solid (Figure 2.1) the magnitude of adhesion is determined from the following equation:

$$W_A = \sigma_{1-g}(1 + \cos\theta), \tag{2.11}$$

the intensity of spreading may also be assessed by the value of wetting angle θ, which is the most important characteristic of the interphase of the three phases.

Value of the equilibrium angle of wetting is determined by the ratio of values of the specific free energies:

$$\cos\theta = (\sigma_{sol-g} - \sigma_{sol-1})/\sigma_{1-g}, \tag{2.12}$$

or

$$\cos\theta = (W_A/\sigma_{1-g}) - 1. \tag{2.13}$$

It is seen that wettability will be good, if $\cos\theta$ is greater than zero, that is in the case, when $\theta < \pi/2$. However, the magnitude of θ depends not only on the ratio of W_A and σ_{1-g}, but also on the condition of the surface of the solid (roughness, presence of contamination, etc.). Therefore, in the determination of θ, it is necessary to take all these factors into account; also, it must be remembered that in spreading, the value of θ usually differs from the equilibrium value.

The spreading of one liquid over the surface of another, similar to the process of a liquid spreading over the surface of a solid, is related to the work of adhesion on the boundary between the liquids and the work of cohesion of the spreading liquid.

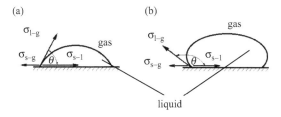

FIGURE 2.1 A drop of liquid over the surface of the solid: (a) good wettability; (b) poor wettability.

A necessary condition for spreading of a liquid (A) over the surface of a liquid (B) is fulfillment of the following inequality:

$$\sigma_B \geq \sigma_A + \sigma_{A-B}. \tag{2.14}$$

Therefore, in a liquid's spreading over the surface of another liquid, the intensity of spreading rises with the increase of the work of adhesion and reduction of the work of cohesion of the spreading liquid.

Difference in the behavior of the wetting and unwetting liquids consists in that the wetting liquid demonstrates a tendency to increasing the area of contact with the contacting body, and an unwetting liquid — to a reduction of this area. This is related to the fact that in a wetting liquid, the energy of the surface layer, contacting the solid, is smaller per unit area, than the energy of the liquid surface layer, not contacting the solid. For the unwetting liquid, this ratio is inverse.

2.1.4 THE ADSORPTION PROCESS

Difference in the magnitude of surface energy per unit surface area of the liquid, is related to the fact that in addition to the attraction forces, acting from the side of the liquid inner layers, the molecule of the liquid surface layer is also under the impact of the forces of attraction or repulsion from the side of the molecules of the solid surface. Therefore, the resultant force acting on the molecules of the liquid surface layer, adjacent to the solid may be either smaller or greater than the forces acting on the molecules of the liquid surface layer, not contacting the solid. However, this depends upon the direction and magnitude of the forces, acting from the side of the molecules of the solid.

If the liquid incorporates molecules of just one kind, then the free surface energy decreases, because of surface reduction to a minimum. However, if several kinds of molecules are present in the liquid, which differ by the force field intensity, then lowering of its surface tension proceeds in a different manner. Molecules with stronger fields are trying to pass from the liquid surface into its volume with a greater force, than the molecules with less–stronger fields. As a result, the content of components characterized by weaker fields of molecular forces of attraction is higher in the surface layer of the liquid than in the solution bulk.

Such an increase of the content of any component of the solution or gas phase in the surface layer is called adsorption.

Adsorption occurs in each case of contact of the surface with gas or liquid. It proceeds at the contact of a gas with a solid or a liquid, at the contact of a liquid with a solid or another liquid, and in some cases even at the contact of two solids. If the component content in the surface layer is greater than in the solution bulk, the adsorption is called positive, and if it is smaller, the adsorption is called negative.

As the value of surface tension of a body is related to the composition of the surface layer, it depends on the magnitude of adsorption. A quantitative relationship between the magnitude of adsorption and change of surface tension yields the known

Gibb's adsorption equation

$$\Gamma = -(c/RT)d\sigma/dc. \qquad (2.15)$$

Value $d\sigma/dc$ is called surface activity of substance. A substance is regarded as surface active, if the surface tension of the liquid decreases with the increase of concentration of this substance in the solution, i.e. at $d\sigma/dc < 0$. In this case, $\Gamma > 0$, i.e. adsorption is positive. If the surface tension of a liquid rises with increase of the substance concentration in the solution ($d\sigma/dc > 0$), then $\Gamma < 0$, i.e. adsorption is negative, and this substance is surface inactive.

As is seen from Equation (2.15), the magnitude of adsorption is related to component concentration in the solution, and, while changing, it passes through an extremum, the adsorption being maximum, if $\Gamma > 0$, and minimum, if $\Gamma < 0$.

Determination of extreme adsorption is of considerable interest. For instance, knowing the magnitude of maximum adsorption Γ_{max}, it is possible to determine the thickness of the adsorption layer

$$\delta = \Gamma_{max} M/\rho_f, \qquad (2.16)$$

as well as calculate the area per one molecule in the surface layer:

$$S = 1/\Gamma_{max} N_o. \qquad (2.17)$$

Here M is the molecular mass of the adsorbed substance; N_o is the Avogadro constant; ρ_f is the density of the substance in the film.

Conducting such calculations allows determination of the form, in which a given component adsorbs on the solution surface.

However, in practice, the relationship between the substance concentration in the solution and the value of surface tension of the latter is often found without determining the value of adsorption. Studying the relationship between surface tension and substance concentration for fatty acids, B. Shishkovsky found that this link is rather accurately expressed by the following equation:

$$\sigma = \sigma_0 - RT \ln[1 + (F - 1)c_N]/S_n, \qquad (2.18)$$

where σ_0 is the surface tension of the pure solvent; F is the value dependent on the coefficient of component activity; S_n is the area taken up by a mole in the surface layer; c_N is the component concentration.

Later on, some researchers noted that for certain metal melts at the change of the alloying substance concentration in them, the change of the value of surface tension can also be defined by Shishkovsky type of equations. Actually, they are not always valid in the entire range of variation of the content of the alloying substance in the melt. However, if Equation (2.18) is valid for a given melt, at extreme adsorption, the surface tension of the solution σ and the surface tensions of the solvent σ_1 and component σ_2 are linked by the following dependence:[63] $\sigma = (\sigma_1 + \sigma_2)/2$, that is at the point of extremum the solution, surface tension is equal to the root mean square value of surface tensions of the solvent and the alloying component.

2.1.5 The Electrochemical Phenomena

Molten slags are solutions consisting of simple and complex ions.[64] Therefore, in welding and surfacing with application of fluxes or electrode coatings containing slag-forming components, the formed slags also are a complex of simple and complex ions. This postulate is confirmed by various facts, and, primarily by electric conductivity and electrolysis, the ability to design galvanic elements, in which the liquid slag is the electrolyte. All this is indicative of the fact, that molten slags are strong electrolytes. It is known however that at the contact of two conducting phases, a difference of electric potentials usually develops between them, which is directly related to the formation of an electric double layer. Three mechanisms of formation of the electric double layer may be active. First, its formation may result from the difference in the speeds of transition of the positively and negatively charged particles from one phase into another. Second, formation of the electric double layer is possible in the case, when the charged particles almost do not go from one phase into another, but have different adsorption ability. In this case, mostly particles of one sign concentrate on the interface which are exactly what determines the sign of the outer half of the double layer. And, finally, an electric double layer may form as a result of orientation on the liquid surface of neutral molecules, which contain electric dipoles.

At the contact of the molten metal and the slag the electric double layer forms as follows. Ions contained in the metal and the slag try to go from one phase into the other at the phases contact, and this transition is determined by the difference of the energies of the particle bond with both the phases. However, the transition of particles will disturb the electric neutrality of both the phases and since in the conductors excess charges are driven to the surface, the electric double layer will form and the associated potential jump will occur on the metal–slag interphase.

Interphase tension in the general case is determined by the energy of the chemical bonds of the surface particles of one phase with those of the other phase, van der Waals bonds, as well as the electrostatic interaction of excess charges in the double layer. Therefore, by varying the metal potential, using an external electric field, it is also possible to change the value of the interphase tension. Change of the value of interphase tension due to polarization of the boundary between the contacting phases is exactly the essence of electrocapillary phenomena. Graphical dependence between interphase tension on the molten metal–electrolyte boundary and the potential of the liquid metal surface is called electrocapillary curve.

Whereas at electrode polarization, the concentration of all the components, except for those, determining the potential, remains practically constant, change of interphase tension related to change of potential, is defined by the known Lippman equation:

$$d\sigma = -\varepsilon d\varphi, \qquad (2.19)$$

where ε is the charge density on the metal; φ is the metal potential.

In this case, the electrocapillary curve usually has the form of a parabola, with its vertex turned upward.

From Equation (2.19) it follows, that the slope of the curve, defining the dependence $\sigma = f(\varphi)$, allows determination of the density of the charge on the metal surface, with $d\sigma/d\varphi = 0$ at the point of maximum of the electrocapillary curve, that is the surface charge being zero. Differentiating Equation (2.19), we can find the capacitance of the electric double layer, C

$$d\varepsilon/d\varphi = d^2\sigma/d\varphi^2 = C. \tag{2.20}$$

It should be taken into account, that Equation (2.19) is only valid for systems which include molten salts or water solutions, characterized by small transition of particles through the interphase. Owing to the polarization of the metal melt, which is in contact with the oxide melt, not only the potential, but also concentration of ions in the near-electrode layers are changed. Since mutual solubility of phases in this case is quite significant, as indicated by large (up to 1000 A/cm^2) values of the exchange currents between the metal and oxide melt, change of ion concentration greatly affects the value of interphase tension in the metal–slag system. This is confirmed, in particular, by a higher sensitivity of the shape of electrocapillary curve to the concentration of particles, characterized by a high surface activity on the interphase.

In this connection, electrocapillary curves in systems with oxide melts should be defined, using the following equation:[65]

$$d\sigma = -\left[\varepsilon + F\sum_{i=1}^{k} n_i(\Gamma_{i\,m} + \Gamma_{i\,el})\right]d\varphi - \sum_{j=k+1}^{m}\Gamma_j\,d\mu_j, \tag{2.21}$$

where m is the total number of components in the system; k is the number of components, the equilibrium of which shifts with the change of potential; $\Gamma_{i\,m}$, $\Gamma_{i\,el}$ is the adsorption of ith component to the interphase from the side of the metal and electrolyte, respectively, F is the Faraday number; μ is the chemical potential of the component.

In Equation (2.21), the expression in the brackets represents adsorption of the components, the equilibrium of which shifts together with the potential, and the latter addend allows for the adsorption of the components, where the equilibrium is not related to polarization. Since in systems with oxide melts, the change of interphase tension is largely dependent on the presence of surfactants in the near-electrode layers, the shape of the electrocapillary curves for such systems usually differs from the parabolic one.[65]

It should be noted that electrocapillary phenomena, probably, have a substantial role in submerged-arc welding, coated-electrode arc welding, and electroslag processes. Therefore, it is necessary to study these phenomena and take them into account when developing the technologies of welding and surfacing, as well as welding consumables.

The above issues are considered in greater detail in the work by V.K. Semenchenko, N.K. Adam, Yu.V. Najdich, and other authors.[48,63,66–68]

2.2 DETERMINATION OF SURFACE PROPERTIES OF MELTS

Methods to determine the surface tension of liquids are divided into two groups: dynamic and static. Static investigation methods became the most widely used, due to a greater accuracy of the experiment.

From a rather large number of the known methods of surface tension measurement,[48,66] the methods of maximal pressure in a bubble, sessile drop, as well as the method of pendant drop and drop weight are the most widely used for high-temperature studies.[48,68,69] Over the last 15–20 yr, the method of a "large drop" and some other methods, the features of which are considered below, began to be used for the determination of surface tension of steel melts.

2.2.1 PENDANT DROP METHOD

This method is most often used in the determination of surface tension of refractory metals and oxides at their melting temperature.[69,70] In this case, a rod of the studied material, heated by various sources (in a furnace[71] or by an electron beam[72–74]), melts, and a drop forms at its tip (Figure 2.2). Determination of the dimensions of this drop allows calculation of the value of surface tension (σ_{m-g}) of the studied material

$$\sigma_{m-g} = \rho g d_1^2 / H_f, \tag{2.22}$$

where H_f is the function of the ratio of the diameter d_s (located at a height equal to maximum drop diameter d_1), to the maximum diameter; ρ is the density of the studied material; g is the acceleration of gravity.

Experiments on the determination of surface tension of some refractory metals and oxides showed, that application of the rod heating by the electron beam allows obtaining accurate results with this method of investigation.[73,74]

2.2.2 DROP WEIGHT METHOD

This method is characterized by the relative simplicity of conducting the experiment and sufficiently easy processing of the obtained results. During the performance of

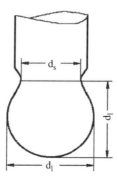

FIGURE 2.2 Measurement of surface tension by the pendant drop method.

investigations by this procedure, the drop detaches from the rod tip at the moment, when the forces of gravity slightly exceed those of surface tension. The value of surface tension is most often calculated by the following formula

$$\sigma_{m-g} = mgF/r, \qquad (2.23)$$

where m is the mass of the detached drop; r is the rod radius; F is the correction factor, which allows for the difference between the weight of the drop, formed at the rod tip, and that of the detached drop.

Drop weight method has been known for a long time, but lately it has been used only in some studies.[70,71,73] Although this method of determination of surface tension allows avoiding contact of the study material with that of the crucible (substrate), its application is limited by its drawbacks. These drawbacks include, first of all, the need to regulate drop formation and its detachment, as relatively accurate results can be obtained only with slow formation of the drop. In addition, accuracy of the results obtained by following this procedure, is influenced even by a slight vibration that occurred during the experiment. A significant disadvantage of the method is also the fact that it allows determination of the value of surface tension only at the temperature of melting of the studied material.

2.2.3 SESSILE DROP METHOD

The shape of the drop, lying freely on a substrate, is determined by the action of the forces of gravity and surface tension. Therefore, knowing the dimensions of the drop, it is possible to find also the value of surface tension. Despite the considerable labor consumption, this method has recently been rather widely used in research to study the surface properties of materials. This is due, primarily, to the following advantages of this method: high accuracy, ability to measure surface tension at temperature much higher than that of material melting; absence of wetting angle in the calculation formulas.

A drop, lying on a substrate of a refractory material, is photographed, and then its dimensions are measured on the negative or photo, namely, maximum diameter d and distance h from the drop plane having the maximum diameter, to the drop top (Figure 2.3). Knowing these dimensions, it is possible by tables or graphs, to find the magnitude of surface tension of the studied material. When the graphical method is

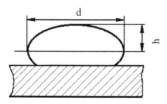

FIGURE 2.3 Measurement of surface tension by the sessile drop method.

used (which is much simpler), discrepancy in the determination of the surface tension value is not more than 1.5–2.5%.

Accuracy of measurement of surface tension of melts by this method becomes higher, if (Auth. Cert. # 1288550, USSR) in addition to maximum diameter, the radius of curvature in the drop top is also found.

Drop dimensions should be determined in the molten condition, as at metal solidification, not only the dimensions, but also the shape are changed, therefore, measurement of the dimensions of solidified drops may result in 20–30% or greater error in the determination of the surface tension. Error in the measurement of solidified drops will be particularly high, if the composition of the atmosphere contacting the molten metal includes gases, dissolved in liquid metal and evolving from it during solidification, for instance, nitrogen. In this case, pores form in the solidified metal drop, which leads to a rather considerable change of the drop dimensions. Dimensions of these pores, and hence, distortion of drop dimensions and possible experimental error (Figure 2.4) are dependent on the composition of the studied metal.

It should be noted that the results obtained depend in this method, also on the drop weight, substrate material, and method of producing it. So, by the data of Muu and Kraus,[75] substrate porosity should not exceed 30–35%, and drop weight, 5 g. In addition, by our data, the measurement accuracy depends on the initial shape of the metal sample and time of soaking of the molten drop at a specified temperature (Figure 2.5). According to Muu and Kraus,[75] it is recommended to use cylindrical samples with the diameter-to-height ratio, equal to unity.

2.2.4 METHOD OF MAXIMUM PRESSURE IN THE BUBBLE

Surface tension of the melt in this case is determined as follows. A capillary tube of a material, inert in relation to the melt, is lowered into the latter. Then the gas is fed into the tube, which results in a bubble being formed at its tip, and the pressure required for the bubble to form is measured. For a small diameter tube, it may be assumed that the forming bubble has a shape, close to a semi-sphere (Figure 2.6). Total pressure, at which the bubble forms at the tube tip, is made up of hydrostatic pressure P_h and pressure P_σ, which is determined by the forces of surface tension. Then,

$$P_h + P_\sigma = gh\rho + 2\sigma_{m-g}/r. \qquad (2.24)$$

(a)   (b)   (c)

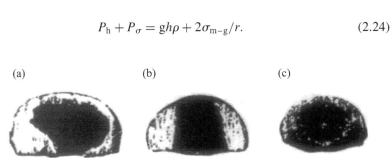

FIGURE 2.4 Appearance of the cross-section of metal drops after their melting and solidification in nitrogen atmosphere: (a) armco-iron; (b) Sv-08 steel; (c) Sv-08G2S steel.

(a)

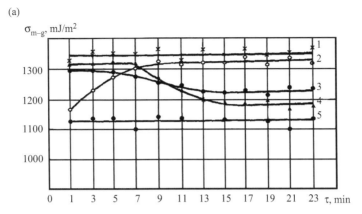

(b)

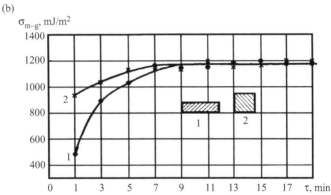

FIGURE 2.5 Influence on the time of soaking the metal melt – hydrogen system (a) and initial shape of the studied sample (steel U8A) in nitrogen (b) on the magnitude of surface tension, determined by the sessile drop method at $T = 1803$ K: [1] 08Kh20N10G6; [2] U8A; [3] Sv-08; [4] armco-iron; [5] Sv-10GS.

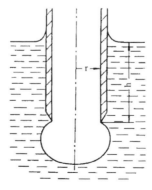

FIGURE 2.6 Measurement of surface tension by the method of maximum pressure in the gas bubble.

Therefore, the value of total pressure depends on the depth of the tube immersion into the melt, melt density, value of its surface tension, and the tube radius. If the liquid wets the tube material, the tube's inner radius is substituted into Equation (2.24), and if it does not wet the tube material, its outer radius is substituted.

The method of maximum pressure in the bubble is often used to measure the surface tension of the metal and the slag melts. The following may be regarded as the disadvantages of this method. With large tube diameters, (and at tube diameter of up to 1 mm, the tubes are clogged with the metal and get melted up), correction coefficients have to be incorporated into the calculation formula, which allow for the deviation of the bubble shape from the spherical one, as well as the impact of the gravity force on the bubble. If therefore becomes more difficult to perform the calculations. As follows from Equation (2.24), during the experiment it is necessary to exactly know the depth of the tube immersion into the melt. In addition, it is necessary to allow for (i) the change of the melt level in the crucible due to the melt ousting from the tube (ii) formation of a gas bubble at the tube tip, and (iii) the depression or capillary rise of the melt, if the distance between the tube and the crucible inner wall is comparatively small.

Neglecting these features in the determination of surface tension may lead to an error of several percent. Some of these factors may be eliminated, by applying crucibles of larger diameter of about 100 mm.[68,76-77] Depth of tube immersion into the melt may be ignored, if two tubes of different diameter are immersed simultaneously into the melt to the same depth (Sugden method). In this case, however, the change of the level of the liquid in the crucible still has to be taken into account.

Despite the above drawbacks, the method of maximum pressure in the bubble is quite often used for the measurement of surface tension of melts.

2.2.5 LARGE DROP METHOD

This method is characterized primarily by a high accuracy of measurement of up to 0.5%. In this case, the melt drop is not on a flat surface, but in a refractory cup (Figure 2.7). Drop symmetry is ensured by the crucible with sharp even edges,

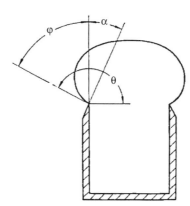

FIGURE 2.7 Measurement of surface tension by the large drop method.

which form a circumference, this being highly important for the accuracy of the experiment. The drop, protruding from the crucible, can be readily photographed, and then also measured. Wetting angle, θ depends on the degree of wetting of the crucible material by the melt; α is the angle of cut of the crucible edge. A stable drop forms in a broad range of values of angle φ, defined by the inequality $\alpha < \varphi < \theta$. And since α can be increased to $10-20°$, this method may be used in the determination of surface properties of melts, which wet the crucible material well.

2.2.6 METHOD OF METAL MELTING IN THE PENDANT STATE

The method may be used only for the determination of surface tension of metal melts, and it is based on the interrelation between the frequency of oscillations of a metal drop and surface tension.[78] Deviation of the drop shape from the equilibrium spherical one is recorded by a high-speed camera. Knowing the drop weight m and frequency of oscillations ω, the value of the metal surface tension is determined by the following formula: $\sigma_{m-g} = 3\ \pi\omega^2 m/8$.

In order to reduce the influence of gravity forces on the drop shape, it is recommended to conduct the tests on drops of $0.5-1$ g weight. This method has the following advantages: (i) absence of molten metal contact with the crucible (substrate), which allows preservation of the initial metal composition during the entire experiment; (ii) absence of metal density or wetting angle in the calculation formula; (iii) possibility to determine the value of surface tension of metals at high temperatures. However, two factors make it more difficult to use this method to determine the surface tension of molten metals they are: (i) the complexity of the unit required for a synchronous filming of the drop in two projections, normal to each other; (ii) the possibility of the drop rotation as a result of non-uniformity of the electromagnetic field of the inductor.

2.2.7 METHOD TO DETERMINE THE SURFACE TENSION OF A METAL IN THE ZONE OF THE ARC ACTIVE SPOT

This method was developed by us (Auth. Cert. # 1583793, USSR) specifically to determine the surface tension of the metal in the zone of the arc active spot. This is important, as it is exactly the part of weld pool, where active interaction of the molten metal with the gas phase occurs, and base-metal penetration depends on the value of surface tension in this zone. The experimental procedure in this case is as follows.

Crucible (2) with the solid prior remelted metal being studied is placed into the working space of furnace (1) (Figure 2.8) under the nonconsumable electrode (4). The nonconsumable electrode is lowered, until it contacts the surface of the metal in the crucible, and then is lifted by a microscrew to the height, H_0, the value of which is registered with the accuracy of hundred fractions of a millimeter, and usually in the range of $3-4$ mm. Then, the working space of the furnace is purged with argon or any other shielding gas. Shielding gas's flow rate remains constant during the experiment. After creation of the specified shielding atmosphere in the working space of the furnace, the oscillator (8) is used to excite an arc discharge

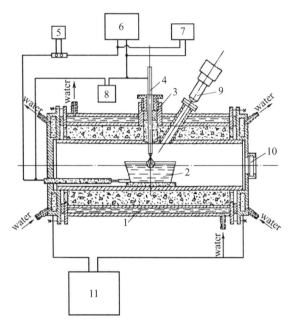

FIGURE 2.8 Schematic of the measurement of the surface tension of metal in the zone of the arc active spot (1) furnace; (2) crulible; (3) welding herd; (4) nanconsumable electrode; (5) ammeter; (6) current source; (7) oscillagraph; (8) oscillator; (9) pyrometer; (10) viewing window; (11) furnace power source.

between the solid metal, which is in the crucible (2), and the nonconsumable electrode. After 0.1s, when the arc voltage becomes constant, the oscillograph (7) records the value of arc voltage U_o at a given welding current. Current value should not be higher than 200 A for an electrode of 2 mm diameter and 250 A for an electrode of 3 mm diameter.

Then the current is switched off and the arc is interrupted. The value of voltage drop per unit arc length is found from U_o/H_o ratio. After this, furnace power source (11) is switched on and the temperature in the furnace rises up to the melting temperature of the studied metal plus $10–20°$. For the equalizing temperature in the metal melt, the system is soaked at this temperature for $30–60$ s. Then at the same value of H_o between the electrode and the melt, an oscillator is used to excite the arc. In this case, the current value remains the same, as in the case of arc excitation between the solid metal and the electrode. Arc pressure, the magnitude of which depends on the established arc current, leads to the formation of a crater in the melt, the depth of which depends on the value of surface tension of the metal melt. Having measured the arc voltage in the presence of the crater, the crater depth may be found from the following expression:

$$h = H_o[(U_1/U_o) - 1],$$

where h is the mark depth; U_1 is the voltage in the arc, running between the electrode and the metal melt.

Surface tension of the metal melt in the zone of the arc active spot is given by the following formula:

$$\sigma_{m-g} = (P_a - \rho g h) r_a^2 / 2h. \tag{2.25}$$

Here P_a is the arc pressure; ρ is the density of the metal melt; g is the acceleration of gravity; r_a is the active spot radius on the metal melt surface.

In addition to the above methods, surface tension of molten metals and slags is sometimes determined, using the methods of (i) capillary rise, (ii) ring break off from the melt surface, (iii) drawing-in of a vertical plate (cylinder), (iv) determination of the meniscus shape, and (v) solid material dispersion under the impact of the high-temperature gas flow (Auth. Cert. #1242767, USSR). However, all these methods are used rather seldom.

It should be noted that optimization of calculations in the determination of surface tension of melts by the methods of pendant or sessile drop allows[79] improving the accuracy of measurement of σ_{m-g} value.

Several works are known, where the authors suggested determination of surface tension of molten metal by the shape of weld pool, which is in the overhead position,[80] in the shape of the deposited bead,[81] or in the shape of a butt-root weld in gravity welding.[82] In these cases, however, in addition to the forces of gravity and surface tension, the molten metal is also under the impact of the forces of arc pressure, electrodynamics, gas flow pressure, etc. It is rather difficult to take all these additional forces into account and to accurately determine their magnitude. In addition, measurement of geometrical dimensions, required for the calculation of surface tension is performed on the solidified metal, which leads to an additional error. Therefore, in the application of these methods, the accuracy of determination of surface tension is very low. This is confirmed, in particular, by a considerable scatter of the experimental data obtained with the proposed procedures, which is equal to 15–20% and more.

2.3 METHODS OF MEASUREMENT OF THE SURFACE TENSION OF SOLIDS METALS

In most of the experimental methods for the determination of the surface tension of solids, the high mobility of atoms, creating the conditions, under which the surface effects may be manifested, is achieved by heating the solids up to their melting T_m and usually point to $(0.8–1.0)T_m$. The thus obtained values of surface tension may be extrapolated to low temperatures due to a constant temperature coefficient of surface tension. Without considering all the methods for the determination of surface tension of solids, let us just dwell on those, which are most often used in practice.

2.3.1 METHOD OF MULTIPHASE EQUILIBRIUM

This method is a modification of the sessile drop method, which is often used to determine the surface tension of metal melts. In the determination of surface tension of solids by this procedure, experiments are conducted as follows.

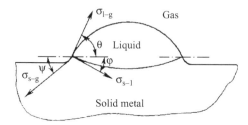

FIGURE 2.9 Schematic of measurement of surface tension of solid metal by the method of multiphase equilibrium.

The studied solid metal acts as a substrate for a drop of substance, chemically neutral in relation to the solid metal. Long-term annealing of the sample at a high enough temperature $(0.6-0.9)$ T_m provides an equilibrium shape of the interfaces. Equilibrium of the surface tension forces (Figure 2.9) for this system will have the following form:

$$\sigma_{1-g} \sin \theta = \sigma_{s-1} \sin \varphi + \sigma_{s-g} \sin \psi,$$
$$\sigma_{1-g} \cos \theta + \sigma_{s-1} \cos \varphi = \sigma_{s-g} \cos \psi. \tag{2.26}$$

Here σ_{1-g}, σ_{s-g} is the surface tension of the drop and the solid, respectively; σ_{s-1} is the interphase tension on the boundary between the liquid and solid.

At a known value of σ_{1-g}, the system described by Equations (2.26) is a closed one and its solution allows finding the values σ_{s-g} and σ_{s-1}. The value of wetting angle, θ may be determined during annealing. The values of the angles, φ and ψ can be measured after the cooling of the system, having first removed the drop from the substrate or a slice of the substrate cut up together with the drop.

This method is quite simple, however, the error in the determination of σ_{s-g} value may reach 30% and even 50%, because of (i) the influence of the liquid drop vapors on the properties of the solid and (ii) the errors related to measurements of small values of the angles, φ and ψ.[83]

2.3.2 Method of "Zero Creep"

This method, proposed by Tamman and essentially improved by Udin,[84] is quite often used for the determination of surface tension of solid metals. It essentially consists in the following: under the impact of forces of surface tension, the area of a solid surface is reduced, as the free surface energy of a system decreases. A similar phenomenon is also observed in the heating of metallic strips or wires of the thickness of up to several tens of micrometers, to temperatures, close to that of meting, when the heated samples become shorter and thicker. If we take several such strips and before heating attach to them, loads of an increasing weight, then upon heating, the samples with smaller loads will contract, and those with greater loads will become longer. By this way, it is possible to select a load, P_0, which will compensate the action of the surface tension forces and to reduce the creep rate of the studied material to zero, i.e. reach the zero creep condition. It is, however, rather difficult to select such a load. Therefore, value P_0 is usually determined by plotting a "deformation rate–load" graph from the results of testing several samples with different loads. Knowing the value of P_0, the value of surface tension of the studied solid is given by the following expression:

$$\sigma_{sol-g} = 2P_o/\pi d, \tag{2.27}$$

where d is the wire diameter.

V.S. Kopan',[85] suggested another variant of the method of zero creep, when the sample is not located vertically, but is lying in an unrestrained horizontal position, resting on two supports. In this case, a dependence exists between the magnitude of sample deflection and tensile load applied on to the sample, which is described by the following equation:

$$P_o = ql^2/8\lambda, \tag{2.28}$$

where q is the weight of unit sample length; l is the sample length between the supports; λ is the magnitude of deflection.

Then, σ_{s-g} is found from the following expression:

$$ql^2/8\lambda = \pi r\sigma_{s-g} - n\pi r^2\sigma', \tag{2.29}$$

where r is the radius of the wire sample σ_{s-g} is the surface tension of the studied metal; σ' is the intergranular surface energy.

The method of zero creep is usually associated with an error of not more than 10% in the determination of surface tension of a solid. In the report by Khokonov, Shebzukhova, and Kokov[86] a new procedure is proposed to measure the surface tension of metals in the solid state, based on the zero creep phenomenon. A feature of this procedure is the use of the compensation principle. This allows applying an additional controllable force to the sample and directly determining the magnitude of critical load P_0, which compensates the action of surface tension. Application of this procedure allows reducing the error in σ_{sol-g} measurement to 1%.

2.3.3 Method of Groove Smoothing

For the performance of experiments by this procedure, one or several grooves, running parallel to each other, are made on the polished surface of a metal sample along its axis. Groove width and their spacing may vary from several microns to several millimeters. After preliminary annealing, the samples are subjected to annealing for several days; the groove profile is determined at the start of the annealing and also during annealing at regular intervals for several hours. In this case, the tendency of the system to reduce the free energy, results in smoothing of the made grooves. The observed smoothing of the grooves at elevated temperatures may proceed due to: surface self-diffusion; bulk self-diffusion; evaporation and deposition of the substance; diffusion of metal vapors through the ambient gaseous medium. Surface self-diffusion prevails for the majority of the metals. If it is ensured that self-diffusion alone is the active mechanism of atoms transfer, then it is possible to determine the constant C_1 by the known equations,[87] its value being related to surface tension by the following dependence:

$$C_1 = \sigma_{s-g} D_1 V_m / kT, \qquad (2.30)$$

where D_1 is the coefficient of surface self-diffusion; V_m is the atomic volume; k is the Boltzmann constant; T is the absolute temperature.

Values of σ_{s-g}, obtained by this method are sufficiently accurate. Calculated formula for the determination of surface tension of solid metals with other mechanisms of groove smoothing are given below, in consideration of the method of thermal etching of grain boundaries.

2.3.4 METHOD OF AUTOELECTRONIC EMISSION

This method may be used to determine the surface tension of solids, capable of emitting electrons upon heating, i.e. metals. Investigations by this procedure are performed, using an electron microscope. The essence of this method consists in the following[83]: on heating a metallic sample, which is an electron emitter, the rounding-off radius of the emitter needle point becomes greater, this resulting in shorter length of the needle point. If high voltage is applied to the cathode, then the field intensity will be maximum at the needle point, decreasing in the direction of the cathode body. Electrostatic forces, arising in this case prevent mass transfer in the direction from the needle point to the body. If minimum stress F_0 is selected, at which no change of the needle point length occurs, then the surface tension of the cathode material can be found from the following equation:

$$\sigma_{s-g} = rF_o^2 / c\pi, \qquad (2.31)$$

where r is the radius of the emitter needle point; c is the coefficient dependent on cathode shape.

This method of determination of the surface tension of solid metals is characterized by the reliability of the obtained results, but requires application of an emitter-sample, which has no grain boundaries in the vicinity of its needle point.

2.3.5 METHOD OF THERMAL ETCHING OF GRAIN BOUNDARIES

In the determination of surface tension of solid metals by this method, a polished polycrystalline metal sample is heated in vacuum or in an inert atmosphere for several tens of hours. Under the impact of the surface tension forces, grooves form along the grain boundaries (Figure 2.10). Groove formation may result from the processes of diffusion, as well as evaporation and condensation. Measured

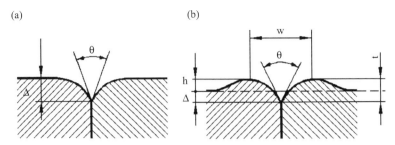

FIGURE 2.10 Appearance of grooves, formed during thermal etching of grain boundaries: (a) in metal evaporation and condensation; (b) as a result of diffusion processes.

groove dimensions are: width, w of forming the groove for diffusion processes; its depth, h; the angle, θ for evaporation and condensation.

In this case, the calculation formulas, depending on the mechanism of groove formation, have the following form[87]:

1. For surface self-diffusion

$$w_1 = 4.6(C_1\tau)^{1/4}; \tag{2.32}$$

2. For bulk self-diffusion

$$w_2 = 5.0(C_2\tau)^{1/3}, \tag{2.33}$$

3. For diffusion of metal vapors through the ambient medium

$$w_3 = 5.0(C_3\tau)^{1/3}, \tag{2.34}$$

4. For evaporation and condensation

$$\Delta = 1.13(C_4\tau)^{1/2}/\mathrm{tg}(\theta/2). \tag{2.35}$$

Processing of the experimental results starts with plotting lgw or $lg\,[\Delta\mathrm{tg}\,(\theta/2)]$ graphs, depending on $lg\,\tau$. If the derived dependencies correspond to the exponent in Equations (2.32) through (2.35), calculations are conducted by the respective formulas, given below:

1. For surface self-diffusion Equation (2.30) is used;
2. For bulk self-diffusion:

$$C_2 = \sigma_{s-g}V_mD_2/kT, \tag{2.36}$$

3. For diffusion of metal vapors through the ambient gaseous medium

$$C_3 = \sigma_{s-g}c_oD_3V_m^2/kT, \tag{2.37}$$

4. For evaporation and condensation

$$C_4 = \sigma_{s-g} P V_m^2 / (2\pi m_a)^{1/2} (kT)^{3/2}. \tag{2.38}$$

Here, C_1, C_2, C_3, and C_4 are the constants which are dependent on the mechanism of groove formation at thermal etching; V_m is the atomic volume; D_1 is the coefficient of surface self-diffusion; D_2 is the coefficient of bulk self-diffusion; D_3 is the coefficient of metal vapors diffusion through the ambient gaseous medium; P is the partial pressure of metal vapor above the flat surface; m_a is the atomic mass; c_o is the number of atoms per unit volume.

Equation (2.35) through (2.38) can be applied in those cases, wherein one of the mechanisms of atom transfer evidently prevails. Influence of individual mechanisms is established from the graphic dependence of w^3 on τ.[87] In case of a linear dependence, it is assumed, that the process proceeds due to bulk self-diffusion. A concave curve is indicative of prevalence of the mechanism of surface self-diffusion, and a convex one indicates the prevalence of the mechanism of evaporation and condensation.

In addition to the above methods of investigation the method of measurement of the rate of pore shrinkage, method of floating wedge, and some other are also used to determine the surface tension of solids.[83,87] These methods, however, are used rather seldom.

2.4 METHODS TO DETERMINE THE SURFACE ENERGY OF THE INTERPHASES

Fusion welding processes are characterized by the presence of systems, including molten metal and slag, as well as solid metal–metal melt. These are exactly the systems for which it is necessary to know the magnitude of surface energy of the interphases.

2.4.1 THE SESSILE DROP METHOD

Measurement of interphase tension in the metal–slag system at high temperatures is performed mainly by the sessile drop method in its two variants. In the first case, X-ray photography of the drop of the molten metal, coated by the slag, is used.[88] This method is used when the densities of the contacting phases differ noticeably from each other. Possibility of long-term soaking of the metal–slag system at a constant temperature allows studying the system in the equilibrium condition at any temperature, assigned in advance. Labor consumption of the method, complexity of selecting the crucible material, which would be equally resistant to the impact of slag and metal melt, are set off by the ability to obtain highly accurate results in the determination of the values of interphase tension.

The measurement on the metal drop in the solidified condition allows eliminating X-ray photography and considerably simplifies the experiment, but leads to significant errors, for the same reasons, as in the σ_{m-g} determination. Value of the interphase tension, determined by this method, is found from tables or graphs, as in the case of determination of surface tension by the sessile drop method, having first measured the required dimensions of the drops.

Another variety of this method consists in that the drop of the liquid slag is placed onto the molten metal surface.[89] The crucible material in this case should be resistant only to the metal, and regular filming may be used to determine the required dimensions of the drop. The disadvantages of this method are: (i) a short time of existence of the slag drop on the surface of the metal melt; (ii) need to measure the wetting angle, which leads to additional error; (iii) the method being of little use in studying interphase tension in systems with well-wetting phases.

2.4.2 THE DROP WEIGHING METHOD

In addition to the sessile drop method the drop weighing method is also used to determine the interphase tension in the metal–slag system.[90] A metal rod attached to scales is placed into a furnace, its lower end being lowered into the crucible with the molten slag. Knowing the weight of the detached drop, Equation (2.23) is used to calculate the value of the interphase tension. We suggested (Auth. Cert. #1571469, USSR), a variant of this method to determine the interphase tension of reactive metals on the boundary with high-active slags. Its essence consists in the following; the lower end of a rod of the studied metal is coated by a layer of flux, on the boundary with which the interphase tension is to be determined. In this case, the flux weight is 5–10% of that of coated part the rod. the thus prepared rod after drying in air is packed into the melting space of a the furnace. The rod end coated by flux is melted-off with formation at the rod end of a metal drop, coated by a thin layer of a flux. After the drop detachment, its weight is determined, and the value of interphase tension is found, using the Equation (2.23).

Application of this method of determination of the interphase tension allows elimination of interaction of the studied metals and slags with the crucible material.

2.4.3 TURNED CAPILLARY METHOD

In investigations of interphase tension of cast iron on the boundary with $CaO-Al_2O_3-SiO_2$ slag[91,92] the method of maximum pressure in the bubble with a turned capillary was used. A sintered corundum tube with a soldered-in capillary of the same material is lowered into a crucible with the molten metal. The molten metal and slag are inside the tube. Rarefaction is slowly created in the tube, which is followed by the metal drop penetrating into the measurement tube, overcoming the interphase tension on the metal–slag boundary. Knowing the value of rarefaction, P and capillary radius, r, the value of interphase tension is found from the following formula:

$$\sigma_{m-sl} = Pr/2. \tag{2.39}$$

This equation does not include the correction for the depth of immersion of the tube into the crucible, which is possible, if the pressure on the metal meniscus in the capillary from the slag side is balanced by the pressure of the metal in the crucible, i.e. $\rho_1 h_1 = \rho_2 h_2$, where h_1, h_2 are respectively the height of the meniscus (in the crucible and of slag in the tube above the metal meniscus in the capillary at maximum rarefaction in the measurement tube); ρ_1, ρ_2 are the densities of the metal and slag, respectively. This method, however, was only used in earlier works[91,92] on the determination of interphase tension in ferrous metal–slag system.

2.4.4 DROPS CONTACT METHOD

A method is proposed (Auth. Cert. # 409115, USSR) to determine the interphase tension, which consists in the following: two drops of the studied liquids with equatorial diameters are brought into contact and the contact angle, β between them is measured. Value of the interphase tension is calculated by formula

$$\sigma_{m-sl} = \sqrt{\sigma_{m-g}^2 + \sigma_{sl-g}^2 + 2\sigma_{m-g}\sigma_{sl-g}\cos\beta}. \qquad (2.40)$$

As is seen from the calculation formula, when interphase tension is determined by this method, it is necessary to first find the values of surface tensions of the contacting liquids, which, certainly, makes this method of measurement of interphase tension more complicated.

It should be noted, that under the conditions of fusion welding, in view of the short time of contact of the molten metal and slag, the equilibrium condition in the metal–slag system, as a rule, is not achieved. Therefore, the welding processes are characterized by dynamic interphase tension, which is smaller, than the static value and is determined by chemical interaction between the metal and slag or transition of the components through the interphase.

2.4.5 X-RAY PHOTOGRAPHY OF A SESSILE DROP

The value of dynamic interphase tension may be determined by the method of x-ray photography of a sessile drop. A simpler method[93] is the one, which allows finding the difference between the static and dynamic interphase tensions $\Delta\sigma_{m-sl}$. In this case, the electric contact method is used to measure the height of metal drop in two successive conditions. Here, $\Delta\sigma_{m-sl} = g\Delta\rho\,(H_1^2 - H_2^2)/2$, where g is the acceleration of gravity; $\Delta\rho$ is the difference of the densities of metal and slag; H_1, H_2 are respectively the height of the metal drop at the initial and final moments. Now the absolute value of the dynamic interphase tension may be determined by the method of a rotating crucible.[93] In this case, filming is used to record the change of the slag lens diameter on the metal-pool surface, and calibration graphs are applied to determine the area of phase contact and interphase tension at each moment of time.

Determination of the interphase surface energy on the boundary of solid phase–native melt is necessary, first of all, to study the process of solidification of the weld-pool metal, as the size of the critical nuclei of the new phase and the progress of the solidification process depends on the value of σ_{sol-sl}.

Certain data has been now obtained on the value of interphase surface energy on the solid metal–melt boundary. In this case, overcooling and different modes of lowering of the melting temperature of fine particles, compared to massive samples, are mainly used to determine the value of σ_{sol-1}. The thus found experimental σ_{sol-1} values are ambiguous, which drawback makes their practical application rather difficult.

2.5 WETTABILITY, SPREADING, AND ELECTROCAPILLARY PHENOMENA

2.5.1 GENERAL CONSIDERATIONS

In the process of liquid spreading over the surface of a solid, wettability of solids by melts is usually studied by applying side filming and vertical recording of the melt drop on the substrate surface. When side filming is used, it is also possible to record the change of contact angles in time, alongside with the change of the drop diameter. In either case, a drop of the melt is deposited on the surface of a preheated solid substrate, both usually having the same temperature.

Sometimes, wettability of the surface of solids is determined by placing a drop of liquid between the plane-parallel surfaces of two solid samples of different compositions. This allows saving the time taken up by the experiment, while the procedure remains the same.

In all the cases, investigations are conducted under the conditions of equal temperature of the solid and liquid phases and absence of temperature gradient in the substrate, i.e. under isothermal conditions. In fusion welding in the presence of a welding heat source (gas flame, electric arc, and electron beam), the processes of wetting and spreading proceed under different conditions. First, the temperature of the substrate (base metal) is below that of the metal melt, that is contacting the base metal. Second, the base metal is non-uniformly heated, that is spreading and wetting occurs at a temperature gradient in the substrate. Finally, the molten metal is affected by the welding heat source, which will also change the conditions of wetting and spreading due to the impact of arc pressure, appearance of reactive forces as a result of metal evaporation in the active spot zone, etc.

2.5.2 A MODERN UNIT FOR STUDYING THE WELDING PHENOMENA

In this connection, it must be mentioned that standard procedures and experimental units, which are used to study the wettability and spreading under isothermal conditions, are unsuitable, as they do not reproduce the conditions, characteristic of the actual welding processes. Therefore, we have developed a unit[94] to study the processes of wetting and spreading under the conditions, close to the actual ones. Schematic of the unit and its appearance are shown in Figure 2.11 and Figure 2.12. The unit chamber consists of a water-cooled case (1), closed from both sides by covers (2) through seals (12). The covers are tightly pressed to the case, using pinch bolts (11), which provides the required air-tightness of the chamber. One of the covers has a viewing window (3), which allows visual

FIGURE 2.11 Appearance of the unit to study the wettability and spreading of metal melts in welding.

observation of the processes of wetting and spreading of the melt over the solid metal, as well as recording these processes by filming.

An electric furnace (4), powered by a transformer is used for independent heating of substrate (6) up to the required temperature. Power is supplied through vacuum electric lead-in (5). The device also permits heating the substrate by passing current, supplied from a controllable current source.

Welding head (9), mounted in a replaceable assembly and fitted with correction device (8), is used to melt metal sample (7). In order to provide a guaranteed gaseous medium, a possibility is envisaged to create vacuum in the chamber, using an inserted nozzle (14), connected to a vacuum system. Temperatures of the melt drop, substrate, and working space of the chamber are monitored, using thermocouples.

The device operates as follows. Before conducting the investigations, a substrate of the studied material is mounted in the furnace (4). The required method of substrate heating is selected, based on the purpose of investigation. After that, the chamber is closed with the covers, which provide the air-tightness of the working space.

Investigations may be conducted either in a controlled atmosphere, or in vacuum. When investigations are performed in a controlled atmosphere, first a vacuum of 10^{-2} mm Hg is induced, and then the required gas is fed (argon, helium, CO_2, air, or a gas mixture).

The process of wetting and spreading of the metal melt over the surface of the solid is photographed using a camera, which is outside the device. Filming is conducted through a viewing window (3). After completion of the experiment, the

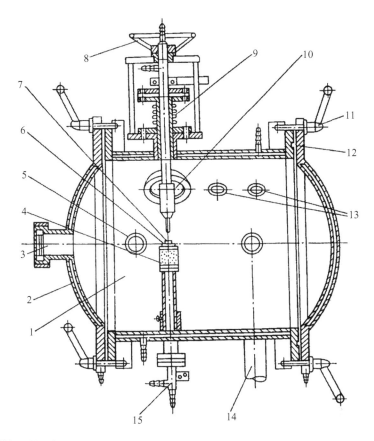

FIGURE 2.12 Diagram of the unit for investigation of wettability in welding: (1) water-cooled case; (2) covers; (3) viewing window; (4) electric furnall; (5) electric lead; (6) substrate; (7) metal sample; (8) correction device; (9) welding herd; (10) viewing window; (11) pinch bolts; (12) seals; (13) input of thermocouple; (14) nozzle; (15) supply of a current.

chamber is cooled to room temperature, then the covers are opened and the studied sample is taken out.

Use of this unit allows establishing the influence of the welding heat sources on the processes of wetting and spreading and on the obtention of data, which may be used for the development of technologies of welding and surfacing of the most diverse materials.

2.5.3 Studying the Electrocapillary Phenomena

When the electrocapillary phenomena in the ferrous metal–slag system were studied, the method of the turned capillary and the sessile drop method, complemented by x-ray photography were used.

In the sessile drop method, the experiment was performed as follows. A metal drop, which was a polarizable electrode, was placed on a substrate with a hole for the electrode, through which the electric current is applied to the drop. An auxiliary electrode was brought into the slag from above, this electrode being enclosed into a corundum tube to avoid contact of the side surface with the slag. The current, flowing through the electrolytic cell, formed by the studied electrode (metal drop), auxiliary electrode, and electrolyte (slag), changed the drop potential.

In this case, the measurement cell can be of any design, but it should provide a uniform distribution of current over the surface of the polarizable electrode. For the electrolytic cell, shown in Figure 2.13, the shape and area of the auxiliary electrode surface provided a uniform polarization of the molten metal drop.

Storage batteries are usually used as the current source. The drop potential is measured with a reference electrode, through which the current does not run. The reference electrode is placed into the same crucible, where the metal drop is, or into an additional compartment. The slag, present in this compartment, should not contain any oxides of the metal being studied.

The first works on electrocapillary phenomena were studied by the sessile drop method.[95,96] The drop potential was measured at the moment of current passage through the electrolytic cell, which led to a certain error in the obtained results. Subsequent studies[65,97–98] used the switching method of drop potential measurement. The switch converts the constant polarizing current into an intermittent current. When the polarizing current is switched off, the switch automatically closes the measurement circuit, and the potential measurement is performed, when there is no polarizing current in the circuit. In this case, the ohmic drop of voltage in the

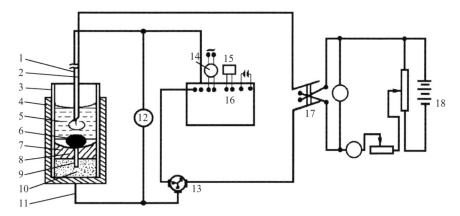

FIGURE 2.13 Schematic of the unit to study electrocapillary phenomena in the melt metal–slag system by the sessile drop method[97]: (1) reference electrode; (2) auxiliary electrode; (3) corundum tube; (4) graphite crucible; (5) slag; (6) molten metal drop; (7) fused magnesium substrate; (8) silver bar; (9) graphite electrode; (10) carbon powder; (11) current supply to the metal; (12) oscillograph; (13) switch; (14) zero-galvanometer; (15) normal element; (16) potentiometer; (17) pole switch; (18) current source.

electrolyte is equal to zero, and does not distort the measurement results. When the switching method of potential measurement is used, the quick-falling fraction of polarization is measured with an oscillograph.

When the electrocapillary phenomena are studied by the turned capillary method (Figure 2.14), the measurement cell differs slightly from the unit for the measurement of interphase tension by this procedure. Just two electrodes (4,5) are added, when the electrocapillary phenomena are studied, the electrodes being used for passing the electric current through the electrolytic cell. One of the electrodes, designed for polarization of the molten metal, which is in the capillary, is immersed into the crucible with the metal (2); the other, protected by the corundum tube from the slag, is connected with the metal inside the tube. After a certain value of polarization has been achieved, rarefaction is gradually created inside the tube. The metal drop penetrates into the measurement tube (1), overcoming the interphase tension on the metal–slag boundary, which results in closing of the circuit. As the metal surface inside the tube is approximately 50 times larger than the metal surface in the capillary, the metal in the capillary is a poralizable electrode. The polarized electrode potential is taken to be the difference between the applied voltage, indicated by the voltmeter readings and the voltage drop across the measurement cell. The latter is equal to the product of the electrolytic cell resistance by the polarizing current. With this method, the role of the reference electrode is played by the metal, which is in the tube.

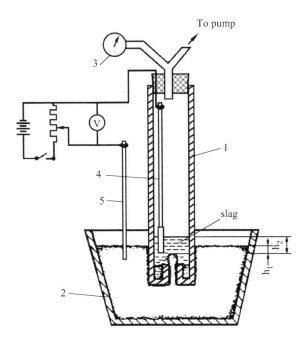

FIGURE 2.14 Investigation of electrocapillary phenomena by the turned capillary method: (1) measurement type; (2) crucible; (3) pressure gauge; (4, 5) electrodes.

2.5.4 A Comparison of the Different Methods

Each of the considered methods of studying the electrocapillary phenomena has its advantages and disadvantages. The turned capillary method is simpler to apply and does not require a sophisticated experimental set up. However, the fact, that the metal drop's surface area changes during the experiment, is a significant drawback of this method. In addition, the reference electrode here is the metal in the tube, which, even though to a small extent, is still subjected to polarization. This can lead to a certain error in the experimental results.

The disadvantage of the sessile drop method, besides the complexity of the experiment, consists in rather large metal drop surface, which demands application of higher current for its polarization. The advantage of the method is that the drop surface practically does not change during the experiment. In addition, when the sessile method is used, the entire system can be maintained until a constant potential is established in the electrode being studied, which is important, as it is generally very difficult to achieve at once a constant value of the potential, corresponding to a certain current density.[66]

3 Results of Studying the Surface Properties and Phenomena

As the surface properties and phenomena have an important role in the processes of welding and surfacing, it is necessary to know what factors and in what way influence their quantitative characteristics. This allows changing the values of surface and interphase tension of metals, as well as controlling the processes of wettability and spreading, which, in turn open up new possibilities in the development of the technologies for welding and surfacing.

3.1 SURFACE TENSION OF MOLTEN AND SOLID METALS

Surface properties of molten and solid metals have attracted the attention of researchers for a long time and a large number of theoretical and experimental works are devoted to their investigation. The first studies on the determination of surface tension of molten low-melting metals were performed more than hundred years ago. Measurements of surface tension of the melts of iron and alloys on its base were conducted by F. Saurwald[99] and his associates in 1929. However, the first study on the measurement of surface tension of molten metals in connection with solution of practical tasks, in particular, with deoxidizing and decarbonizing of metals, was performed later by 1944.[100–102] Since then a rather large number of works have been performed over a relatively short period, and the interest to studying this issue remains to be high.

3.1.1 INFLUENCE OF ALLOY COMPOSITION

In practical welding, one often has to deal not with pure metals, but with different alloys; therefore, the attention is now focused on the investigation of the influence of individual components on the surface tension of the melts. Experiments, performed by various researchers showed that the surface tension of metal melts depends largely on what elements and in what amount are present in them. Such research has been conducted for the most diverse metals. In this work, we will consider only the surface properties of those metals, which are most often used in metal-structure fabrication, more specifically, the influence of components on surface tension of iron, copper, aluminum, titanium, and nickel. Most of the results, given below, are obtained at the melt contact with inert gases (argon, helium) and less often at contact with hydrogen or under vacuum.

Several investigators have studied the influence of carbon on the surface tension of molten iron and have produced significantly different results. According to Esche and Peter, carbon is a highly surface-active element in Fe–C alloys,[103] and surface tension of iron is noticeably lowered with the increase of carbon concentration in the melt. Unlike these data, in other works,[104–107] it is found that carbon has an insignificant surface activity in iron, and with the increase of its content, the surface tension of iron indeed decreases, but only slightly. In this case, the value σ_{m-g} decreases monotonically.

Finally, the data of Halden and Kindgery showed that carbon does not show any surface activity in Fe–C alloys,[108] and increase of its concentration in the melt up to 4% does not change the value of surface tension of molten iron.

Our studies showed that carbon has an insignificant surface activity in iron, which coincides with the data of the early works.[104–107]

Thus there are different opinions on the surface activity of carbon in iron. However, the surface activity of sulfur in iron and iron-based alloys has not been refated. All the investigators who studied the influence of sulfur on the surface tension of molten iron, have noted its high surface activity.[106,107,109] However, the values of surface tension at the same concentrations of sulfur, derived by different researchers, are somewhat differ from each other. Dependence of sulfur surface activity in iron-based alloys on the presence of various impurities, for instance oxygen, in the melt has been found.[110]

Studies on the influence of phosphorus on the surface tension of iron showed it to be a surfactant, both in commercial grade iron[111,112] and in Fe–C[112] and Fe–O[113] alloys. Dependence of phosphorus surface activity on carbon and oxygen content in iron has been found. Increase of carbon content in the Fe–C melt lowers the surface activity, and increase of oxygen content in iron, contrarily, enhances the phosphorus surface activity.

A large number of studies are devoted to the investigation of the influence of various alloying additives (Si, Mn, Ti, Cr, etc.) on the surface tension of iron. Investigation of silicon influence on the surface tension of iron showed that silicon is a surfactant in binary Fe–Si alloys.[107,114,115] Manganese has a somewhat higher surface activity in molten iron.[107,109,115] Here both silicon and manganese surface activity changes in the presence of a third component in the melt.[109,110] Study of iron–nickel alloys showed[107,119,120] that increase of Ni content in the alloy enhances the surface tension of the melt. In the Ni–C–Fe ternary system, however, with a carbon content of approximately 1.6%, the increase of nickel content lowers the value of σ_{m-g}.[121] From the same data, in vacuum and on the boundary with the gas mixture of hydrogen and helium, cobalt addition to the Fe–C alloy decreases the surface tension of the melt at low cobalt concentrations; at cobalt concentrations above 12 wt.%, an increase of the alloy surface tension is found. In the ternary Fe–Ni–Co system without C, by the data of Dzhemilev, Popel', and Tsarevsky,[122] the increase of nickel and cobalt content in iron enhances the value of σ_{m-g}.

Investigation of the influence of titanium and vanadium on surface tension showed these elements to be inactive in iron and iron–carbon alloys.[123–125]

Vanadium addition to low-carbon steel also increases its surface tension.[71] Increase of tungsten and molybdenum content in iron, similar to titanium and vanadium, also results in a higher surface tension of the iron melt.[126,127]

Unlike Ti, V, Mo, and W, chromium is a surfactant in pure iron.[107,123,127] Addition of chromium to low-carbon steel, as shown by the data of Yakobashvili,[71] increases its surface tension. However, from the work of Baum, Gel'd, and Akshentsev it follows that chromium is a surfactant in an iron–carbon alloy with about 1% C.[128] Here, the surface activity of chromium in Fe–C–Cr system is higher than in Fe–Cr system. According to Ayushina, Levin, and Gel'd, aluminum also shows surface activity in iron melts.[129]

The surface activity of rare-earth elements in iron and steels was studied in connection with their use for steel alloying. It is found[130,131,132] that cerium and lanthanum are surfactants both in pure iron and in steels. Boron, by the data of Bobkova et al. and Pirogov et al. at up to 0.5 wt% content in steel and in pure iron, does not influence the melt's surface tension.[133,134] Surface tension of iron markedly drops only at a boron content of about 50 wt%.[135] In 12 KhMF steel, however, boron surface activity was noted[133] at its content of about 0.01 wt% in steel. Selenium and tellurium also have considerable surface activity in iron.[136,137]

3.1.2 THE INFLUENCE OF GASES

3.1.2.1 Hydrogen, Nitrogen, and Oxygen

Several investigations have been made on the influence of gases, soluble in metals (H_2, N_2, O_2), on the surface tension of the melts of iron or iron-based alloys. All these studies have indicated high surface activity of oxygen,[104,106–108,110] being the highest of all the elements studied. Actually, Halden and Kindgery found that at low concentrations, sulfur has a higher surface activity than oxygen.[108] On the contrary, at high concentrations, its surface activity is lower. This conclusion, however, is not confirmed by the work of other researchers. Besides, in view of the low electronegativity of sulfur, compared with that of oxygen, the sulfur atoms, penetrating into iron, should deform the electron cloud of the iron atoms to a smaller degree and, hence, should also cause a weakening of the interaction among the iron atoms than caused by oxygen. Therefore, oxygen should lower the iron surface tension more than sulfur does. In other words, at equal concentrations, sulfur should have a higher surface activity in the iron melt.

Studies on the influence of nitrogen showed nitrogen to be a surfactant both in pure iron and in binary iron-based alloys.[10,108,138,139] Here, somewhat differing results were derived. By the data of another study,[10] increase of Si, Mn, and C content in the iron melt lowers the surface activity of nitrogen. However, the data of Borodulin, Kurochkin, and Umrikhin[138] indicates that the presence of carbon in the melt not only does not lower, but even does increase somewhat the surface activity of nitrogen, compared to its surface activity in pure iron. Nitrogen surface activity is practically independent on carbon concentration.

Using the sessile drop method, the surface tension of Fe–O, Fe–N, and Fe–O–N melts at the temperature of 1823 K was studied.[140] In this work, it is established that both nitrogen and oxygen are surfactants in molten iron. Surface activity of nitrogen is reduced with the increase of oxygen content in the metal. The change of the value of surface tension of Fe–O–N melt with the change in content of nitrogen (c_N) and oxygen (c_o) in it is given by the following expression:

$$\sigma = 1990 - 318 \ln{(1 + 130c_o)} - (282 - 763c_o - 83152c_o) \ln{(1 + 45c_N)}.$$

As the dependence of nitrogen surface activity on the presence of a second component in iron[10,138,140] is established, investigation of nitrogen surface activity in multicomponent systems, encountered in practice, acquires special importance. However, there are only a few such works.[141–143] According to the data of Kreshchanovski, Prosvirin, and Zaletaeva,[142] increase of nitrogen concentration in Kh15N25M steel enhances its surface tension.

Information on hydrogen influence on surface tension of iron is contradictory. For instance, hydrogen does not demonstrate any surface activity in pure iron and in Fe–Mn alloy,[144] while being a surfactant in Fe–Si, Fe–S, and Fe–P alloys. In Fe–C alloys, hydrogen will be surface-active only at the temperatures below 1783–1823 K.[144] At higher temperatures, no influence of hydrogen on the surface tension of an iron–carbon melt was observed.

On the other hand, another group[145] found hydrogen to be a surfactant in armco-iron and Fe–C alloys at a temperature of 1848 K, and its surface activity to depend on carbon content in the melt. Increase of carbon concentration in iron lowers the hydrogen surface activity, and at a C content of 1.5%, hydrogen becomes inactive in iron. Similar to carbon, presence of silicon in iron also lowers the surface activity of hydrogen, and it is known already that at 0.5% Si in the melt, hydrogen does not change the melt's surface tension.[145]

In addition to the metal composition, the value of surface tension of the metal melt is essentially influenced by the composition of the gaseous medium contacting the metal. Studying the influence of the gaseous medium on the value of σ_{m-g} is particularly important for welders, as most diverse shielding gases are used in welding and surfacing, namely, inert gases (argon, helium, and their mixtures) and active gases (carbon dioxide, air, water vapor, $Ar + CO_2$, etc.). Determination of the values of surface tension of metal melts in different gaseous media may provide a better insight into the process of pore formation. In addition, such data are required in the development of technologies for welding and surfacing, as well as advanced welding consumables.

In this connection, we have used the sessile drop method to determine the values of surface tension of some steels, often applied in welding fabrication, in argon and helium. It appeared that argon replacement by helium practically does not influence the value of surface tension of the metal melt. Table 3.2 gives the values of σ_{m-g} for the studied materials only in argon.

Investigation of surface properties of metal melts in nitrogen and hydrogen showed that argon replacement by hydrogen led to lowering of σ_{m-g} value for armco-iron and Sv-08 and St3 steels. For Sv-10GS and Sv-08G2S steels, the surface tension was approximately the same both in argon and in hydrogen. With replacement of argon by hydrogen the value of surface tension increased for U8A and 08Kh20N10G6 steels.

Determination of surface tension of armco-iron and the above steels in nitrogen atmosphere and comparison of the obtained data with the value of surface tension of the same metals on the boundary with argon showed, that σ_{m-g} value decreased in contact with nitrogen for all the studied metals, except for 08Kh20N10G6 steel.

3.1.2.2 The Behavior of CO

In addition to studying the influence of nitrogen and hydrogen on the surface tension of steels, we have also conducted experiments to determine σ_{m-g} value in CO atmosphere. Such investigations are of considerable interest for welders, as this gas participates in the process of pore formation in welds. Furthermore, in submerged-arc and covered–electrode welding, the composition of the arc atmosphere consists of CO by 89–93%,[146] that is, this gas may influence electrode-metal transfer and other processes, proceeding at the stage of the drop or the weld pool. However, such an investigation has not been conducted, as far as we know.

During the experiments with CO to create an atmosphere consisting of CO, the gas was passed through graphite powder, heated to 1823 K. A medium, consisting of just CO is provided.[147] Determination of surface tension of armco-iron and Sv-08, Sv-08G2S, Sv-10GS, St3, U8A, and 08Kh20N10G6 steels in CO atmosphere showed that value of σ_{m-g} is lower in carbon monoxide, than in argon for all the metals. This can be explained as follows.

It is known[148] that carbon monooxide diffuses into iron, as carbon and oxygen, which form as a result of dissociation of a CO molecule. The dissociation occurs because the iron crystal lattice cannot accommodate the larger CO molecule. However, only such gases may diffuse through the metal, which are adsorbed by the metal through activated adsorption, that is adsorption which is irreversible. The activated nature of carbon monoxide adsorption by the metal is confirmed by the increase of carbon content in the metal during the experiment (Table 3.1) at unchanged content of other elements (Mn, Cr, Si, etc.). In the study of Mazanek, an increase of carbon content in the metal, contacting was noted with CO at temperatures above 1073 K.[149]

Simultaneous consideration of Table 3.1 and Table 3.2 shows that lowering of the value of surface tension of the studied metals is most probably related to the change of carbon concentration in the melt. More intensive lowering of the surface tension of Sv-08 and St3 steels, compared to armco-iron, is attributable to a higher content of Si and Mn in these steels, where the surface activity grows with the increase of carbon content in the metal. Lowering of σ_{m-g} of the steels, Sv-08G2S and Sv-10GS in CO atmosphere is attributable to the increase of surface activity of S, Mn, and Si, and, possibly, also to the formation of a certain amount of silicon carbides. For U8A steel, in which the initial concentration of carbon is quite high, a certain increase of C content, due to carbonization of metal in contact with CO, does not significantly change the properties of the melt or the σ_{m-g} value.

TABLE 3.1
Initial and Final Concentrations of Carbon in Metals, Contacting CO (T = 1803 K; τ = 15 min)

No	Metal	$[C]_{in}$, wt%	$[C]_{fin}$, wt%
1	Armco-iron	0.05	0.29
2	Sv-08	0.09	0.34
3	Sv-08G2S	0.08	0.41
4	Sv-10GS	0.15	0.38
5	08Kh20N10G6	0.13	0.69
6	U8A	0.83	0.89
7	St3	0.18	0.38

Of all the studied metals, the value of surface tension of 08Kh20N10G6 steel changes in the most pronounced manner. The observed lowering of σ_{m-g} value at the increase of carbon content in steel is attributable to the formation of chromium carbides, which are driven to the metal surface due to a weaker bond with the melt, thus reducing the value of surface tension. This is confirmed by the data of Baum, Gel'd, and Akshentsev[128] where in the case of addition of 0.1% Cr to Fe–C alloy, containing 0.4% C, the surface tension of the melt, contacting hydrogen at 1843 K, was lowered by 160 mJ/m^2.

3.1.2.3 The Behavior of CO_2/Air

Since in welding fabrication, the shielding gas often is CO_2 and sometimes, also air,[1,23,150] we have studied the influence of these gases on surface tension of metal melts at short-time interaction of the metal and gas phase. We are not aware of any work to determine the surface tension of metals in CO_2, and the change of σ_{m-g} value of ferrous alloys containing silicon and phosphorus was studied, by feeding small portions of air into the chamber.[151] In this case, it was found that at feeding of the first portion of air, σ_{m-g} immediately dropped from 1180 to 700–800 mJ/m^2. Further loading of air did not lead to any change of the surface tension of a metal melt. Authors of Esche and Peter[151] attributed it to the formation of a slag film on the melt surface, which prevented new portions of oxygen from approaching the metal.

Formation of a slag film on the molten melt surface, contacting the oxidizing gas medium is noted in several works[152,153] In this case, it turned out that formation of such a film depends on carbon content in the metal. Experiments[153] and calculations[154] were performed to determine the critical concentration of carbon in the metal, at which no slag film forms on the oxidizing metal surface. Temperature influence on the value of critical concentration of carbon is shown in Figure 3.1. For armco-iron and the steels, namely, Sv-08, St3, Sv-10GS, and Sv-08G2S, the carbon content was not higher than 0.18%, which is approximately three times

TABLE 3.2
Average Values of Surface Tension and Temperature Coefficient of Metal Melts

Metal	Gas Medium	T, K	σ_{m-g}, mJ/m^2	$d\sigma_{m-g}/dJ$, mJ/m^2 K	Measurement Method
Cv-08	Ar	1803–2003	1279	−0.68	2
	He	1803–2003	1260	−0.64	2
	N$_2$	1803–2003	1039	+0.15	2
	CO	1803–2003	1170	−0.46	2
	H$_2$	1803–2003	1210	−0.43	2
Sv-08G2S	Ar	1803–2003	1120	−0.69	2
	He	1803–2003	1126	−0.78	2
	N$_2$	1803	1076	—	2
		1853	1041	—	2
		1903	1056	—	2
		1953	1040	—	2
		2003	1034	—	2
	CO	1803–2003	1022	−0.69	2
	H$_2$	1803–2003	1140	−0.125	2
Sv-10GS	Ar	1803–2003	1125	−0.7	2
	He	1803–2003	1133	−0.56	2
	N$_2$	1803	1097	—	2
		1853	979	—	2
		1903	1027	—	2
		1953	1007	—	2
		2003	986	—	2
	CO	1803–2003	1097	−0.56	2
	H$_2$	1803–2003	1120	−0.24	2
08Kh20N10G6	Ar	1803–2003	1234	−0.48	2
	He	1803–2003	1230	−0.5	2
	N$_2$	1803–2003	1272	−0.76	2
	CO	1803–2003	961	−0.75	
	H$_2$	1803–2003	1359	+0.145	2
U8A	Ar	1803–2003	1186	+0.25	2
	He	1803–2003	1176	+0.33	2
	N$_2$	1803–2003	1111	−0.41	2
	CO	1803–2003	1153	+0.35	2
	H$_2$	1803–2003	1314	+0.07	2
St3	Ar	1803–2003	1264	−0.63	2
	He	1803–2003	1261	−0.67	2
	N$_2$	1803–2003	1016	+0.14	2
	CO	1803–2003	1148	−0.52	2
	H$_2$	1803–2003	1201	−0.41	2

(*Table continued*)

TABLE 3.2 *Continued*

Metal	Gas Medium	T, K	σ_{m-g}, mJ/m^2	$d\sigma_{m-g}/dJ$, mJ/m^2 K	Measurement Method
08Yu	He	1823–1843	1395	—	—
10KhSND	He	1823–1843	1380	—	—
Steel 50	He	1823–1843	1280	—	—
ShKh-15	Ar	1823	1630	—	—
Steel 20	He	1823	1340	—	5
40Kh	He	1823	1370	—	5
18KhN3MA	He	1823	1470	—	5
30KhGSA	He	1823	1310	—	5
06Kh18N12	He	1823	1250–1440	—	5
G13L	He	1823	1130–1200	—	5
G13FL	He	1823	1250	—	5
VT1-00	Ar	1973	1521	—	3
	H$_2$	1973	1482	—	3
VT6	Ar	1973	1459	—	3
	H$_2$	1973	1394	—	3
VT14	Ar	1973	1462	—	3
	H$_2$	1973	1386	—	3
VT20	Ar	1973	1496	—	3
	H$_2$	1973	1428	—	3
Titanium(99)	Vacuum	1953	1588 ± 6	—	4
Aluminium	Ar	973–1573	867	−0.125	2
(99.99%)	He	973–1573	871	−0.123	2
	H$_2$	973–1573	866	−0.195	2
	Vacuum	973–1373	883	−0.127	2
Alloy AD-33	Ar	973–1573	851	−0.13	2
	He	973–1573	851	−0.142	2
	H$_2$	973–1573	840	−0.133	2
	Vacuum	973–1373	874	−0.12	2
Alloy AK5	Ar	973–1573	842	−0.13	2
	He	973–1573	848	−0.152	2
	H$_2$	973–1573	843	−0.157	2
	Vacuum	973–1373	868	−0.135	2
Copper M1	Ar	1373–2073	1258	−0.175	2
	N$_2$	1373–2073	1204	−0.206	2
BrKMts3-1	Ar	1373–2073	1126	−0.465	2
	N$_2$	1373–2073	1093	−0.52	2
Brass L90	Ar	1373	1310	—	2

Note: In the measurement of surface tension in the temperature range, σ_{m-g} values are given for the minimum temperature in the range. Methods of surface tension measurement: (1) maximum pressure in the bubble; (2) sessile drop; (3) drop weight; (4) pendant drop; (5) large-sized drop.

[C]cr, wt%

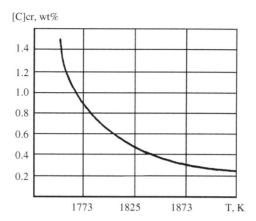

FIGURE 3.1 Temperature influence on the value of critical concentration of carbon in the metal. (From Stretstov, F.N., Perevalov, N.N., and Travin, O.V. *Theory of Metallurgical Processes*, Metallurgiya, Moscow, 1967, p. 45.)

smaller than the critical concentration of carbon, corresponding to a temperature of 1823 K.[154] Therefore, the slag film will form on the surface of these metals in contact with the oxidizing gas.

According to Medzhibozhsky,[153] in iron oxidation by oxygen or air, first iron oxide forms on the metal surface, which is then reduced to ferrous oxide, which is the base of the slag film. High capillary activity of oxygen on the boundary between metal and slag is noted in work,[155] and is confirmed by our studies, the results of which are given in Section 3.2. According to Popel'[155] FeO content in the slag of 84% is sufficient for a drop of 11 g weight to spread completely.

At the contact of metals with small amounts of C, Mn, and Si (armco-iron, Sv-08 and St3 steels) with CO_2 or air, a slag crust, almost completely consisting of ferrous oxide, forms on the drop surface. This is exactly the reason for the lowering of σ_{m-g} value for these metals in their interaction with an oxidizing gaseous medium. Change of σ_{m-g} value depends (Figure 3.2 through Figure 3.5) on the metal composition, oxidizing gas flow (Q_g), and temperature. For U8A steel, in view of the high content of carbon in the metal, such a shell does not form at short-time contacts with CO_2 or air. However, oxygen transition from the gas phase into the metal is also not observed which is also associated with a high content of carbon in the metal. For these reasons, the value of surface tension of U8A steel does not change but interaction with CO_2 and air.

Surface tension of Sv-08G2S and Sc-10GS steels, in which the total content of Mn and Si is 2.5–3.0%, decreases in contact with an oxidizing gaseous medium, but only slightly. This is, apparently, related primarily to the presence of silicon in the steels, which shows some surface activity in iron-based alloys. This is confirmed by the results of the measurement of surface tension of Fe–Si alloys and E3A steel with 2.88% Si. As is shown already (Figure 3.5), at a silicon content in the metal of about

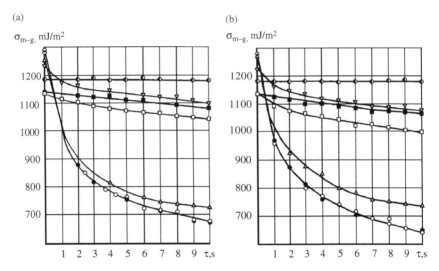

FIGURE 3.2   Change of surface tension of metals depending on the time of contact with CO_2 (a) and air (b) at T = 1803 K and $Q = 84 \cdot 10^{-7}$ m^3/s: ($\bullet$) armco-iron; ($\circ$) Sv-08; ($\triangle$) St.3; ($\square$) Sv-08G2S; ($\blacksquare$) Sv-10GS; ($\triangledown$) 08Kh20N10G6; ($\odot$) U8A.

1.2% the surface tension of Fe−Si alloy, which is in contact with CO_2, practically does not change. Further increase of silicon content in the metal leads even to a rise of its surface tension on the boundary with the oxidizing gaseous medium. The cause for the observed phenomenon may be silicon adsorption at the metal−gas

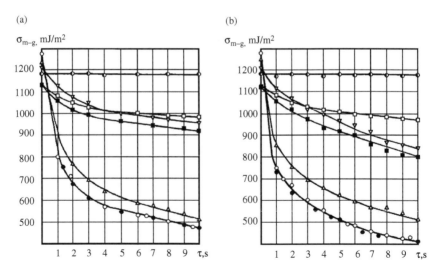

FIGURE 3.3   Change of surface tension of metals with the time of contact with CO_2 (a) and air (b) at T = 1803 K and $Q = 140 \cdot 10^{-7}$ m^3/s: ($\bullet$) armco-iron; ($\circ$) Sv-08; ($\triangle$) St.3; ($\square$) Sv-08G2S; ($\blacksquare$) Sv-10GS; ($\triangledown$) 08Kh20N10G6; ($\odot$) U8A.

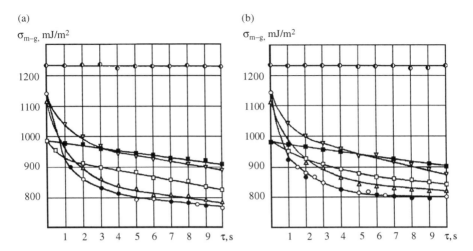

FIGURE 3.4 Change of surface tension of metals depending on the time of contact with CO_2 (a) and air (b) at T = 2003 K and $Q = 84 \cdot 10^{-7}$ m³/s: (•) armco-iron; (○) Sv-08; (Δ) St.3; (□) Sv-08G2S; (■) Sv-10GS; (▽) 08Kh20N10G6; (◉) U8A.

boundary and formation of a slag film, containing a significant amount of SiO_2. Slag films of such a composition readily wet the metal,[4] while providing its reliable protection from the contact of the gaseous medium. In addition, if SiO_2 is the basis of the slag shell, transfer of oxygen from the slag into the metal may be difficult.

A slight decrease of the value of surface tension of 08Kh20N10G6 steel at short-time contact of the melt with air or CO_2 is attributable primarily to a high content of Mn and Cr in the steel.

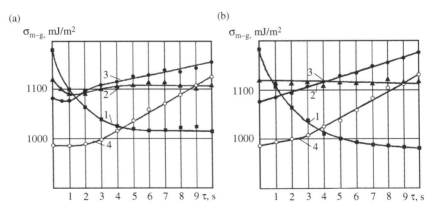

FIGURE 3.5 Change of surface tension of Fe–Si alloys with the time of contact with CO_2 (a) and air (b) at T = 1803 K and Q = $84 \cdot 10^{-7}$ m³/s. Si (wt%): [1] 0.51; [2] 1.21; [3] 1.95. E3A steel is shown in [4].

3.1.2.4 The Behavior of Binary Gas Mixtures

For practical purposes, especially in the study of the process of pore formation, it is important to know the influence of binary gas mixtures on the value of surface tension of the metal melts. Moreover, the composition of the atmosphere, in which the arc runs, also consists of several gases.

3.1.2.4.1 Armco-Iron and Steels

We have determined the value of surface tension of armco-iron and steels on the boundary with gas mixtures of $N_2 + H_2$, $N_2 + CO$, $H_2 + CO$. Results of this research are shown in Figure 3.6.

3.1.2.4.2 Aluminum and its Alloys

Aluminum and its alloys are widely applied in the fabrication of the most diverse welded structures. The main alloying components in aluminum-based alloys are magnesium, manganese, silicon, and iron. In this case, as a rule, the total content of alloying components in the alloys, which are used in welded structures, does not exceed 8–9%.

Values of surface tension of molten aluminum, according to the data of different authors, somewhat differ in their value. The most probable valid values of σ_{m-g} are given in some publications,[156–158] according to which the magnitude of surface tension of molten aluminum at 933 K is 865–870 mJ/m².

Investigations of the influence of alloying additives on the surface tension of molten aluminum demonstrated that iron is inactive on the surface of liquid aluminum.[159,160] According to published data,[160] σ_{Al-Fe} isotherm at the temperature of 973 K in the range of variation of iron concentration in aluminum from 0 to 20% is given by a linear equation:

$$\sigma_{Al-Fe} = 851 + 0.62C_{Fe},$$

where C_{Fe} is the iron content, in wt%.

Studying the influence of silicon additives on the surface tension of aluminum melt showed[159,160] that the value of the melt interphase tension practically does not change with the increase of silicon content up to 18%. Addition of Mn into the aluminum melt, similar to silicon, does not lead to a noticeable change of σ_{m-g} value.[159] Unlike these elements, magnesium is a surfactant in liquid aluminum.[160–162] Increase of Mg content in the aluminum melt up to 10.6 wt% lowers σ_{Mg-Al} value from 850 to 660 mJ/m² at 973 K.

Thus, among the alloying elements, which are most often present in aluminum alloys applied in welded structure fabrication, only magnesium demonstrates surface activity in molten aluminum. It should be noted that experiments to study the surface properties of the melts of aluminum and alloys on its base were conducted in the atmosphere of argon, hydrogen, or in vacuum. And this did not actually influence the obtained value of σ_{m-g}.

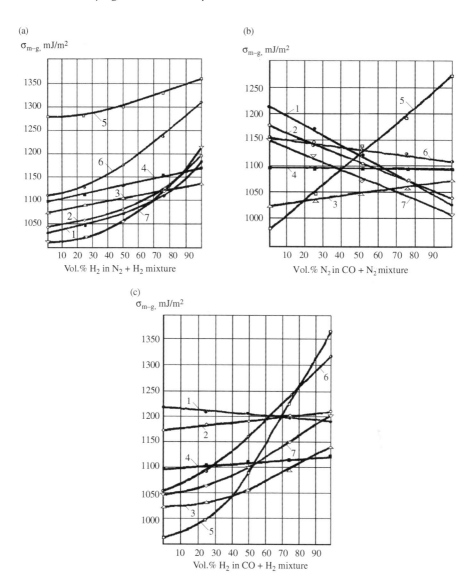

FIGURE 3.6 Change of surface tension of metals depending on the gas composition (T = 1803 K): (1) armco-iron; (2) Sv-08; (3) Sv-08G2S; (4) Sv-10 GS; (5) 08Kh20N10G6; (6) U8A; (7) St.3.

3.1.2.4.3 Copper and its Alloys

Copper and alloys on its base are applied in engineering to manufacture heat exchangers, switch-gear devices, etc. In copper-based alloys the main alloying components are Mn, Al, Fe, Si, and Zn. Therefore, let us consider the influence of these components on surface tension of copper melts, as well as the influence of oxygen, sulfur, and lead, as the presence of the latter elements in copper essentially influences its properties.

According to some investigations,[163,164] the surface tension of liquid copper at temperatures close to 1373 K is equal to 1360 mJ/m^2. Addition of silicon to copper leads[165] to lowering of the value of copper surface tension. We have obtained the same results, which are presented in Figure 3.7.[166] In argon and in nitrogen, σ_{m-g} values of copper M1 and Cu–Si alloys were similar, and decreased with temperature increase. According to our data,[166] manganese also exhibits surface activity in molten copper. An investigation of the influence of aluminum additives on the surface tension of copper[167] showed, that aluminum demonstrates a weak surface activity in liquid copper. At the temperature of 1373 K, addition of 11.0 wt% Al to the copper melt, led to σ_{m-g} lowering by 110 mJ/m^2. Iron does not have any surface activity in copper melts.[168] Increase of iron content results in a higher surface tension of liquid copper. In Cu–Zn alloys, with an increase of Zn content from 0% up to 10%, the value of surface tension of the melt increases by approximately 60 mJ/m^2.[168] Thus, zinc is not a surfactant in molten copper. Sulfur[169] and oxygen[169,170] have considerable surface activity in copper melts. It was shown already that at oxygen content of 0.04 wt% in copper, the surface tension of the copper melt drops from 1370 to 1270 mJ/m^2 ($T = 1423$ K). Increase of oxygen concentration up to 0.88 wt% leads to σ_{m-g} dropping to 530 mJ/m^2. The influence of the change of sulfur content in the copper melt on its surface tension is shown in Figure 3.8.[169] Presence of lead in copper also leads to lowering of surface tension of the copper melt.[168,171] Addition of approximately 4.0 wt% of lead to copper leads to lowering of σ_{m-g} value of the copper melt by 150 mJ/m^2.[168]

Thus, silicon, aluminum, and manganese are surface-active components for copper. However, oxygen and sulfur have the highest surface activity in copper melts.

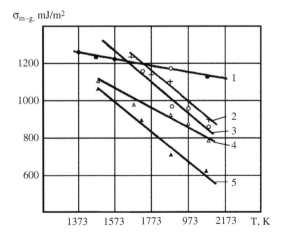

FIGURE 3.7 Change of surface tension of copper M1 and Cu–Si alloys depending on the temperature in argon. Silicon content (wt%): (1) 0; (2) 0.52; (3) 2.20; (4) 3.15; (5) 4.68.

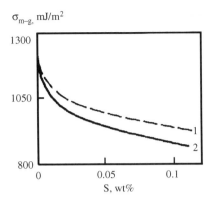

FIGURE 3.8 Influence of sulfur content on surface tension of a copper melt: T(K): (1) 1573; (2) 1373.

3.1.2.4.4 Titanium and its Alloys

Welded structures of titanium and its alloys are becoming ever wider applied in chemical engineering, aircraft construction, and other sectors of industry. Surface tension of molten titanium, determined by different authors, rather noticeably differs in its value. By the results of various studies[172–174] σ_{Ti} value at the melting point is equal to approximately 1550–1580 mJ/m^2. Unfortunately, there is not so much data on the influence of alloying additives on titanium surface tension. It is known,[175] for instance, that presence of aluminum in liquid titanium leads (Figure 3.9) to lowering of the surface tension value of the melt. Iron also is a surfactant in titanium, but tin (Figure 3.10) has a much higher surface activity in molten Ti.[175] Change of surface tension of Ti–C alloy at the change of C content in the melt from 0 to 1.0 wt%,[176] is given by the following

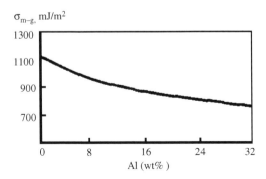

FIGURE 3.9 Influence of aluminum content on surface tension of a titanium melt (T = 2023 K).

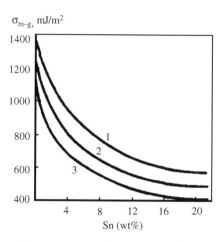

FIGURE 3.10 Influence of tin content on surface tension of a titanium melt. T(K): (1) 2023; (2) 2123; (3) 2243.

equation at 1623 K:

$$\sigma_{Ti-C} = 1360 + 80C_C,$$

where C_C is the carbon concentration in Ti.

Thus, carbon is inactive on liquid titanium surface. Methods of drop weight and pendant drop were used by us to determine the values of surface tension of some titanium-based alloys, which are often applied in the fabrication of welded structures, at their melting temperature, in argon and hydrogen atmosphere. The results of these investigations are given in Table 3.2.

3.1.2.4.5 Nickel and its Alloys

Nickel and its alloys are being applied in chemical engineering, electrical engineering, and other productions. Nickel of N0 grade, in which the total content of impurities does not exceed 0.07%, is often used in fabrication of weldments in the chemical industry. Copper, chromium, aluminum, or manganese are the usual alloying elements in nickel-based alloys.

Nickel is characterized by a high surface tension. By published data[177,178] the value σ_{m-g} is equal to approximately 1770 mJ/m^2 for pure nickel. When nickel is added to aluminum, the value of surface tension of Ni melt decreases (Figure 3.11).[179] Surface-active components for nickel are silicon[180] and copper.[181] Contradictory results are obtained on the surface activity of chromium in molten Ni. According to the results of Kurkjian and Kingery[182] chromium is inactive in Ni–Cr alloys. However, by the data of reported studies,[183–185] with an increase of Cr concentration in nickel, a slight monotonic decrease of surface tension of the latter is observed. The results of the latter works are, probably, more accurate.

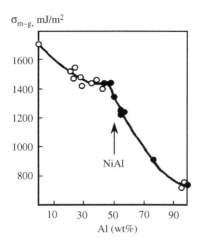

FIGURE 3.11 Influence of aluminum content on surface tension of a nickel melt (T = 1913 K).

Selenium, tellurium, and sulfur demonstrate a considerable surface activity in nickel.[185-188] At the change of selenium concentration in nickel from 0 to 0.3 wt% at T = 1823 K, the σ_{Ni-Se} isotherm is defined by the following equation:

$$\sigma_{Ni-Se} = 1770 - 700lg(1 + 3000C_{Se}),$$

where C_{se} is the mole concentration of selenium in nickel.

When nickel contained tellurium, the concentration of which varied in the range from 0 to 0.2 wt%, the alloy isotherm was given by the following equation:

$$\sigma_{Ni-Te} = 1770 - 700lg(1 + 1200CTe),$$

where C_{te} is the Te mole fraction.

The influence of sulfur on surface tension of nickel is shown in Figure 3.12.[188]

3.1.2.4.6 The Role of Metal Temperature

It should be noted that the value of surface tension of metals depends not only on the composition of the metal and gas phase, but also on the melt temperature. The magnitude and sign of the temperature coefficient (dσ/dT) are determined by the composition of the metal and gaseous medium,[104,109,138,144,189,190] and it is not always a negative value. For instance, by the data of a study[138] the surface tension of pure iron in inert gas atmosphere decreases with temperature rise, and in nitrogen atmosphere it increases. For Fe-C alloy, which is in contact with inert gas, an increase of surface tension is observed with temperature rise up to 1823 K. Further increase of the melt temperature to 1923 K leads to lowering of the surface tension of the metal, and at temperatures above 1923-1943 K, the temperature coefficient of the melt becomes

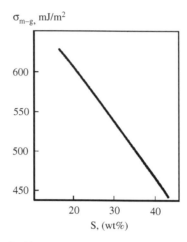

FIGURE 3.12 Influence of sulfur content on surface tension of a nickel melt (T = 1573 K).

positive again[104,190]. On the boundary with nitrogen, the temperature coefficient for iron–carbon melts does not change its sign with temperature rise and remains positive all the time.[138]

To solve many practical problems, it is often necessary to know the values of surface tension of steels and alloys, used for the fabrication of various weldments. Data on the surface properties of some steels and alloys, most often used in welded structure fabrication, are given in Table 3.2, which is based on our research[191–194] and studies.[76,141,143]

Unlike investigations to determine the surface tension of metal melts, much fewer studies are devoted to the measurement of surface tension of solid metals. This is obviously related both to the difficulty of conducting such experiments, and to insufficient reliability of the applied procedures. Table 3.3, based on a review, made by Jones,[87] gives the values of surface tension of a number of metals in the solid state, using the most reliable experimental methods.

As is seen from the data given in Table 3.3, σ_{s-g} values in vacuum and in inert atmosphere (Ar, He) are in excellent agreement. Therefore, the inert atmosphere has little effect on the surface tension of the solid metal. In hydrogen atmosphere, σ_{s-g} value is somewhat higher than in vacuum.

3.2 INTERPHASE SURFACE ENERGIES IN METAL–SLAG AND SOLID METAL–MELT SYSTEMS

In submerged-arc or covered-electrode welding and surfacing, as well as in electro-slag processes, the molten metal is in contact with the slag. Interphase tension on metal–slag boundary, has a substantial influence on many processes, running in welding. That is why the interest of researchers to studying the interphase tension in metal–slag system (σ_{m-s}) is understandable. However, only development of

TABLE 3.3
Surface Tension of Some Metals in the Solid State

Metal	T, K	Atmosphere	σ_{s-g}, mJ/m^2	Method of Measurement
Aluminum	453	Vacuum	1140 ± 200	1
Titanium	1873	Vacuum	1700	2
Iron	1733	Argon	1910 ± 190	2
	1723	Ar + 10%H$_2$	2090 ± 100	2
	1683	Hydrogen	2320 ± 80	2
Copper	1006	Helium	1720	2
	1173	Vacuum	1750 ± 89	2
	1323	—	1735 ± 45	2
Chromium	1823	Vacuum	2200 ± 250	2
	1873	Argon	2390 ± 500	4
	1823	—	2400	2
Nickel	1616	Argon	1820 ± 180	2
	1492	Vacuum	1860 ± 190	5
Niobium	2523	—	2100	2
Molybdenum	1700	Vacuum	2200 ± 200	6
	2773	Argon	1920 ± 200	2
Tungsten	2030	Vacuum	2900 ± 290	6
Zinc	753	Helium	830	2
Tin	490	Vacuum	673	7
	490	Argon	668	7
Lead	583	Vacuum	560	7

Note: Method of σ_{s-g} measurement: (1) rate of pore shrinkage; (2) zero creep method, (3) thermal etching of grain boundaries by evaporation and condensation of metal vapors; (4) thermal etching of grain boundaries by diffusion of metal vapors into the gas phase; (5) rate of groove smoothing; (6) emission ion microscope; (7) compensation method of zero creep.

procedures, allowing determination of σ_{m-s} values for metals in the molten state,[88–90] permitted conducting several interesting studies in this field. The first experimental work, which yielded valid results, was conducted by O.A. Esin, S.I. Popel', and Yu.P. Nikitin with their associates in 1950–1951. In earlier studies, for instance by Leontieva in 1949,[195] the interphase tension was determined by measuring the dimensions of metal drops, solidified in the slag. This resulted in considerable errors and detracted the application from the practical merit.

3.2.1 Composition of The System

Already, the first studies showed that the value of interphase tension depends both on the metal composition and that of the slag, and is determined by the forces of interaction of the contacting phases. In this connection, many investigations are devoted to studying the influence of various components, added to the metal or slag, on σ_{m-s}

value. This section deals with the influence of various factors on the value of inter-phase tension in the systems, where the metal phase was iron or iron-based alloys, as steels are exactly the material having the biggest application in welded structure fabrication.

3.2.1.1 Carbon

According to published work[196,197] carbon addition to iron leads to lowering of the value of interphase tension of the melt on its boundary with multicomponent $CaO-SiO_2-MgO-Al_2O_3-CaF_2$ slag and other slags consisting of three components like $CaO-SiO_2-Al_2O_3$ and $Al_2O_3-SiO_2-MnO$. In this case, the influence of carbon (added to the metal) on the interphase tension is more pronounced than its influence on the surface tension.

3.2.1.2 Sulfur

Sulfur, which is a surfactant on the metal–gas interphase, demonstrates considerable surface activity in the metal–slag system also. Addition of sulfur to iron and iron-based alloys markedly reduces the interphase tension of the metal melt on the boundary with various slag systems.[93,198,199] By the data of Popel'[198] addition of sulfur to the slag alone or its simultaneous addition to the slag and the metal, also lowers the inter-phase tension. Decrease of σ_{m-s} value was observed with the at addition of calcium or manganese sulfides to $CaO-SiO_2-MnO$ slag.[199] Only by the data of Bobkova and Petukhov,[200] increase of sulfur content in the slag, consisting of CaO and Al_2O_3, did not lower the σ_{m-s} value, but even led to a certain increase of this value, and another study[201] reports a certain inactivity of sulfur on cast iron–slag interphase.

3.2.1.3 Phosphorus

Similar to carbon, phosphorus is also a capillary-active element on the metal–slag interphase. Phosphorus addition to iron leads to lowering of its interphase tension on the boundary with $CaO-Al_2O_3-SiO_2$ slag melt.[196] A study of the interphase of Fe–P alloy with $MnO-SiO_2$ slag also revealed the dependence of the interphase tension on phosphorus content in the metal. The interphase tension dropped with the increase of phosphorus concentration in iron to 1%. Further rise of P content had only a slight influence on the interphase tension. Addition of P_2O_5 to the slag, con-sisting of MnO and SiO_2 at constant content of phosphorus in the metal, also lowered the interphase tension.[199]

3.2.1.4 Oxygen

Oxygen has a significant influence on the interphase tension.[202–204] This is due to the high surface activity of oxygen, dissolved in the metal. Even at a comparatively low concentration of O_2 in the metal volume, its content in the surface layer is consider-able. Enrichment of the metal-melt surface layer with oxygen reduces the difference in the structure of the surface layers of the metal and the oxide phase, thus enhancing the bond between the contacting phases and lowering the interphase tension. It is interesting that presence of nitrogen in the metal changes the oxygen capillary activity.[205] Therefore, both on the boundary with slag and on the boundary with the gaseous medium, simultaneous presence of oxygen and nitrogen in the metal melt has a mutual influence on the surface activity of each of them.

3.2.1.5 Cr, Mo, W, and Si

The same components exerts different influences on surface tension and interphase tension. Thus while addition of titanium and vanadium to iron and Fe–C alloys improves the surface tension of metal melts, presence of the same components in the melt contacting the slag, lowers the σ_{m-s} value.[71]

Investigation of the influence of chromium, molybdenum, and tungsten dissolved in the metal, on the interphase tension value, showed that the increase of Cr and Mo content in the metal lowers the interphase tension of the metal melt on the boundary with white steel–melting slag and binary $CaO–Al_2O_3$ slag.[206] Tungsten addition to the metal, contacting the same slags, led to an increase of the σ_{m-s} value.[207]

A considerable number of works are devoted to the study of the influence of manganese and silicon addition to the metal, on the interphase tension.[207–210] In the binary Fe–Si and Fe–Mn alloys, it was found, that Mn and Si lower the interphase tension and have a higher capillary activity here than on the interphase with gas.[207] Some ternary metallic systems, which included silicon or manganese, were also studied on the interphase with various slags. It was found that iron replacement by chromium in ternary Fe–Si–Cr system at a constant content of Si resulted in an insignificant rise of the interphase tension on the boundary with $SiO_2–Al_2O_3–CaO$ slag melt,[205] whereas the increase of Si content in Fe–Cr alloy lowered the interphase tension.[205] Decrease of the latter with the increase of Si concentration in the metal, was also noted for Fe–C–Si alloy on the interphase with $SiO_2–MnO–Al_2O_3$ slag.[209]

3.2.1.6 Individual Slag Components

Many studies are devoted to the investigation of the influence of individual slag components on the interphase tension. The most detailed studies were performed on the change of σ_{m-s} value with change in the component and their ratio with the same components in $CaO–Al_2O_3$, $CaO–SiO_2$, $CaO–Al_2O_3–SiO_2$, $Al_2O_3–Mn–SiO_2$, and $MnO–SiO_2–CaO$ slag systems on the interphase with iron, cast iron, and some steels.

Investigation of the change in the ratio of individual components in $CaO–Al_2O_3–SiO_2$ slag on the interphase with iron showed that with SiO_2 replacement by CaO and Al_2O_3, the interphase tension somewhat increases.[196] Here, the observed increase is more significant in the case of SiO_2 replacement by CaO. SiO_2 replacement by CaO causes an increase of the interphase tension; this observation is also true in the contact of the binary $CaO–Al_2O_3$ slag system with cast iron. The same results were obtained with $CaO–Al_2O_3$ slag–melt interphase with 30KhGSA, 40KhNMA steels.[211] In this case, SiO_2 addition to the slag also led to lowering of interphase tension while MgO addition caused a certain increase of its value.

Addition of ferrous oxide to the slag has a more noticeable influence on the inter-phase tension of iron and its alloys on the boundary with various slags.[202,203,212] Substituting any of the oxides by ferrous oxide lowers the interphase tension, although to varying degrees. This effect is manifested even at small additives of FeO to the slag. With FeO content of 1.3 wt% in the slag, consisting mostly of CaO and SiO_2, the interphase tension is equal to 1060 mJ/m^2. Increase of FeO concentration by just 2.9wt% lowers the interphase tension by 200 mJ/m^2.[203]

Manganous oxide demonstrates considerable capillary activity on the metal–slag interphase. Investigation of interphase tension in systems, in which the slags consisted of MnO, SiO_2, and Al_2O_3, and the metal phase was iron and Fe–Si alloy, revealed that replacement by MnO of both SiO_2 and Al_2O_3 in the slag lowers the interphase tension.[89,209] Its value is also decreased when MnO is substituted for CaO in the slag consisting of MnO–SiO_2–CaO.[209] It was observed that influence on σ_{m-s} by MnO is somewhat weaker than that by FeO. The higher capillary activity of ferrous oxide is confirmed by the rise of interphase tension between iron and slag, when ferrous oxide is replaced by MnO in FeO–MnO–Fe_2O_3 slag. Addition of MnO instead of FeO to the silicate melt of SiO_2–FeO–MnO–MgO also improves the interphase tension in the metal melt–slag system.[213]

Several works were devoted to the study of influence of sodium oxide on the interphase tension.[203,210,214,215] Na_2O substitution for SiO_2 in SiO_2–Na_2O slag melt lowers σ_{m-s} value of the Fe–Si melt. A similar influence of Na_2O on the interphase tension was also noted in the work of Popel', Esin, and Gel'd,[210] wherein CaO was replaced by sodium oxide in the cast iron–(CaO–Al_2O_3–SiO_2) slag system. However, the data of other groups[214,216] indicate that Na_2O does not have high capillary activity on the metal–slag boundary, and the observed lowering of interphase tension with increase of Na_2O content in the slag is attributable to the increase of FeO concentration in it.

TiO_2 influence on the interphase tension, studied by A.A. Deryabin and L.N. Saburov, showed that enrichment of CaO, Al_2O_3, and SiO_2 slag with titanium dioxide, results in a slight lowering of the interphase tension between iron and slag.

3.2.2 Temperature of the System

Besides the composition of the contacting phases, the interphase tension value depends on the temperature of metal–slag system. Comparatively few works are devoted to the study of the influence of temperature on σ_{m-s} value.[211,212,217,218] Most of the researchers note that the interphase tension on the metal melt–slag boundary decreases with the rise of temperature.[211,212,218] However, in the work of Georgiev, Yavojsky, and Bobkova[217] an increase of interphase tension of Fe–Mn–P melt on the boundary with SiO_2–MnO–CaO–MgO slag was found with temperature rise.

3.2.3 Experimental Determination of Interphase Tension

We are now aware that the value of interphase tension in the metal–slag system depends on the composition of the contacting phases and system temperature.

Unfortunately, the above results of studying the interphase tension often cannot be completely applied to the welding processes, primarily because of the difference of the welding slag systems from the metallurgical systems. While the metallurgical slags mostly consist of SiO_2, CaO, Al_2O_3, and MgO, for the majority of welding fluxes, used in electric-arc welding, as well as slags, formed in covered-electrode welding, the main components are SiO_2 and MnO. These components often are responsible for the non-metallic inclusions, present in welds. Therefore, from a large number of investigations, carried out by the metallurgists, only a few are known,[89,204,208,209,213] in which the systems close to the welding systems in their composition, have been studied. Besides, the majority of studies of interphase tension in the metal–slag system were done at temperatures close to the metal-melting temperature. Now under the conditions of welding, the metal–slag system temperature is much higher. For instance, in arc welding processes, it reaches 2473–2573 K at the drop stage.[219] In the weld pool in the case of submerged-arc welding of low-carbon steel, it is close to 2043 K.[220]

3.2.3.1 The Available Information on Welding Processes

In terms of welding processes, only a few studies have been performed so far of interphase tension on the metal–slag boundary.[90,221–224] Interphase tension of steels (Sv-08, Sv-08G2S, Sv-08KhGSMF, 10KhF, 10F13) on the boundary with some industrial fluxes[90] was determined. In addition, the interphase tension of armco-iron on the interphase with fluxes (OSTs-45, AN-42, OF-6) was studied.[90,222] Interphase tension was also studied on the boundary of metal and slag, which was a binary alloy, based on CaF_2.[90,225]

It should be noted that due to the features of the procedure used in the work of Yakobashvili,[90] when the interphase was determined, it varied simultaneously with the change of slag composition. This does not permit to state categorically the cause for the variation of interphase tension as being the change in the composition of slag or the temperature. Nonetheless, these experiments allowed not only studying the interphase tension, but also correlating σ_{m-s} value with the slag-crust detachment from the deposited bead surface.

The study of Tarlinskii and Yatsenko[223] is concerned with the interphase tension on the boundary of the metal with the slag, formed at melting of coatings of UONI 13/55, ANO-9, and VSTs-4A electrodes. A relationship is established between σ_{m-s} value and the possibility to use these electrodes to perform work at downhill welding.

Interphase tension in the system of (CaF_2–SiO_2–Al_2O_3–MgO) slag–EP690 type steel was studied by Slivinsky, Kopersak, and Solokha,[224] and for fluxes AN-26, AN-18, and ANF-14 by the same authors.[221] In addition, we studied[212,226] the interphase tension of low-carbon steel Sv-08 on the boundary with binary SiO_2-based slags and ternary SiO_2–MnO slags. It was found that FeO, MnO, CaO, and MgO have the strongest influence on σ_{m-s} value in systems with binary slags. CaO and MgO increase the interphase tension, and FeO and MnO lower it. Replacement of SiO_2 by Na_2O, K_2O, Al_2O_3, or TiO_2 changes the σ_{m-s} value only slightly.

In systems with ternary slags, addition of Al_2O_3, MgO, and CaO oxides instead of SiO_2 increases the interphase tension; alumina changing the σ_{m-s} value to the least and MgO to the most. Addition of Na_2O, K_2O, TiO_2, and FeO to a ternary slag lowers the σ_{m-s} value. In this case, the farther is the oxide located in the above sequence, the more marked is its influence on the σ_{m-s} value. Temperature increase lowers the value of interphase tension for all the studied metal–slag systems. Lowering of interphase tension value with temperature rise is more pronounced, compared to the change of interphase tension.

Our data on the influence of slag components and temperature on interphase tension in Sv-08 steel–slag system, determined by the sessile drop method, are given in Table 3.4.

3.2.3.2 Solidification Processes

Studying the solidification process requires knowledge of the value of interphase energy (σ_{sol-l}) on the solid phase–own melt interphase. However, studies, devoted to determination of σ_{sol-l} value are rather few. Moreover, experimentally found values of this quantity often do not coincide, which makes their practical application difficult. Turnbull together with Cech, used a microscope with a built-in heating stage, to determine σ_{sol-l} values for many metals, by measuring the over-cooling of metal solidification point.[43] As the used droplets of metal melt were very small ($10-100$ μm diameter), the possibility of heterogeneous nucleation on accidental non-metallic inclusions was eliminated. Results of some of these experiments are given in Table 3.5.

Comparing the results of calculation of σ_{sol-l} value (see Table 1.2) with experimental data in Table 3.5, it is seen that there is a good agreement of the experimental data with the calculated data, when Equation (1.15) through Equation (1.18) and Equation (1.20), and Equation (1.21) are used.

3.3 WETTABILITY AND SPREADING OF METAL MELTS OVER THE SURFACE OF SOLIDS

Wettability and spreading, primarily, in solid metal–metal melt systems have an important role in fusion welding processes. As the value of the equilibrium wetting angle, θ (see Section 2.1) is determined by the ratio of the values of specific surface energies, then all the factors, affecting the values of specific surface energies of the contacting phases, should also influence the wettability of the solid by the melt. Such factors include, primarily, the composition of the solid substrate and the melt, composition of the gas atmosphere, system temperature, as well as the presence of chemical interaction between the solid and the melt. The last factor influences the value of the interphase tension on the solid–melt boundary, and also the surface tension of the melt due to the change of the composition of the melt. Let us consider the influence of the above-mentioned and other factors on the wettability of a solid by melts and their spreading over a solid surface.

TABLE 3.4

Values of Interphase Tension and Temperature Coefficient for Sv-08 Steel–Slag System

Slag	Component Content, wt%	T, K	σ_{m-s}, mJ/m^2	$d\sigma_{m-s}/dT$, mJ/(m^2.K)
SiO_2	40	1823–1923	1150[a]	−3.34
CaO	60			
SiO_2	50	1803–1903	1086[a]	−3.1
CaO	50			
SiO_2	60	1803–1853	1042[a]	−3.02
CaO	40			
SiO_2	60	1853–1903	1198[b]	−3.41
MgO	40			
SiO_2	92	1853–1903	979[b]	−1.01
Al_2O_3	8			
SiO_2	50	1803–1903	763	−1.91
MnO	50			
SiO_2	30	1803–1903	648	−1.5
MnO	70			
SiO_2	38	1803–1903	414	−0.95
FeO	62			
SiO_2	89.5	1853–1903	966[b]	−1.38
TiO_2	10.5			
SiO_2	92	1803–1903	950	−1.2
K_2O	8			
SiO_2	83	1803–1853	935	−1.3
K_2O	17			
SiO_2	75	1803–1853	924	−1.06
K_2O	25			
SiO_2	92	1803–1853	945	−1.3
Na_2O	8			
SiO_2	83	1803–1853	940	−1.14
Na_2O	17			
SiO_2	75	1803–1903	933	−1.17
Na_2O	25			
SiO_2	50	1803–1903	770	−1.4
MnO	40			
Na_2O	10			
SiO_2	50	1803–1903	755	−1.64
MnO	40			
K_2O	10			
SiO_2	50	1803–1903	785	−1.11
MnO	40			
TiO_2	10			

(*Table continued*)

TABLE 3.4 *Continued*

Slag	Component Content, wt%	T, K	σ_{m-s}, mJ/m²	$d\sigma_{m-s}/dT$, mJ/(m².K)
SiO₂	50	1803–1903	833	−1.62
MnO	40			
Al₂O₃	10			
SiO₂	50	1803–1903	907	−2.85
MnO	40			
MgO	10			
SiO₂	50	1803–1903	863	−2.45
MnO	40			
CaO	10			
SiO₂	50	1803–1903	633	−0.72
MnO	40			
FeO	10			

[a]Measurement at 1823 K.
[b]Measurement at 1853 K.

3.3.1 INFLUENCE OF MATERIAL COMPOSITION

Influence of the composition of the material of the substrate and the melt on wettability is confirmed by numerous experiments. Already in the first experiments to study the wettability of solid substrates by molten metals it was observed[227] that transition metals, not forming compounds with carbon, do not wet graphite. The transition metals, as well as carbide-forming metals (Al, Si), wet graphite well. Change of the composition of the metal melt, for instance, addition of chromium and titanium to tin and copper, markedly improves graphite wettability. Saturation of metal (Fe, Ni, Co, Pd) by carbon leads to a marked increase of the wetting angle. An influence of the compositions of the substrate and molten metal on wettability was also observed in some studies.[228,229]

TABLE 3.5
Value of Surface Tension on Crystal–Own Melt Interphase

Metal	σ_{sol-l}, mJ/m²
Al	93.0
Cu	177.0
Ni	255.0
Co	234.0
Fe	204.0

It should be noted that change of the wetting angle with the change of the com-position of the substrate or the metal melt largely depends on the change of σ_{sol-l} value. In this case, the chemical interaction of the contacting phases has a significant influence on the value of interphase energy of the phase boundary and solid metal wettability by the melt.[230–232]

It is known,[231] that lowering of the interphase tension in a non-equilibrium system is a function of the difference in the chemical potentials of the system com-ponents. This follows from the expression for determination of interphase energy on the solid–liquid interphase[233]:

$$\sigma_{sol-l} = \sigma_1(P_{sol} - P_l) - \sigma_2(\mu_{sol} - \mu_l), \tag{3.1}$$

where σ_1, σ_2 are the contributions to the value of interphase energy, determined by the difference in the properties of the solid and liquid phases, conditionally desig-nated as $P_{sol} - P_l$, and difference of chemical potentials of the component, μ.

Both the factors depend on temperature, and have an opposite influence on σ_{sol-l} value with temperature change.

As the value of θ depends on σ_{sol-l} value, change of graphite wettability with the change of carbon concentration in Fe–C, Ni–C, CO–C, and Pd–C melts becomes understandable.

Thus, at a low content of carbon in the melts, there is high degree of wetting which is due to the process of carbon dissolution in the melt. However, at the contact of the melt with the solid phase, wettability depends also on the chemical activity of the contacting phases. For instance, when wettability of refractory metals (W, Mo) by tellurium, antimony, and their melts is studied, it is noted[234] that at the same temperatures (723–923 K) tellurium wets molybdenum readily, but does not wet tungsten. The found difference is attributable to the different chemi-cal activity of tellurium in relation to these metals. So reaction of tellurium with molybdenum starts in the solid phase at 703 K,[235] and that with tungsten only at 903–923 K.[236] As the values of σ_{sol-l} σ_{l-g}, and σ_{sol-g} depend on the temperature, the angle, θ is also related to change of temperature of the melt–solid system.

As a rule, temperature increase is accompanied by lowering of wetting angle.[237,238] In some, cases, however, particularly, when the melt interaction with a solid substrate results in the formation of intermetallics, the value of θ may increase, starting from a certain temperature. So, in steel wetting by zinc melt, reduction of θ to zero with time occurs in the temperature range of 748–873 K. At further increase of temperature up to 923 K, formation of intermetallics resulted in a final value of θ equal to $10°$.

Gas atmosphere strongly influences solids' wettability with molten metals. This influence is particularly noticeable when an essential lowering of the value of surface tension of the metal melt occurs on the interphase with the gas being applied. For instance, in our studies,[239] it was noted that replacement of Ar by CO_2 or air markedly lowers the value of θ on the metal melt–solid oxide interphase.

In the wetting of oxide substrates (Al_2O_3), by melts of armco-iron and St3 steel ($T = 1803K$), after 5 sec of the melt contact with the oxidizing gas, the value of θ

drops from 130 and 124° to 60 and 72°, respectively. An improvement of wettability, when argon was replaced by CO_2 was also found in the cast iron–Sv-08 wire system.

The wettability by molten metals of solid substrates can also be changed by creating a vacuum. Wettability is generally believed to be improved according to the vacuum level. This is attributed to the removal of the oxide film from the solid metal surface, the presence of which impairs spreading and wetting of the solid metal by the metal melts. The research, however, provides evidence to the effect[240] that with the higher level of rarefaction, wettability first becomes better, and then starts deteriorating. Similar dependencies were established for different systems (Fe–Ag, Cu–Ga, Cu–In, etc.). The effect found is probably attributable to the change of the reactivity of the solid-phase oxides with the change of the vacuum level.

3.3.2 INFLUENCE OF SURFACE ROUGHNESS

In addition to the presence of the oxide film, the wettability is also influenced by the roughness of the substrate surface. Influence of microrelief on the wettability is noted by many researchers.[66,241,242] This dependence is attributed to the presence of microprotrusions and microdepressions on the solid surface, leading to the increase of the actual surface of the solid, compared to the ideal smooth surface. In this case, it is rational to differentiate between the macroscopic contact angle, θ and microcontact angle θ_0 formed by the melt on the ideal smooth surface.

According to Deryagin,[242] the following dependence exists between angles θ and θ_0:

$$\cos \theta = k \cos \theta_0 = (S/S_0) \cos \theta_0, \tag{3.2}$$

where k is the coefficient of roughness; S and S_0 are the actual and apparent surfaces of the solid. The k value is sometimes found from the following expression: $k = 1/\cos\beta$, where $\cos\beta$ is the average value of cosine of the microrelief slope.

The above equation is valid when the height and the distance between the adjacent ridges of the microrelief are small, compared to the capillary constant and radius of microcurvature of the meniscus near the walls. From this equation it follows that, if $\theta < 90°$, the presence of the roughness weakens the wettability, and if $\theta > 90°$, it, by contrast, enhances the wettability.

Later, B.V. Deryagin and L.M. Sherbakov[243] showed that the contact angle depends not only on the value of the free surface energies and microrelief of the solid surface, but also on the drop weight. The results of studying the wettability under the zero gravity conditions, however, are indicative of the fact,[244] that in this case the contact angle values are the same as on the ground. Gravitational forces obviously do not make any significant influence on θ value.

3.3.3 INFLUENCE OF EXTERNAL FIELDS

It should be noted that in the solid substrate–melt–electrolyte system, θ value is influenced by the external field, superposed onto the system. By the results of investigations,[245,246] changing of the value of the electrochemical polarization allows changing the rate of melt spreading over the solid surface by tens of times. Melt spreading may also be accelerated due to the presence of a temperature gradient in a solid. In this case, the spreading process is called thermal spreading.[245] It is found[245,247] that during the thermal spreading of a liquid chemically interacting with the substrate, the rate of the liquid spreading may be varied by changing the temperature gradient in the substrate.

Simultaneous and separate influence of electrochemical polarization and temperature gradient on spreading of the metal melt was studied by Bykhovsky.[248] Spreading of aluminum alloy AK5 over steel VNS-9 was investigated in a melt of KCl + NaCl chlorides of equimolar composition with addition of 5 wt% $PbCl_2$. Investigation results are shown in Figure 3.13. As is seen, at $T = 973$ K spreading of aluminum melt proceeded only at anode polarization, and the spreading rates were very small. Temperature increase up to 1023 K, improved the wettability and spreading of the aluminum melt both at anode and cathode polarization. Spreading rate rose abruptly at the increase of melt temperature up to 1073 K. Spreading rate maximums were shifted to the anode side, compared to the maximum of the curve, derived at 1023 K.

Presence of a temperature gradient in the substrate (Figure 3.13) increases the rate of the alloy spreading: six times at temperature gradient $\Delta T = 5°/cm$ and 8 times at $\Delta T = 9°/cm$, compared to the isothermal spreading.

Change of wettability and spreading of copper melts over cast iron surface at superposition of an external field on the melt–solid metal system is also reported.[249]

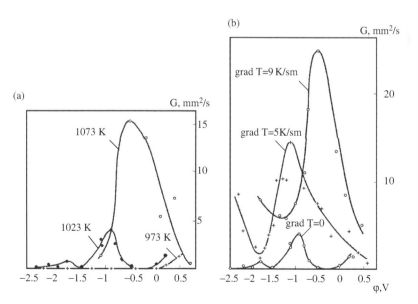

FIGURE 3.13 Dependence of the rate of spreading of aluminum alloy AK5 over VNS-9 steel on the magnitude of electric potential (a) and temperature gradient in the substrate (b).

For all the studied alloys, based on copper (brass L63, bronze BrKMts3-1, BrAMts9-2) an improvement of melt spreading on current passage through the melt–solid metal interphase is noted. Current density should be greater than $1-1.5$ A/mm^2. For BrAMts9-2 bronze, spreading depends also on current polarity. Without current application and at reverse-polarity current (melt is the anode), bronze spreading over the surface of the solid cast iron is much worse, than when direct-polarity current (melt is the cathode) is used.

Melt spreading over the solid surface is related to a certain degree to wettability. Good wetting, however, is a necessary yet an insufficient condition for the liquid to spread over the solid surface. A purely thermodynamic consideration of the spreading process, even with the known values of free surface energies, does not provide a reply to the question of whether this metal melt will spread over the solid metal surface or not. This is related to the fact, that at the contact of the molten and solid metals, mutual diffusion of components and running of chemical processes on the interphase may result in an essential change of the composition of the contacting phases, and, therefore, of the values of σ_{sol-g}, σ_{l-g}, and σ_{sol-l} and their ratio.

The molten metal spreading over the surface of solid metals can occur as a result of flowing of the liquid-phase layer, due to surface diffusion (migration) of atoms. Further causes are: evaporation and subsequent condensation of particles on the solid metal surface and dissolution of the melt in the solid metal; in bulk diffusion of its atoms in the direction of smaller concentrations and their subsequent adsorption on its surface. Prevalence of a particular mechanism in the overall process of the melt propagation, depends on the properties of the contacting phase, temperature, pressure, contact time, etc.

For welding processes, in view of the short time of their running, the most characteristic feature is spreading of comparatively thick layers of molten metal over the surface of the solid metal. Let us, therefore, consider in more detail the influence of various factors solely for this case.

The spreading process can be conditionally subdivided into three consecutive stages. This first stage, which usually lasts for fractions of a second, leads to the metal-drop flattening under the action of the capillary forces and the gravitational force. In this stage, spreading is mostly impeded by the forces of inertia. The second stage, which is the main one, is characterized by involvement of the major amount of the melt in the movement. In this stage, the role of viscous forces becomes more important, and the size of the drop for circular and linear spreading is defined from expressions,[250] respectively: $r = (6 \text{ m } \Delta\sigma/\pi\eta\gamma\chi)^{1/4} \tau^{1/4}$; $x = (3m \, \Delta\sigma/2a\eta\gamma\chi)^{1/3} \tau^{1/3}$, where r, x are the distances from the center of the drop, placed on the solid surface; m is the drop weight; χ is the dimensionless coefficient, allowing for the influence of the curvature of the liquid surface on its viscous resistance, $\chi \approx 10$; a is the width of a rectilinear path, along which the liquid metal spreads; $\Delta\sigma$ is the resultant force of tensions of the interfaces, referred to unit perimeter length, $\Delta\sigma = \sigma_{s-g} - \sigma_{s-l} - \sigma_{m-g} \cos\theta$. The $\Delta\sigma$ value can vary from the maximum value, equal to $2\sigma_{m-g}$ at $\theta = \pi$, to zero. Taking into account the surface roughness, $\Delta\sigma = k \, (\sigma_{s-g} - \sigma_{s-l}) - \sigma_{m-g} \cos\theta$. In this case, $k = 1$, when spreading over a liquid, and when spreading over a solid, $k > 1$. At the third stage, the spreading rate rapidly drops and the spreading process stops.

Thus, the process of molten metal spreading over the surface of the solid metal is influenced, besides the surface forces, by melt viscosity, its density, drop weight, and the state of the solid surface, and also, as shown above, by the presence of a temperature gradient in the solid metal and an external electric field, applied onto the melt – solid metal system. In the actual welding process, the number of factors influencing the spreading process is much higher. Let us first of all consider spreading and wettability under isothermal conditions.

3.3.4 ISOTHERMAL CONDITIONS

Most of the studies performed so far are devoted to the investigation of wettability and spreading of molten metals over the surface of solids under isothermal conditions. Although these results cannot be transferred directly to the welding processes, they still permit establishing the influence of various factors on the processes of wettability and spreading.

3.3.4.1 Steel/Copper and Copper-Based alloys

Let us consider the practically important systems for welding fabrication, and, first of all, the system of solid steel–melts of copper and copper-based alloys. In the study of Bogatyrenko et al.[251] wettability and spreading of copper M1 and bronzes BrAZhNMts 8.5-4-5-1.5, Br KMts 3-1, BrON 8.5-3, and Br MN40 over substrates of 12Kh18N10T and 45 steels were investigated. Wettability was studied in a high-temperature vacuum unit in a vacuum of $(2-4) \cdot 10^{-3}$ Pa at temperature of 1573 K. Heating of the substrate and the drop was conducted separately, and then they were brought into contact at the experimental temperature. Substrates were subjected to thermo-vacuum treatment for 30 min at 1573 K, during which the sample of the wetting metal was outside the heating zone. Time of contact of the melt with the substrate under isothermal conditions was 1.6 and 300s. Then sample was cooled to room temperature. Wetting angles were measured by the shadow image of a cold drop in a measuring microscope UIM-21. Data scatter was equal to 2–3%.

Experiments showed that all the studied melts readily wetted the steel substrates. Complete wetting of steel 45 by all the bronzes and copper, and of 12Kh18N10T steel by BrMN40 bronze was found already after 1 s. For other systems with 12Kh18N10T steel, the contact angle was not more than 20° (for M1 copper). Spreading process began at the moment the drop touched the substrate surface. High initial rates of spreading (0.3–0.6 m/s) and small wetting angles are indicative of establishment of a good contact on the interphase of metal melt–solid metal for these systems.

Investigation of wettability of austenite (steel 12Kh18N9T) and ferrite (steel 08Kh17T) by a copper melt showed,[252] that at 1373 K values of the wetting angle were 28–44° for austenite, and 92–100° for ferrite.

In the study of Popel'[253] the rates of liquid copper spreading over the surface of iron of different degrees of oxidation are determined at 1393 K. It is noted that the melts spread with varying contact angles. This is indicative of the influence of the degree of metal oxidation on its wettability by the copper melt.

Investigations, devoted to studying the wettability and spreading of melts of tin–lead and aluminum–manganese bronzes over steel 20 and armco-iron, were performed by Sivkov et al.[254] Experiments were conducted in an atmosphere of high-purity helium 1423 K. Drop weight was 0.14 g. Three types of metal substrates were used in the experiments: first group was not preheated; second group was preheated by the auto-vacuum surfacing mode; third was held (oxidized) in air at 1273 K for 20 min. Samples of armco-iron were not preheated.

Bronzes of grades BrOS8-12 and BrAMts9-2 spread with practically identical initial rates of 0.75 m/s. After $1.5 \cdot 10^{-2}$ s from the moment of the start of the drop contact with the substrate, the spreading rates were reduced to 0.17 m/s for BrOS8-12 bronze and 0.10 m/s for BrAMts9-2. After $(3.5-4) \cdot 10^{-2}$ s, a drop of bronze BrAMts9-2 stops spreading, having reached a radius of 5.1 mm at a wetting angle of $10°$. Bronze of BrOS8-12 grade completely spreads over a steel substrate. Initial spreading rate along a steel substrate is maximum for bronze BrOS8-12 and is equal to 1.0 m/s.

Initial rate of spreading of BrOS8-12 bronze on samples of the second group, heated in a container up to 873 K, was equal to 0.85 m/s, and for those, heated up to 1273 K, it was 1.0 m/s. Thus, an increase of the temperature of substrate heating leads to an increase of the initial spreading rates.

Wetting angle of the third group samples, oxidized in air at 1273 K, was equal to $110°$, and phases adhesion was 650 mJ/m². Poor wetting and low adhesion of phases are attributable to the presence of oxide film on the steel surface. Therefore, bronze of BrOS8-12 grade provides better wetting and spreading over steel, than bronze of BrAMts9-2 grade, as well as a high adhesion of phases.

We studied the wettability of steels St3 and 12Kh18N10T by melts of copper M1 and copper-based alloys: BrKMts3-1, MNZhKT5-1-0.2-0.2, and LK62-0.5 at 1373 K. As is seen in Figure 3.14, all the studied copper alloys readily wet steel St3, but do not wet steel 12Kh1810T so well.

3.3.4.2 Steel Aluminum

Welded structures for most diverse purposes often require joining aluminum to steel and other metals. Let us first consider the processes of wetting in aluminum–steel system.

Information on iron (steel) wetting by an aluminum melt is limited.[255–257] Spreading of liquid aluminum or alloys on its base over the solid metal surface differs from spreading of other metals. This is because the aluminum melt is covered by a highly stable oxide film which has a high melting temperature (about 2323 K). Therefore, we actually have a system, consisting not of three, but of four phases: gas phase–oxide–aluminum melt–solid metal. In should further be taken into account that a strong chemical interaction of the contacting phases proceeds in systems with aluminum melts. This leads to a noticeable change of the properties of contacting phases, as their composition is changed.

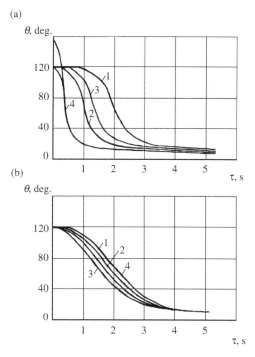

FIGURE 3.14 Charge in time of the angle of wetting of St3 (a) and 12Kh18N10T (b) steels by copper melts (T = 1373 K): [1] M1 copper; [2] BrKMts 3-1; [3] MNZhKT5-1-0,2-0,2; [4] LK62-0,5.

Kinetics of aluminum spreading over the iron surface was studied under vacuum at the pressure of $(1-3) \cdot 10^{-3}$ MPa in a unit, described by Eremenko, Natanzon, and Ryabov.[258] As the iron–aluminum system features a considerable reactivity, characterization of contact interaction requires obtaining data on spreading at a short time of contact. Therefore, experiments were conducted, with preheating the substrate and the aluminum sample separately, and bringing them into contact only after the required vacuum and temperature have been reached. AV000 aluminum (99.99% Al) and armco–iron were used in the experiments. Before the experiment, the substrate surface was ground, polished, washed in benzene and alcohol, annealed in vacuum at 1273 K, and cooled with the furnace.

The derived dependencies of the change of wetting angle in time τ at temperatures of 973, 1023, 1073, and 1173 K are shown in Figure 3.15. As is seen, iron–aluminum system is characterized by a satisfactory wettability ($\theta < 90°$). Change of the drop diameter in time follows a parabolic law $d^2 = k\tau$, where k is the coefficient of proportionality, characterizing the rate of aluminum spreading over iron under isothermal conditions. Spreading rate rises with temperature. The work of aluminum adhesion to iron has a considerable magnitude ($1300-1500$ mJ/m^2) in the temperature range of $973-1173$ K.

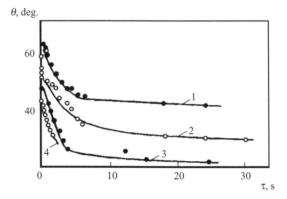

FIGURE 3.15 Charge in time of the angle of wetting of iron by aluminum melt. T(K): (1) 973; (2) 1023; (3) 1073; (4) 1173.

3.3.4.3 Copper/Aluminum

Aluminum melts wet copper substrates better, than iron does. Temperature increase from 1073 to 1273 K only slightly influences the value of the contact angle.[67] As is seen from Figure 3.16, the contact angle decreases abruptly in the first 1 to 1.5 min, and then changes only slightly. At copper contact with aluminum melt, copper diffuses into aluminum and aluminum into copper. Copper diffusion into the aluminum melt should prevail, as diffusion into the liquid proceeds much faster than that into the solid. Copper wetting by the aluminum melt may be improved, if a coating is first applied onto the surface of the copper substrate.

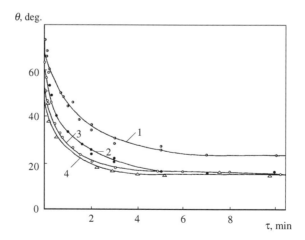

FIGURE 3.16 Change in time of the angle of wetting of copper by aluminum melt. T(K): (1) 973; (2) 1073; (3) 1173; (4) 1273.

Coating influence was studied[67] in vacuum at a temperature range of 973–1073 K. All the coatings, excepting zinc, improve the wetting and spreading. However, their influence is different and greatly dependends on the temperature. Zinc coatings have a small effect on wettability, probably because zinc quickly evaporates in vacuum. Substrate coating with only tin slightly reduced the value of the contact angle. Wettability and spreading improved in the most dramatic manner, when silver was deposited onto the substrate. Now, if the experiment temperature does not exceed 1023 K, aluminum spreads over the copper substrate with a silver coating in 4–5 min, forming an initial contact angle of about 40°, and a final angle of about 10°. The situation is different when the system is at higher temperatures. So, at 1073 K, the initial contact angle is equal to 8°, and quickly (within 2 min) drops to zero. This is, apparently, related to the formation of an eutectic in copper–silver system at the temperature of 1053 K. This is confirmed by the results of the experiments conducted to study silver wettability by aluminum melt. In this case, at 1073 K, wettability and spreading are worse than in Cu–Ag–Al system.

3.3.4.4 Titanium and Titanium Alloys/Aluminum

Several reports have been published, which are devoted to studying the wettability and spreading of aluminum melts over substrates of titanium alloys, molybdenum, niobium, and nickel.[94,259–261]

Wettability of titanium alloys VT-1, VT-20, VT-14, VT6S, OT4-1, and of titanium by molten aluminum was studied in the temperature range of 943–1143 K in helium atmosphere. It is established that in the first 20 s, the wetting angle decreases quickly enough. At system soaking for approximately 1 min, aluminum spreading stops, and the contact angle remains unchanged. For titanium, this angle is zero (temperature of 1183 and 1393 K), and for the VT-20 alloy, it is close to 15° (temperature of 1223 and 1323 K). According to our data, titanium wettability by an aluminum melt also depends on the drop weight m_d (Figure 3.17). Apparently, in this case, a change of the drop weight leads to a change of the conditions of oxide film removal from the aluminum melt surface. In all the cases, spreading rates became higher with temperature rise.

Studying the wettability of molybdenum and niobium by liquid aluminum and alloys on its base showed,[94,261] that Mo and Nb are readily wetted by aluminum melts. Wettability improves with temperature, and is independent of the melt composition. Increase of the total content of alloying components improves the wetting both of molybdenum and niobium by molten aluminum.

In the work by Sebo and Havalda,[261] the temperature and time dependence of the contact angle and the work of adhesion were studied in molybdenum wetting by the aluminum melt. As is seen from Figure 3.18, below 1173 K, aluminum poorly wets molybdenum, and the value of the contact angle practically does not change with time. Work of aluminum adhesion to molybdenum increases with temperature, particularly at temperatures above 1123 K. Temperature dependence of the work of aluminum adhesion to molybdenum shows, that wetting in this system may be regarded as that of the chemical type.

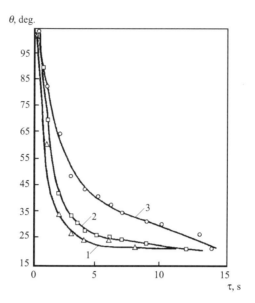

FIGURE 3.17 Change in time of angle of wetting of titanium by aluminum melt. Drop weight, m_d (mg): (1) 190; (2) 392; (3) 95.

3.3.5 BEHAVIOR UNDER NON-ISOTHERMAL PROCESS

As was noted above, in fusion welding, spreading of molten metal and solid metal wettability by the molten metal proceed under non-isothermal conditions. Non-isothermal processes of spreading and wetting, occurring in welding, braze-welding, and brazing have been insufficiently studied so far. This hinders selection of the optimal conditions for implementation of the technological process.

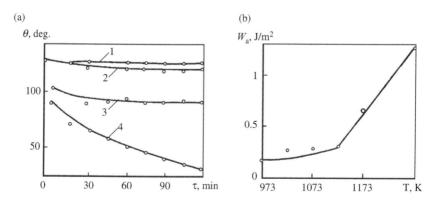

FIGURE 3.18 Change in time of the angle of wetting in aluminum–molybdenum system (a) and of work of aluminum adhesion to molybdenum, W_a depending on the temperature (K) (b) [1] 1023; [2] 1123; [3] 1173; [4] 1273.

3.3.5.1 Influence of Arc Parameters

One of the features of wetting and spreading in fusion welding is the presence of a welding heat source. In particular, in arc welding processes, the molten metal is exposed to the welding arc pressure which cannot be neglected if the welding current exceeds 200 A,[262] photocapillary effect (resulting from radiation), and possibly also the manifestation of electrocapillary phenomena. In application of electron beam and laser processes, the high concentration of heat gives rise to the reactive forces, which may also influence the metal melt spreading and change of wetting angle.

Influence of arc pressure on molten metal spreading has also been considered in.[94] In order to assess this phenomenon, the shape of the surface of spreading liquid metal was calculated, which is exposed to the gravitational forces, surface tension forces, and force of the impact of the arc. Making the assumptions, which are usually taken in such calculations,[263] and assigning the arc pressure distribution by a normal law[264]:

$$p = p_0 \exp(-kx^2),$$

where p_0 is the maximum pressure on the arc axis, and k is the coefficient of concentration. A differential equation was derived for the equilibrium of the liquid-phase surface in the form of a plane (two-dimensional) drop:

$$\frac{\sigma z''}{(1 + z'^2)^{3/2}} = \rho z + \frac{\sigma}{R_0} + p_0[1 - \exp(-kx^2)], \qquad (3.3)$$

where R_0 is the radius of curvature of weld convexity on the axis of symmetry at $x = 0$.

Adding dimensionless coordinates:

$$x = \frac{X}{a_c}; \quad z = \frac{Z}{a_c};$$

and dimensionless parameters

$$\text{æ} = ka_c^2; \quad z_0 = \frac{a_c^2}{R_0}; \quad p^* = \frac{p_0}{\sqrt{\sigma\varphi}}, \quad f_0 = \frac{F_d}{a_c^2}; \quad b_0 = \frac{B}{a_c};$$

(where a_c is the capillary constant; F_d is the area of the deposited metal; B is the weld width), a differential equation is derived:

$$\frac{z''}{(1 + z'^2)^{3/2}} = z + z_0 + p^*[1 - \exp(-\text{æ}x^2)]. \qquad (3.4)$$

The latter equation was solved by Runge–Kutta method. In this case, the normality condition

$$f_0 = 2 \int_0^{b_0/2} z(x)\mathrm{d}x,$$

and value of contact angle

$$z' = tg\theta \quad \text{at} \quad x = b_0/2,$$

were assigned, where b_0 is the dimensionless width of the convexity; f_0 is the dimensionless area of the deposit.

Calculations performed using Equation (3.4), showed (Figure 3.19), that spreading increases with the increase of the liquid-phase size (this is attributable to the impact of the gravitational forces) and increase of the axial pressure of the arc. Therefore, increase of welding current and reduction of electrode diameter, which result in higher axial pressure of the arc should improve the melt spreading over the solid metal surface.

3.3.5.2 Gas-Shielded Welding

In gas-shielded welding, the wettability of solids by molten metals is strongly affected by the gaseous medium. This influence is particularly noticeable when an essential lowering of the value of surface tension of the metal melt occurs on the boundary with the applied gas. This is confirmed, in particular, by the results of Popel'.[253] For instance, upon addition to the aluminum melt of Si, Mn, Cu, and Zn, which only slightly influence the value of surface tension of aluminum, no significant change in aluminum spreading over steel was observed. Aluminum alloying with Mg, Pb, Bi, and Cd noticeably decreased the value of σ_{1-g} and improved the melt spreading.

3.3.5.3 The Fluoc Effect

In welding with fluxes, the wettability and spreading will further depend on the flux composition, as in this case, the value of interphase tension on metal–slag boundary will also change. Addition to the flux of components, lower the interphase tension, should improve the wettability of the solid metal by the molten metal, and addition of components that increase the interphase tension, should result in

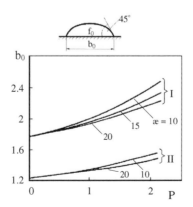

FIGURE 3.19 Influence of arc pressure on the width of molten metal spreading. f_0: (I) 0.5; (II) 0.25.

deterioration of the wettability. In welding of low-carbon steels, FeO and MnO should be added to the slag in order to decrease σ_{m-s}, and CaO, MgO, or SiO_2 should be added to increase the interphase tension, as these components have the strongest influence on σ_{m-s} value (see Table 3.4). In addition, the value of interphase tension may be changed by applying an external electric field to the metal–slag system. which will be shown below.

3.3.5.4 The Role of Preheading

With fusion welding processes, using a filler metal, wetting and spreading of the molten electrode metal proceeds under the conditions that are different from the isothermal conditions. In this case, the electrode-metal temperature is much higher than that of the weld-pool metal and of the solid (base) metal, and, moreover, the base metal is non-uniformly heated. Temperature distribution in the solid (base) metal depends, primarily, on the applied heat source, thermophysical properties of the base metal, and the welding modes.

As is seen in Figure 3.20, the base metal should be preheated in the most uniform manner, when gas flame is used as the heat source. The greatest temperature gradient in the solid metal will be found, when concentrated heat sources (electron, laser beam) are applied. Influence of the thermo-physical properties of the base metal and welding modes on temperature distribution in the base metal is readily seen from Figure 3.21.[1] However, as this will be shown further on, spreading of the metal melt over the surface of the solid metal proceeds only after the solid metal has reached a certain temperature. Therefore, it is anticipated, that the metal melt will stop spreading, when it has reached the isotherms with a lower temperature.

In view of the presence of a temperature gradient in the base metal, the value of the driving force of the melt spreading over the surface of the part, $\Delta\sigma$ under the condition, that the driving force is concentrated on the wetting perimeter and is consumed in overcoming the kinetic resistance, according to,[265] is equal to:

$$\Delta\sigma = \sigma_{1-g}(\cos\theta_0 - \cos\theta),\qquad(3.5)$$

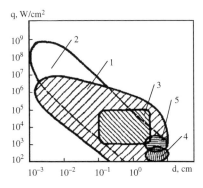

FIGURE 3.20   Concentration of energy of welding heat sources: (1) electron beam; (2) laser beam; (3) welding arc; (4) gas flame; (5) arc plasma.

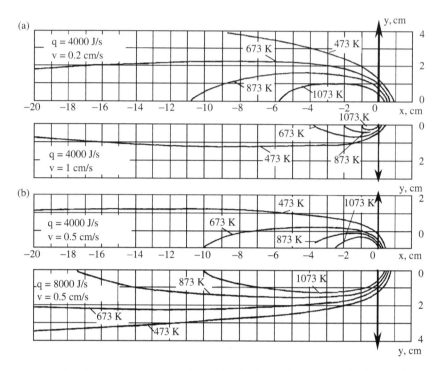

FIGURE 3.21 Appearance of temperature fields in the steel plate 1-cm thick ($\lambda = 0.38$ W/cm; $c\gamma = 4.8$ J/cm$^3 \cdot$ deg; a = 0.08 cm^2/sec) under the impact of an arc discharge on it: (a) change of welding speed; (b) change of the heat source power.

where θ_0 is the equilibrium value of the wetting angle; θ is the current value of the wetting angle.

Proceeding from Equation (3.5), the change in the dimensions of the drop, spreading under the impact of the surface tension forces, is given by the following expression:

$$\frac{dr}{d\tau} = K\sigma_{1-g}(\cos \theta_0 - \cos \theta), \tag{3.6}$$

where K is the constant of the spreading rate, determined by calculations or by experiment. Under these conditions, three cases are possible:

First case: When $dr/d\tau > 0$ that is liquid spreading over the surface of the solid proceeds. This is possible, as is seen from Equation (3.6), only, if the difference, $(\cos\theta_0 - \cos\theta) > 0$, which corresponds to the positive gradient of temperatures in a solid.

Second case: Value $dr/d\tau = 0$, that is no spreading proceeds and the liquid on the solid surface is in the static condition. This will occur if $(\cos\theta_0 - \cos\theta) = 0$, which corresponds to the absence of a temperature gradient in the solid and equality of angles θ_0 and θ.

Third case: Value $dr/d\tau < 0$, that is, contraction of the liquid present on the
solid surface is observed. This will occur if $(\cos \theta_0 - \cos \theta) < 0$, that is, at
a negative temperature gradient in the solid.

The third case is the most typical for the conditions of fusion welding, when
lowering of the base metal temperature is found with increase of distance from
the heat source. As the value of wetting angle decreases with the increase of the
temperature of the solid, the farther from the heat source is the zone of the solid
and melt contact, the greater will be the value of θ for a given zone. Therefore,
melt contraction proceeds in the direction from large θ values to smaller θ values.
This is particularly characteristic of cases of surfacing by an indirect arc, when
the base metal is insufficiently heated, but it may be also observed with other processes of welding and surfacing.

Thus, in fusion welding, the processes of wetting and spreading differ from
wetting and spreading, proceeding under isothermal conditions. This is primarily
related to the presence of a welding heat source and non-uniform heating of the metal.

So far, there is no work devoted to studying the wettability and spreading
of metal melts over the solid metal under impact of a welding heat source on the
melt. Such studies are important, however, as wettability of solid metals by the
melts is responsible for the formation of contact between the parts being welded,
and, hence, the strength of the welded joint and development of various weld
defects. Moreover, there is no data on the wettability of solid metals by iron
sulfide melt, although the latter has the determinant role in the process of formation
of solidification cracks of sulfide origin. Therefore, let us consider these issues in
greater detail, presenting them in separate paragraphs.

3.4 INFLUENCE OF WELDING HEAT SOURCES ON WETTABILITY AND SPREADING

Arc welding and surfacing are the most often employed techniques in the fabrication
of welded structures of most diverse applications. Let us, therefore, first of all consider the influence of the arc discharge on the processes of wetting and spreading in
the metal melt–solid metal system.

Standard procedures and installations, which are used to study the wettability
and spreading under isothermal conditions, are unsuitable, as they do not reproduce
the conditions characteristic of the actual welding processes. In the latter case, the
molten metal is influenced by the pressure of the arc (beam, gas), the base metal
is non-uniformly heated, and the electrode-metal drop temperature is much higher
than that of the weld pool and the base metal.

3.4.1 INFLUENCE OF ARC DISCHARGE

Using the unit developed by us, which was described in Section 2.5, we studied the
influence of the arc discharge on the wettability and spreading of aluminum melts
over the surface of boron, steel, molybdenum, tungsten, niobium, zirconium, and

titanium substrates, as well as of copper melts over the surface of copper and steel substrates. Experiments were conducted in argon atmosphere, by varying the welding current, arcing time, and the initial temperature of the substrate. In experiments with aluminum melts, the drop weight and weight of flux applied onto the welding were also varied.

As is seen from Figure 3.22, when the substrate is of steel VNS-9 (18Kh15N5AM3), which is used for the fabrication of fibers in composite materials, the wettability of the substrate by pure aluminum melt is improved with the increase of welding current. Silicon addition to the aluminum melt promotes better wettability. This effect is particularly pronounced at low currents (10–30 A). In both the cases, the drop spreads ($\theta < 50°$) practically in the first few seconds.

Value of the wetting angle depends on the drop weight and with its increase, the effectiveness of the arc discharge influence becomes smaller. Addition of Si to the melt in this case also improves the steel wettability by the aluminum melt.

Change in wetting angle, depending on the time of the arc impact on the molten metal, current value, and drop weight is rather accurately defined by the following formula:

$$\theta = 2.8\, m_\mathrm{d}[(0.9\, m_\mathrm{d}/I_\mathrm{w}) - 0.1]^{\tau-1}(m_\mathrm{d}/25)^{-\tau} + 3\tau, \tag{3.7}$$

which is valid for $0 < \tau < 4$ sec, 40 mg $< m_\mathrm{d} <$ 60 mg, 30 A $I_\mathrm{w} <$ 50 A. The above formula is readily usable in the selection of the welding modes.

Thus, increasing the welding current and reducing the drop weight improves the wettability of VNS-9 steel by the aluminum melt. Increase of substrate temperature from 973 to 1273 K did not have any noticeable influence on the wettability of this steel by aluminum melts. Although the melt temperature rises to 1273–1472 K under the arc impact with any weight of the drop, the main cause for wettability improvement in this case, probably, is the removal of the oxide film under the impact of the arc. This is, in particular, indicated by the appearance of aluminum

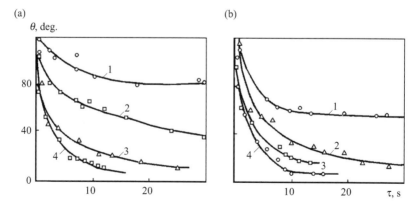

FIGURE 3.22 Change in time of the angle of wetting of VNS-9 steel by melts of aluminum (a) and AK5 alloy (b) I_w(A): (1) 10; (2) 30; (3) 45; (4) 75.

(a)   (b)   (c)

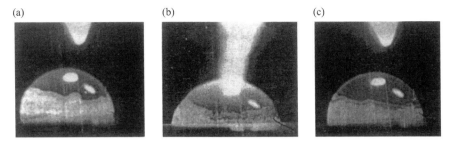

FIGURE 3.23   Appearance of aluminum drops cleaned from the oxide film under the action of the electric arc at $I_w = 45$ A and after a time interval(s) of (a) 2; (b) 3; (c) 4.

drops, shown in Figure 3.23. So, with the increase of the weight of the drop, accompanied by the increase of its surface, more time is required (at unchanged current) to remove the oxide film, and deterioration of steel wettability by aluminum melt is found. Therefore, wettability of steel by the aluminum melt can be improved by using various techniques, which allow removing the oxide film from the aluminum melt surface, for instance, by using a flux.

3.4.1.1 Direct Arc

Investigation showed that aluminum and AK5 alloy do not wet the steel substrate under the regular conditions. In this case, the values of wetting angles are equal to $120-146°$, and increase of substrate temperature from 973 to 1273 K does not have any noticeable influence on the value of these angles. A change in θ, which is caused by interaction of the melt with the substrate, is found only at long-term contact of the aluminum melt with the substrate (approximately $10-15$ min). Application of a small amount of AF-4a flux (5–7 mg) on the aluminum melt with unchanged initial temperature of the substrate (923 K) leads to an abrupt reduction of the wetting angle with the increase of welding current from 20 to 95 A, particularly in the first $3-5$ sec. Influence of drop weight in this case is manifested to a smaller degree than in the experiments, without flux, which is also indicative of a decisive influence of the oxide film on the process of spreading of the aluminum melt over the steel substrate.

Thus, wettability of steel VNS-9 by the melts of aluminum and aluminum–silicon alloy AK-5 is improved under the influence of a direct arc. In this case, however, the steel substrate is preheated to $1200-1500$ K. This leads to formation of intermetallics on the steel–aluminum boundary, and, hence, to lowering of the mechanical properties of the welded joint. Overheating of the steel substrate can be avoided if an indirect arc in used in welding. However, there is no data now on the features of steel wetting by aluminum melts with the application of an indirect arc.

3.4.1.2 Indirect Arc

We studied the wettability of VNS-9 steel by aluminum melt at the melting of a wire of A99 of 1.2 mm diameter by an indirect arc. Two schematics of electrode wire melting were used. In the first case, the arc was running between the aluminum wire, fed automatically vertically down and the tungsten electrode, placed at an angle of 45° to the wire; AC was used in this case. In the second case, the nonconsumable electrode was a copper ring, which enclosed the aluminum wire; DC was applied, and the electrode wire was the cathode. All the experiments were conducted in a unit, the design of which was described above. The gaseous medium was argon. Change of the wetting angle in time, depending on substrate temperature T_s, welding current I_w, arcing time τ_a, and time τ of system soaking are given in Figure 3.24. As is seen, change of the wetting angle in this case depends primarily on arcing time τ_a, which changed in the experiments with the change of the electrode wire feed rate. The longer the duration of the arc running between the aluminum wire and the additional electrode, the better is the steel substrate wetted by the aluminum melt. This is, apparently, related to a more complete removal of the oxide film from the drop surface. However, as the time of the drop existence at the electrode wire tip, and, hence, the arcing time, are short, the value of θ remains to be rather high (50–60°). Similar dependencies are also observed during the performance of experiments by the second schematic. In this case, the values of the wetting angle also remain high. It should be noted that with an indirect arc, the weight of the drop, which transfers to the steel substrate surface, is great (250–320 mg).

Wetting of a steel substrate by an aluminum melt by an indirect arc is improved with deposition of AF-4a flux on the aluminum wire (Figure 3.25). At the same temperature of the substrate (1023 K) and current (20 A), addition of flux at of 4–8% of the wire weight leads to reduction of the wetting angle approximately to 20°. No significant overheating of the steel plate is observed in this case. Substrate

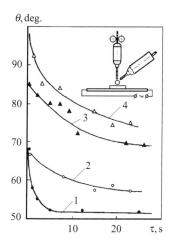

FIGURE 3.24 Change in time of the angle of wetting of VNS-9 steel by aluminum melt under the action of the indirect arc. T(K), I_W(A), τ(s) are: (1) 1033, 60, 4; (2) 1123, 60, 2.3; (3) 1073, 60, 1; (4) 1123, 75, 0.6s.

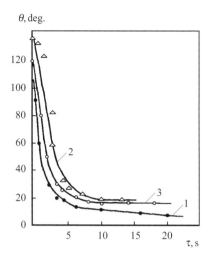

FIGURE 3.25 Change in time of the angle of wetting with deposition of AF-4a, Flux on the aluminum wire. m_f (mg): (1) 19.8; (2) 11.44; (3) 7.

temperature usually does not exceed 1023 K and can be easily adjusted by varying the distance from the electrodes to the plate.

Similar to welding by a direct arc, increase of welding current, substrate preheating, and use of flux improve the wettability of VNS-9 steel by the aluminum melt. Applying the flux on the aluminum wire surface reduces the weight of the electrode drop (m_d). Without the flux, at $I_w = 50$ A, $m_d = 263$ mg; at the same value of current, but with 5 mg of flux, $m_d = 149$ mg. The latter, apparently, is attributable to the fact, that the interphase tension on the molten aluminum–slag boundary is lower than the aluminum surface tension, and also that a certain amount of Na, which is a surfactant in aluminum, may pass from the flux into the metal.[266]

3.4.1.3 The Role of Substrate Composition

In addition to VNS-9 steel, wettability of armco-iron, St3, 45, and U10 steels, cast iron, as well as B, Nb, Ti, Zr, W, and Mo by aluminum melts was also studied at the impact of a direct arc on the melt. Experimental procedure and the unit, in which experiments were performed, remained the same. During the experiments, the temperature of substrate preheating was constant at 933–953 K. Investigation of wettability of armco-iron, steels, and cast iron is indicative of the fact that the wetting angle is influenced by the substrate material. The higher the carbon content in iron (Figure 3.26), the better is this metal wetted by the aluminum melt. The rate of growth and structure of the intermetallic phase zone also depend on the content of carbon in the metal. Aluminum contact with armco-iron results in the formation of a diffusion layer with deep protrusions inside the iron. In the contact with steels St3 and steel 45, the height of the protrusions decreases. Steel U10

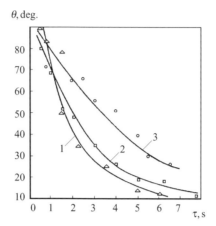

FIGURE 3.26 Change in time of the angle of wetting steels 45 (1), St3 (2), and armco-iron (3) aluminum melt, ($m_d = 170$ mg; $I_w = 75$ A; $T = 933$ K).

develops a comparatively even diffusion layer with small protrusions, growing toward aluminum. A thinner diffusion layer forms at contact of the aluminum melt with cast iron. Thus, the nature of the solid metal interaction with the aluminum melt, and, therefore, the wettability, are related to carbon content in the solid metal.

Carbon solubility in molten aluminum is low and is equal to less than 0.05 wt% at 1573–1773 K. In this case, the probability of formation of Al_4C_3 carbide (Al_3C by other data) is high.[267] In keeping with the published data,[268] at the contact of aluminum melt, with carbon-containing solid metal, the carbide phase grows in the direction of the metal melt. Such an interaction is observed, when the melt contains transition metals or other elements, forming strong carbides with carbon (for instance, Si, Al, B). In this case, an intensive wetting is observed (Figure 3.26), but without any noticeable damage of the solid metal surface, as it is preserved by the carbide layer.

3.4.1.4 Composite Materials

Composite materials, where boron fibers are used, and aluminum is the matrix, are applied in industry. Average strength of boron fibers of 100 μm diameter is up to 4600 MPa, and in some batches the maximal strength is 8460 MPa. Such a high strength of the fibers in combination with a low weight, make them a promising material, which may become applied in fabrication of composite materials. Therefore, we conducted experiments to study the wettability of boron plates by an aluminum melt under the impact of an arc discharge on the melt. The contact time was limited to 10 sec, as at more than 5 sec duration of the contact of the aluminum melt with boron fibers, the fiber strength does not exceed 0.76 of the initial value.[269] This effect is attributable to chemical interaction of boron with molten aluminum.

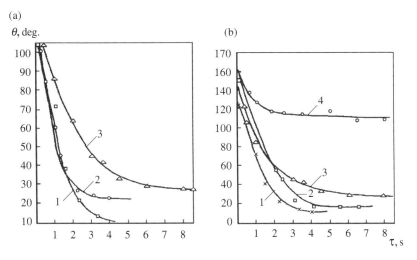

FIGURE 3.27 Change in time of the angle of wetting of boron by aluminum melt. (a) at different values of I_w (A) and $m_d = 125$ mg: [1] 45, [2] 30, [3] 10; (b) at different values of I_w (A) and m_d (mg): [1] 45, 125; [2] 45, 258; [3] 10, 125; [4] 10, 325.

As is seen from Figure 3.27, similar to wettability of VHS-9 steel, boron wettability by aluminum melt improves with the increase of current and reduction of drop weight. A similar influence of welding current and drop weight was demonstrated for Nb–Al systems (Figure 3.28). An essential influence of the substrate material is readily seen in comparison of the data on wettability of Ti, Zr, and W

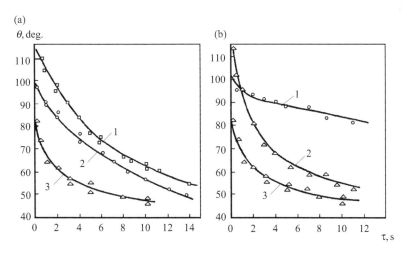

FIGURE 3.28 Change in time of the angle of wetting of niobium by aluminum melt. (a) at different values of m_d (mg) and $I_w = 75$ A: [1] 250; [2] 180; [3] 125; (b) at different values of I_w (A) and $m_d = 125$ mg: [1] 30; [2] 45; [3] 75.

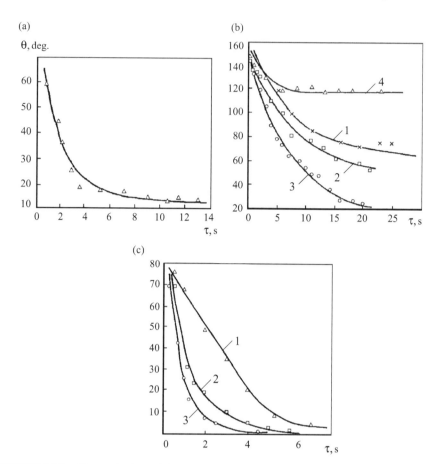

FIGURE 3.29 Change in time of the angle of wetting by aluminum melt of zirconium (a) at $m_d = 118$ mg; $I_w = 30$ A, tungsten; (b) at $m_d = 170$ mg and different I(A): [1] 30; [2] 45; [3] 75; [4] 10 and titanium; (c) at $m_d = 170$ mg; [1] 10; [2] 20; [3] 45.

by aluminum (Figure 3.29). Under the same experimental conditions, Ti substrates are wetted the best and W substrates, the worst.

Chemical interaction of the substrate and aluminum melt proceeds in all the studied systems. The intensity of this interaction depends on welding current and time of contact of the melt and substrate, and increases with the rise of these parameters. Change of wetting angle also depends on the intensity of the physicochemical processes proceeding on the solid metal–melt interphase, as this is accompanied by a change of values σ_{1-g} and σ_{sol-1}. As was noted above, at solid metal contact with aluminum melts, the wettability is largely related to the removal of oxide film from liquid aluminum surface. Oxide film removal is not only the result of the impact of the electric arc, but it also depends on the chemical affinity of the substrate material to oxygen.

Apparently, this must be the reason for a slight improvement of the wettability in the presence of carbon in the solid metal, as well as when titanium is used as the substrate. The aluminum melt wets titanium well (Figure 3.17) even without the impact of the arc discharge on the melt.

We also studied the wettability of copper and steel substrates by melts of M3r, M1 copper, and copper-based alloys MNZhKT-5-1-0.2-0.2, BrKMts3-1, and LK62-0.5 under the impact of an arc discharge on the melt. Experiments were conducted in argon, and the change of wetting angle was recorded by a camera. Experimental results, given in Figure 3.30, are indicative of the fact, that wetting of steel substrates by copper melts improves with the increase of welding current and arcing time.

It is noted that at the arc impact on the copper melt, the latter starts wetting the solid metals only after a certain interval of time after the arc ignition. The latter is related to heating of solid metal up to a certain temperature, at which wetting starts, which was discussed in Chapter 1. This is clear from Figure 3.31, where the data on the influence of arc discharge on solid copper wettability by the melt of M1 copper is given. As is seen, wettability and spreading of the copper melt over the surface of solid copper starts when the substrate temperature reaches 950–1050 K. Low values of wetting angle (10° down to 8°) are characteristic for temperatures of 1350–1400 K.

3.4.2 GENERAL MECHANISTIC CONSIDERATIONS

The mechanism of the influence of arc discharge on wetting processes may be different. At spreading of aluminum melts, the presence of the arc promotes removal of oxide film from the aluminum melt surface and increase of the temperature of the substrate and the melt. At spreading of copper melts, the influence of the arc discharge is primarily related to substrate preheating. This, in particular, is confirmed by the data, given in Figure 3.32. As is seen, for each melt–solid metal pair there

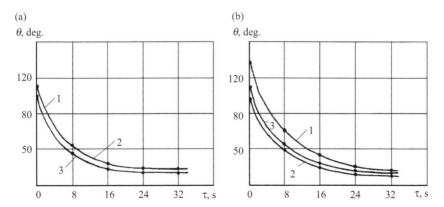

FIGURE 3.30 Change in time of the angle of wetting of St3 (a) and 12Kh18N10T (b) steels by copper melts ($I_w = 30$ A; $U_a = 12$ V): [1] M3r copper; [2] BrKMts3-1 bronze; [3] MNGKT5-1-0,2-0,2 alloy.

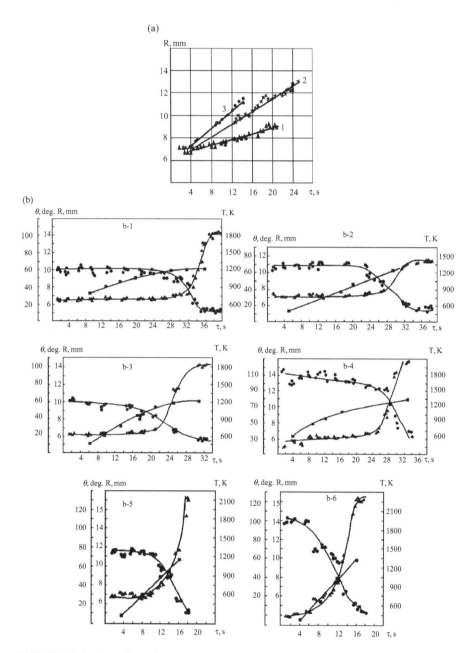

FIGURE 3.31 Spreading of a copper drop. Over a steel (a) at different I_w (A): [1] 73; [2] 83; [3] 93. Over copper substrate (b) at different I_w (A): [1] 30; [2] 53; [3] 63; [4] 73; [5] 83; [6] 93. Electric arc parameters: (•) (θ) = f(τ); (▲) R = f(τ); (■) T = f(τ).

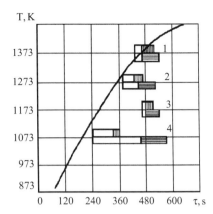

FIGURE 3.32 Temperature-time intervals of melting and spreading of copper alloys over steel substrates: (1) M3r; (2) BrKMts3-1; (3) NZhKT5-1-0,2-0,2; (4) LK62-0,5; (▥) in the interval of spreading over St3 steel, (▤) in the interval of spreading over 12Kh18N10T steel, (▢) in the alloy in the liquid state.

exists a certain temperature–time interval of wetting and spreading. The quality of the welded joint in electron beam welding with filler metal or electron beam surfacing with powder materials or filler metal largely depends on the wettability of the solid metal by the melt. In this connection, investigation of wetting and spreading of the metal melt over the surface of solid metal under the impact of the electron beam on the melt, is of considerable interest. More so, since such research was not conducted earlier.

Experiments were performed as follows. A drop of metal (copper M1) was applied onto a substrate, which was of copper M1 and steel St3. Then the substrate with the drop was placed into the chamber of the electron beam unit, where after establishing a vacuum, the drop was exposed to the electron beam. Drop spreading under the impact of the electron beam was recorded with a camera. Filming frequency was 48 frames per second. Values of wetting angle and change of the drop diameter in time were measured from the film in a measuring microscope.

As is seen from Figure 3.33, spreading of a copper melt over the surface of solid copper starts, as in the case of the impact of an arc discharge, after a certain interval of time, but this period is shorter. On a steel substrate, spreading starts practically immediately and proceeds faster with current rise.

Our results on wetting and spreading under the impact of the electron beam on the melt are significantly different from the known theoretical models and experimental data, obtained when studying the process of copper spreading over steels under isothermal conditions. This may result from either metal overheating under the impact of the electron beam, or appearance of additional forces, which primarily, are the reactive forces arising due to evaporation of a part of the metal. Appearance of these forces is caused by the flying out particles transferring their pulse to the liquid metal, which should promote the melt spreading. Here, the magnitude of

(a)

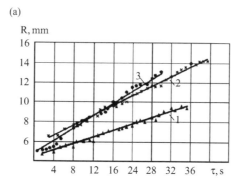

(b)

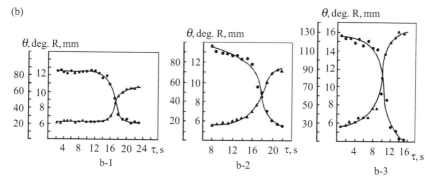

FIGURE 3.33 Spreading of molten copper over (a) steel at different I_w (mA): [1] 20; [2] 25; [3] 30 and (b) copper substrate at different I_w (mA): [1] 20; [2] 25; [3] 30 under the impact of the electron beam.

the reactive force is equal to

$$F_r = dP/dt \approx \Delta P/\Delta t, \tag{3.8}$$

where ΔP is the pulse, transferred by flying out particles to the molten metal during time Δt.

For numerical assessment of these forces, it may be assumed that the velocity of particles of evaporating metal is equal to[270]:

$$V_{av} = \sqrt{8kT/\pi m}, \tag{3.9}$$

where k is the Boltzmann constant; T is the metal boiling temperature; m is the mass of the flying-out particle.

Then the magnitude of the pulse is equal to:

$$\Delta P = \Delta m V_{av}, \tag{3.10}$$

where Δm is the loss of metal mass.

3.4.2.1 Experimental Determination of Δ_m

The value of Δm can be determined experimentally. For this purpose, a sample, which was a copper or steel substrate of $5 \times 30 \times 30$ mm^3 size with a drop of metal (copper) of a registered weight was exposed to the thermal impact of the electron beam and weighed in an analytical balance before and after the experiment. The difference of the two measurements yielded the Δm value. For comparison purposes, experiments were also conducted to determine the reactive forces under the impact on the metal of the electric arc, running between the tungsten electrode and the copper drop in argon atmosphere. The tungsten electrode was the cathode, and the copper drop, the anode. Time of the heat source impact on the metal was evaluated with great accuracy by the experimental filmogram. Filming was conducted with a frequency of 48 frames per second.

Results of the conducted experiments are given in Table 3.6. Obtained values of reactive forces under the impact of the arc on the metal are in good agreement with the theoretical estimates, derived by G.I. Leskov and V.I. Dyatlov. This leads to the conclusion on the validity of this procedure for the determination of the reactive force magnitude, and allows F_r value to be estimated under the impact of the electron beam on the metal.

As is seen from Table 3.6 in the melting of a drop of metal under the impact of both the electric arc and the electron beam, the reactive forces essentially depend on welding current and become greater with current rise. On the other hand, the reactive forces, arising in evaporation of a copper drop from a steel substrate are approximately 1.5 times greater than that in the evaporation of a similar drop from a copper substrate, both under the impact of the electric arc and the electron beam. Apparently, this is attributable to different thermo-physical properties of the substrate.

Obtained results indicate that for the studied systems under the impact of the electron beam on the metal, the reactive forces attain considerable magnitude and are approximately 2–3 times greater than under the impact of the arc. It should

TABLE 3.6
Magnitudes of Reactive Forces under the Impact of the Electric Arc and the Electron Beam on the Metal

Method of Action	Substrate Material	Initial Weight of the Drop (g)	Change of Drop Weight (g)	Current Applied	Magnitude of Reactive Force, (N)
Electric arc	Copper M1	1.143	0.0111	30 A	208×10^{-5}
	Steel St3	1.335	0.025	30 A	310×10^{-5}
Electron beam	Copper M1	0.520	0.0433	30 mA	420×10^{-5}
	Copper M1	0.520	0.0283	25 mA	310×10^{-5}
	Copper M1	0.520	0.0264	20 mA	120×10^{-5}
Electron beam	Steel St3	0.520	0.1172	20 mA	310×10^{-5}
	Steel St3	0.520	0.1758	25 mA	660×10^{-5}
	Steel St3	0.520	0.1976	30 mA	750×10^{-5}

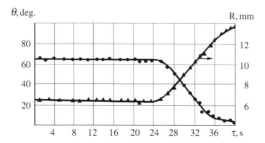

FIGURE 3.34 Change in time of the angle of wetting and R of steel St 3 by copper melt under the effect of the reducing flame.

be noted that when the electron beam is used, the reactive forces are concentrated, which is confirmed by the formation of a crater in the zone of the most intensive evaporation of the material.

3.4.2.2 The Nature of the Gas Flame

Unlike the electron beam. the gas flame belongs to heat sources with a low concentration of heat (Figure 3.20). Even though gas welding and surfacing are comparatively seldom used in modern fabrication, it is still interesting to establish the influence of this heat source on the processes of wetting and spreading in the solid metal–melt system.

As is known, the gas flame is characterized by comparatively low temperatures (2700–3500 K). This is, however, much higher than the melting temperature of many metals, for instance, low-carbon steels, copper-based alloys, etc. Therefore, the gas flame may influence the spreading and wetting of the solid metal by metal melts.

The results of investigations to study the impact of the gas flame on the wettability of steel St3 by a melt of M1 copper are given in Figure 3.34 and Figure 3.35. It is seen in Figure 3.34 that when a reducing flame is used, copper spreading over steel starts after a longer time, compared to the arc impact, but the magnitude of the contact wetting angle is reduced to smaller values with time: θ is reduced somewhat slower than under the impact of the arc.

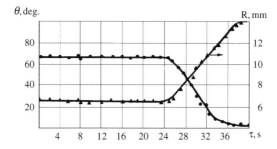

FIGURE 3.35 Change in time of the angle of wetting and R of St3 steel by copper melt under the effect of the oxidizing flame.

The observed features of spreading and wetting are attributable to the difference of the thermal characteristics of the gas flame, compared to the arc discharge. Due to a lower thermal power of the gas flame more time is required for melting the copper drop and heating the steel plate, leading to extension of the period, when the melt drop does not wet the solid metal. However, as the gas flame is a less-concentrated heat source than the arc is, with application of the gas flame, the steel plate is heated more uniformly. Consequently, the isotherms limiting the area of spreading of the copper melt, shift to a greater distance. Therefore, the melt spreads more readily, and the wetting angle has lower values, compared to the impact of the arc discharge on the melt. When an oxidizing flame is used (Figure 3.35), wettability and spreading are somewhat better than when a reducing gas flame is used. This is accounted for by lowering of surface tension of copper as a result of its oxidation, as oxygen is a surfactant for copper. This observation is a further confirmation of the statement on interrelation of the magnitudes of the melt surface tension, wetting angle, and melt spreading.

Thus, wetting of the solid metal by the molten metal changes under the impact of a welding heat source on the metal melt, and this change depends on the kind of the applied heat source.

3.5 WETTABILITY OF THE SOLID METAL BY NONMETAL MELTS

For fusion welding processes, the most important is to study the wetting and spreading behavior for the following systems: solid metal–ferrous sulfide melt and solid metal–melt of fluorides of alkali and alkaline-earth metals. Investigation of the first system will permit analyzing the process of solidification cracking in welds. The second system is important, as in this case, the protective properties of the slag and the depth of metal penetration depend on the liquid-phase wetting and spreading.

3.5.1 THE INFLUENCE OF SULFUR/SULFIDES

Investigations of wetting and spreading of ferrous sulfides over the solid metal surface have not been conducted earlier, as far as we know. On the other hand, it is known,[271,272] that improvement of wettability of the metal grains by the liquid phase increases the probability of solidification cracking in welds. It is further known,[273] that in the welding of structural steels, the metal susceptibility to solidification cracking rises with sulfur content in the metal. This is attributable to the ability of sulfur to form low-melting sulfide inclusions, primarily Fe–FeS, which remain molten at the moment, when the tensile stresses, arising at cooling of the weld metal reach a significant value.

Presence of the melt in the metal influences even the location of the initiating crack. The energy required for the formation of such a crack, depending on the fracture mode, is defined as follows[274]:

1. Transcrystalline fracture in the absence of the melt

$$A_1 = 2\sigma_{\text{sol-g}}, \tag{3.11}$$

2. Intercrystalline fracture in the absence of the melt

$$A_2 = 2\sigma_{\text{sol}-\text{g}} - \sigma_{\text{sol}-\text{sol}} \tag{3.12}$$

3. Transcrystalline fracture in the presence of the melt

$$A_3 = 2\sigma_{\text{sol}-\text{m}}, \tag{3.13}$$

4. Intercrystalline fracture in the presence of the melt

$$A_4 = 2\sigma_{\text{sol}-\text{m}} - \sigma_{\text{sol}-\text{sol}} = \sigma_{\text{sol}-\text{sol}}\left[\left(\frac{1}{\cos\frac{\theta}{2}}\right) - 1\right], \tag{3.14}$$

5. Fracture in the melt

$$A_5 = 2\sigma_{\text{m}-\text{g}} \tag{3.15}$$

Calculations, performed by Equation (3.11) through Equation (3.15), indicate, that energy consumption in metal fracture is greatly reduced in the presence of the melt on grain boundaries, and this reduction is the greater, the smaller the magnitude of angle, θ. For low-carbon steel (0.09% C), when the melt and crystals are of the same composition, $A_1 = 2766$ mJ/m^2, $A_2 = 2305$ mJ/m^2, $A_5 = 2558$ mJ/m^2, while $A_3 = 408$ mJ/m^2, $A_4 = 461$ mJ/m^2 at $\theta = 120°$, and $A_4 = 69$ mJ/m^2 at $\theta = 60°$.[275]

Therefore, studying the wettability of a solid metal by FeS melt and the features of ferrous sulfide spreading over the surface of the solid metal may allow establishing the mechanism of solidification cracking in welds. In addition, wettability is important in the formation of nonmetallic inclusions. We studied[276] the wettability of substrates of various metals, as well as of Al_2O_3, SiO_2, and FeO oxides, by ferrous sulfide.

Experiments on studying the wettability of solids by ferrous sulfide melt were conducted in a unit, which included Tamman furnace, and the welding transformer TDF-1601 was used as the current source. Before the experiment, the substrates were ground and rinsed with alcohol. Substrates of ferrous oxide were in the form of a tablet, compacted under a pressure of 4.443 GPa.

FeS melt was fed onto the substrate from a steel tube, located at the distance of 3–4 mm from the substrate surface. Formation of a drop at the tube tip, its transfer to the substrate, and spreading were recorded with a "Krasnogorsk-3" camera, in which the standard lens was replaced by the "Industar-22" lens. Filming was usually performed with the frequency of 48 frames per second. The moment of the drop of FeS melt touching the tube and substrate was taken to be zero. During the entire experiment, argon $(5–7) \cdot 10^{-5}$ m^3/sec, removed of oxygen and moisture, was fed into the furnace. Temperature in the furnace was constant in all the experiments, and was equal to 1373 K. Wetting angles were measured in the photographs in a measuring microscope.

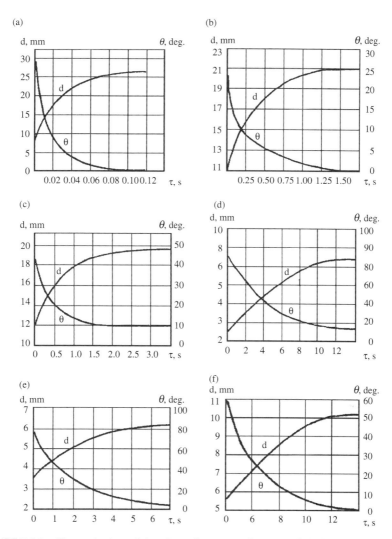

FIGURE 3.36 Change in time of drop base diameter and contact abgle in wetting of metals by (a) FeS melt: armco-iron; (b) 40Kh steel; (c) U10; (d) 1Kh18N9T; (e) 30KhGSA; (f) E3A.

As is seen from Figure 3.36, FeS melt readily wets all the studied materials. Particularly good wetting was observed in those cases, when the substrates were of armco-iron and 40 Kh steel. Spreading of a drop of ferrous sulfide over these metals proceeded at a higher rate. For armco-iron–FeS system, it was 238 mm/sec. Presence of carbon, chromium, molybdenum, nickel, silicon, vanadium, manganese, and aluminum in the metal leads to lowering of the rate of spreading of FeS melt over the metal surface. For instance, the rate of spreading of FeS melt over the surface of a substrate of steel 9 KhF is 24 mm/s, and for steel 08Yu, it is 6.6 mm/sec.

Great magnitudes of wetting angles and low rates of spreading are characteristic for the cases of spreading of FeS melt over the surface of corundum and quartz. In FeS spreading over the surface of corundum, the spreading rate is equal to 0.3–0.4 mm/sec, and in spreading over quartz, it is 0.6–0.7 mm/sec. Ferrous sulfide melt flows readily over the substrate of ferrous oxide, just as over armco-iron.

Thus, ferrous sulfide melt readily wets both solid metals and oxides, and, hence, can easily precipitate in the form of thin interlayers on the boundaries of metal grains and on the surface of non-metallic oxide inclusions, particularly, if the latter contain a large amount of ferrous oxide.

3.5.2 INFLUENCE OF CHLORIDES/FLUORIDES OF FLUXES

Various paste-like fluxes, applied onto the part prior to welding, are often used[1] in the welding of items of titanium and titanium-based alloys. This allows achieving the required depth of penetration with a smaller magnitude of heat input (Figure 3.37). Such an influence of paste-like fluxes is usually attributed to two causes: change of the characteristics of the arc discharge due to appearance of fluoride or chloride ions in the arc (depending on the composition of the applied paste-like fluxes) or good wetting of titanium by the flux melt. The first cause can hardly account for the observed effect, as the presence of fluoride or chloride ions in the arc discharge lowers the arcing stability. Therefore, the arc discharge concentration should not occur here, and hence, the penetration depth should not increase. It is obvious that the second explanation is more likely. We conducted investigations[277] to study wetting of titanium by molten single-component fluoride fluxes at the temperature of 1723 K in argon atmosphere. The experimental set up, used in this case, allows performing separate heating of solid titanium and flux. This permits eliminating the interaction between them during heating and bringing the drop of the molten flux into contact with the surface of the titanium substrate at the specified temperature of solid metal.

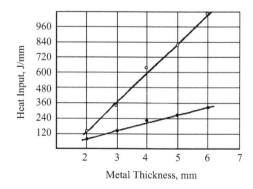

FIGURE 3.37 Values of heat input, providing complete penetration of titanium: (○) without pastelike flux; (●) with pastelike flux.

3.5.2.1 The Experimental Procedure

Experiments were conducted as follows. A graphite support with a cleaned and degreased plate of VT1 titanium of $25 \times 25 \times 5$ mm^3 size was placed in the central part of a horizontal tubular graphite heater. The studied flux was poured into a low-carbon steel tube of 3 mm diameter, which was placed into a corundum guide tube with a packer, mounted outside the heating zone.

After sealing the unit, it was heated, while argon was continuously fed into the working space of the frunace. Argon was thoroughly purified by removing oxygen and moisture, before introduction in to the furnace. The temperature in the furnace was controlled by a KSP-4 potentiometer with a VP $5/20$ tungsten–rhenium thermocouple, connected to it. The thermocouple hot junction was located in the opening of the graphite support directly under the titanium plate.

After attaining 1773 K, the tube with the flux was lowered into the melting space and the tube lower edge was positioned at a distance of 5–6 mm from the substrate. The flux powder in the tube melted, a drop formed on the edge, which transferred to the titanium substrate. The process of transfer of the drop of molten flux to the substrate and its spreading over solid titanium were recorded with a Krasnogorsk-3 camera. Filming speed was 48 frames per second.

Investigation of wetting of solid titanium by molten fluorides of sodium, lithium, calcium, magnesium, barium, and strontium showed that all of them wet the metal very well at 1773 K. Wetting angles for all the studied single-component fluxes are zero after a short time interval. Therefore, the rate of change of θ was taken as the wettability criterion. Experimental results, presented in Table 3.7, are indicative of the fact, that fluorides of alkali metals wet titanium better than fluorides of alkaline-earth metals.

Obtained data correlate well with the results of investigations on the influence of flux composition on the dimensions of welds.[278] According to these data, fluoride fluxes that more readily wet titanium promote a deeper penetration of the metal due to less-stronger driving away of the molten flux by the arc discharge.

3.6 ELECTROCAPILLARY PHENOMENA IN A METAL–SLAG SYSTEM

The ion theory of slag structure is universally recognized now, as it provides a better explanation for different experimental data, than the molecular theory does. The ion

---

TABLE 3.7
Wetting of Titanium by Single-Component Fluoride Fluxes

Flux Composition	NaF	LiF	MgF$_2$	StF$_2$	CaF$_2$	BaF$_2$
Change of angle θ in time, deg/sec	124	83	44	40	80	19
Time for complete spreading ($\theta = 0^0$), sec	0.25	0.3	0.5	0.375	0.5	0.8

theory recognizes the existence of both free and bound oxides, but it denies their molecular base and states, that the structural units are not the electrically neutral molecules, but charged ions.

3.6.1 ELECTROCHEMICAL NATURE OF INTERACTIONS

Various known facts and primarily electrolysis and electrical conductivity of slags has shown that molten slags are electrolytes and are in fact stronger electrolytes.[64] In this connection, the interaction of the metal melt and slag should be regarded as an electrochemical interaction. This is confirmed by a number of experimental data,[241] in particular, occurrence of electrocapillary phenomena in the system of molten metal–slag.

Presence of a potential jump on the molten metal–slag interphase led to the assumption (which was later confirmed experimentally) that superposition of an external electric field on this system changes the magnitude of interphase tension on metal–slag boundary.[95,96]

3.6.2 THE APPROACH

In the first studies of electrocapillary phenomena in the system of molten ferrous metal–slag, the sessile drop method complemented by x-ray photography was used. Electrocapillary phenomena were studied in the system of cast iron–slag, consisting of SiO_2, Na_2O, CaO, and Al_2O_3. In this case, just the cathode branches of the electrocapillary curves were obtained, and the influence of the metal and slag composition on the change of the magnitude of interphase tension with the change of metal potential was established. Cathode branches of electrocapillary curve were also obtained for Fe–P alloys, with phosphorus content of 10.6 and 19.8% on the boundary with $CaO–Al_2O_3–SiO_2$ and $CaO–Al_2O_3–SiO_2–Na_2O$ slags.[96]

B.V. Patrov, studying the electrocapillary phenomena in cast iron–slag system ($CaO–SiO_2–Al_2O_3$) by the method of maximal pressure with reverse capillary, obtained[91,92] both the anode and cathode branches of the electrocapillary curve. Consequently, Patrov raised suspicion about the reliability of the procedure, applied in the early work.[96] The relatively large area of the metal drop and, hence, a considerable magnitude of polarizing current, were regarded as the disadvantages of the sessile drop method.

However, new experiments,[97] conducted by A.A. Deryabin and S.I. Popel' and complemented, unlike previous work,[95,96] by switching measurement of the drop potential, also revealed just the cathode branches of electrocapillary curve for cast iron–slag system ($Al_2O_3–SiO_2–CaO$). A maximum and anode branch were found only for ShKh-15 steel and Fe–C alloy, containing 1.3% carbon. Deryabin and Popel' attributed the existence of anode branches of electrocapillary curve to polarization of these metals at, lower carbon contents,[97] and, as a result, to the possibility of oxygen dissolution in them. In order to confirm this assumption, experiments were conducted to study the electrocapillary phenomena on the boundary of copper, silver, and gold with $Al_2O_3–CaO–SiO_2$ slag.[65] Selection of metals was dictated by different solubility of oxygen in them. In this case, no anode branch was obtained for gold, which almost does not dissolve oxygen, while for copper, which dissolves oxygen, both the anode and cathode branches of the electrocapillary curve were derived. The results of these investigations permitted the investigatons[65,97] to state that the method of x-ray photography of the sessile drop is suitable for studying the electrocapillary phenomena, and the appearance of the derived curves is attributable not to the disadvantages of the procedure, but to the features of the studied systems.

In addition to studying the influence of D.C.,[98] the influence of A.C. on the value of the interphase tension in cast iron–slag system was investigated. It turned out that for this system, a.c. flowing through the metal–slag boundary practically does not affect the magnitude of the interphase tension.

Most of the investigations to study the electrocapillary phenomena were thus conducted for cast iron–slag systems; in systems with ShKh-15 steel and iron–carbon alloy, containing 1.3% C the metal phase have been reported in only one publication.[97] However, low-carbon steels are often used in welding, and the slags mainly contain SiO_2 or SiO_2–MnO. On the other hand, investigations to study the nature of electrocapillary phenomena were conducted at temperatures not exceeding 1773 K, which is below the temperatures, characteristic for welding processes.

Under the conditions of arc and electroslag welding processes, an external electric field is always superposed on the metal–slag system. Therefore, a study of the influence of electric current, flowing through the metal–slag boundary on the magnitude of interphase tension is of special interest for welders. It is obvious that such investigations will allow us to understand better and interpret many processes that occur in welding, for instance, influence of current polarity on the dimensions of electrode drops and weld porosity, formation of undercuts, etc.

However, in order to study the electrocapillary phenomena in welding processes, it is necessary to know the density of current flowing through the metal–slag boundary in its different sections.

3.6.3 EXPERIMENTAL METHODS

Experimental methods, used to find the value of current flowing through the slag in submerged-arc welding, are rather complicated and do not allow finding the current density in different sections of the boundary between the metal and slag. Therefore, in order to determine the current density in this case, it is rational to apply the calculation methods.

Assuming that the area existence of the molten slag is limited by points with the temperature of 1773 and 2573 K, we obtain that at the welding modes of $I_w = 750$ A, $U_a = 30$ V, $V_w = 0.0056$ m/s, $d_e = 5$ mm. The electrolytic cell (Figure 3.38) will have approximately the following dimensions: $R_2 = 3$ mm; $H_1 = 2$ mm; $H_2 = 10$ mm.[219] Let us assume that the unmolten flux is an insulator, the molten electrode-metal has the shape of a cylinder, surfaces of electrode metal and weld pool are the surfaces of equal potentials, and no electrode polarization proceeds during the passage of current. We further assume that the change of temperature on the surfaces of the electrode and the weld pool is from 2573 to 1773 K and the change of specific electric conductivity of the slag in this temperature range is linear, and the arc column has the shape of a cylinder with radius R_1, equal to the radius of the electrode wire.

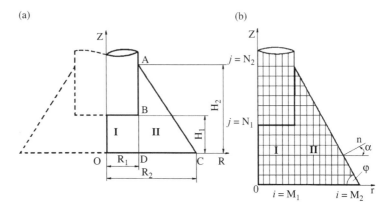

FIGURE 3.38 Schematic of the considered electrolytic cell: (a) general view; (b) with the applied net.

The problem of finding the current distribution on the boundaries of a slag with the molten metal of the electrode and the weld pool is reduced to integration of the equation:

$$\frac{1}{r}\frac{\partial}{\partial r}\left[r\chi(r; z)\frac{\partial U}{\partial z}\right] + \frac{\partial}{\partial r}\left[\chi(r; z)\frac{\partial U}{\partial z}\right] = 0 \tag{3.16}$$

in regions 1 and 11 (Figure 4.5) at the following boundary conditions:

$$U_{s\ cyl} = U_a, \tag{3.17}$$

$$U_{z=0} = 0; \quad 0 < r < R_2, \tag{3.18}$$

$$\frac{\partial U}{\partial n_{s\ cone}} = 0, \tag{3.19}$$

$$\frac{\partial U}{\partial r_{r=0}} = 0; \quad 0 < z < H_1, \tag{3.20}$$

$$U_{11} - U_1 = 0$$

$$\chi_{11}\frac{\partial U_{11}}{\partial n} = \chi_1\frac{\partial U_1}{\partial n}. \tag{3.21}$$

As a rigorous analytical solution of problem given by Equation (3.16) through Equation (3.21) is difficult, we will solve the defined problem by an approximate method, namely the method of finite differences. For this purpose, let us apply a net (Figure 3.38) to the considered area, which will be rectangular matched, that is, all the boundary points fall into the grid nodes. Let us write Equation (3.16) as follows:

$$\frac{1}{r}\chi(r; z)\frac{\partial U}{\partial r} + \frac{\partial U}{\partial r}\left[\chi(r; z)\frac{\partial U}{\partial r}\right] + \frac{\partial U}{\partial z}\left[\chi(r; z)\frac{\partial U}{\partial z}\right] = 0.$$

Approximating this equation[279,280] yields

$$\frac{\chi_{ij}(U_{i+1,j} - U_{i,j})}{r_i h_j} + \frac{1}{a}\left[\frac{a_{i+1,j}(U_{i+1,j} - U_{i,j})}{h_i} - \frac{a_{i,j}(U_{i,j} - U_{i-1,j})}{h_{i-1}}\right]$$

$$+ \frac{1}{b}\left[\frac{b_{i,j+1}(U_{i,j+1} - U_{i,j})}{l_i} - \frac{b_{i,j}(U_{i,j} - U_{i,j-1})}{l_{j-l}}\right] = 0$$

where h_i is the ith step on axis r, l_j is the jth step on axis z;

$$a_{ij} = (\chi_{i,j} + \chi_{i-1,j})/2; \quad b_{ij} = (\chi_{ij} + \chi_{i,j-1})/2,$$

$$a = (h_i - h_{i-1})/2; \quad b = (l_j + l_{j-1})/2.$$

Designating

$$\rho = \frac{\chi_{ij}}{(r_i - h_i)}; \qquad \rho_1 = \frac{(\chi_{i+1,j} + \chi_{ij})}{h_i \, hh_i};$$

$$\rho_2 = \frac{(\chi_{ij} + \chi_{i-1,j})}{h_{i-1} \, hh_i}; \qquad \rho_3 = \frac{(\chi_{i,j+1} + \chi_{ij})}{l_j \, ll_j};$$

$$\rho_4 = \frac{(\chi_{ij} + \chi_{i,j+1})}{l_{j-1} \, ll_j}; \qquad hh_i = h_i + h_{i-1}; \quad ll_j = l_j + l_{j-1};$$

after transformations, we have

$$U_{ij} = \frac{U_{i+1,j}(\rho + \rho_1) + U_{i-1,j}\rho_2 + U_{ij}\rho_3 + U_{i,j-1}\rho_4}{\rho + \rho_1 + \rho_2 + \rho_3 + \rho_4}.$$

$$(3.22)$$

We will use Seidel method[281] to solve the system of algebraic equations, found through approximation. Then, Equation (3.22) becomes:

$$U_{ij}^v = \frac{U_{i+1,j}^v(\rho + \rho_1) + U_{i-1,j}^{v+1}\rho_2 + U_{ij}^v\rho_3 + U_{i,j-1}^{v+1}\rho_4}{\rho + \rho_1 + \rho_2 + \rho_3 + \rho_4},$$

$$(3.23)$$

where v is the iteration number.

Approximation of the boundary conditions yields[279,280]: from the condition (Equation (3.17))

$$U_{ij} = 0; \quad i = 1, \, M; \quad j = 1, \tag{3.24}$$

from the condition (Equation (3.20))

$$U_{ij} = U_{i+1}; \quad 0 < z < H; \quad i = 1; \quad j = 1, N, \tag{3.25}$$

from the condition (Equation (3.18))

$$U_{ij} = U_a; \quad i = M_1; \quad j = N_1 N_2;$$
$$U_{ij} = U_a; \quad j = N_1; \quad i = 1, M_1; \tag{3.26}$$

from the condition (Equation (3.19))

$$\frac{\partial U}{\partial n} = \frac{\partial U}{\partial r} \cos (n, {}^\wedge r) + \frac{\partial U}{\partial z} \cos (n, {}^\wedge z) = 0,$$

where $(n, {}^\wedge r) = \alpha = 90° - \varphi; \quad (n, {}^\wedge z) = \varphi$, (Figure 3.38)
Then,

$$U_{ij} = \frac{\left[\frac{U_{i-1,j} \sin \varphi}{h_{i-1}} + \frac{U_{i,j-1} \cos \varphi}{l_{j-1}} \right]}{\frac{\sin \varphi}{h_{i-1}} + \frac{\cos \varphi}{l_{j-1}}}. \tag{3.27}$$

Then, finally, from matched condition (Equation (3.21)) on the boundary of regions 1 and 11 we have:

$$\frac{\chi_{11}(U_{i+1,j} - U_{ij})}{h_i} = \frac{\chi_1(U_{ij} - U_{i-1,j})}{h_{i-1}},$$

where χ_1, χ_{11} is the specific electric conductivity of the medium in regions 1 and 11, respectively.
Then,

$$U_{ij} = \frac{\left[(\chi_{11} U_{i+1,j}/h_i) + (\chi_1 U_{i-1,j}/h_{i-1}) \right]}{(\chi_{11}/h_i + \chi_1/h_{i-1})}.$$

Considering, that $\chi_{11} \neq$ const, we get

$$U_{ij} = \frac{\left[((\chi_{11ij} + \chi_{11i+1j})/2h_i) + (\chi_1 U_{i-1,j}/h_{i-1}) \right]}{\left[((\chi_{11ij} + \chi_{11i+1j})/2h_i) \right] + (\chi_1/h_{i-1})}. \tag{3.28}$$

Thus, solution of the problem (Equation (3.16) through Equation (3.21)) is reduced to solving the difference problem (Equation (3.22) through Equation (3.28)), which may be implemented, using a computer.

3.6.4 THE INFLUENCE OF CELL DIMENSIONS

Calculations to determine the density of current on the interphases of slag with the molten electrode metal and weld-pool metal showed (Figure 3.39 through Figure 3.42), that current distribution on the above interphases depends on the cell dimensions H_1, H_2, R_1, and R_2, as well as on specific electric conductivity of arc column, χ_1. Values χ_1 of 20 and 30 S/mm are assumed in calculations, which corresponds to the reported data.[282] It was further assumed, than χ_1 is the same in different points of the region 1. It is seen that current density in the low-temperature

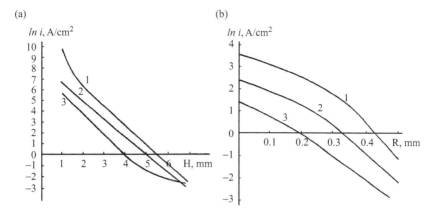

FIGURE 3.39 Influence of the change of the radius of distance between the base metal and electrode on current distribution on the interphases of slag–electrode metal (a) slag–weld pool (b) at $\chi_1 = 20\,S/mm$. The cell dimensions (mm); $R_1 = 2.5$; $R_2 = 3$; $H_2 = 10$; [1] $H_1 = 4$; [2] $H_1 = 3$; [3] $H_1 = 2$.

parts of metal–slag boundary in submerged-arc welding is equal to thousand fractions of an ampere per square millimeter. This is exactly the current density, which was assumed, when studying the electrocapillary phenomena by the sessile drop method.[283]

The diagram of an electrolytic cell, used when studying the electrocapillary phenomena, and of the electric circuit is shown in Figure 3.43.

The electrode used for current supply to the molten metal, was a rod of zirconium diboride, which has a high melting temperature, quite good electric conductivity, and almost does not interact with the metal melt. The auxiliary electrode was a tungsten

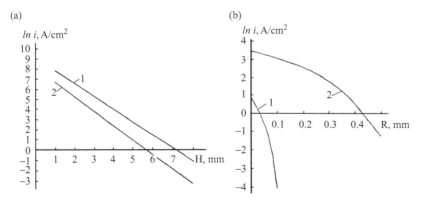

FIGURE 3.40 Influence of the change of the thickness of the molten slag layer on current distribution on the interphases of slag–electrode metal (a) slag–weld-pool metal (b) at $\chi_1 = 20\,S/mm$. The cell dimensions (mm): $R_1 = 2.5$; $H_1 = 2$; $H_2 = 10$: [1] $R_1 = 2.6$ mm; [2] $R_2 = 3$ mm.

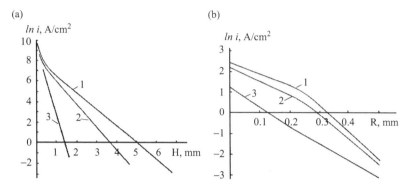

FIGURE 3.41 Influence of the change of the height of the molten slag on current distribution on the boundaries of slag–electrode metal (a) and slag–weld-pool metal (b) at $\chi_1 - 20$ S/mm. The cell dimensions, (mm): $R_1 = 2.5$; $R_2 = 3$, $H_1 = 3$. H_2 (mm): [1] 10; [2] 8; [3] 5.

rod of 6 mm diameter, enclosed into a corundum tube. Absence of visible distortions of the drop profile at current passage through the electrolytic cell is indicative of its uniform polarization, when the rod is used as the auxiliary electrode. The auxiliary electrode was located in the crucible center directly above the drop, and the distance from the drop top to the electrode tip was kept approximately the same in different experiments, as it may affect the accuracy of the experiment, and it varied between 6 and 8 mm.[284] In this case, the dependence of interphase tension not on the magnitude of the drop potential, but on the current density flowing through the metal–slag interphase, was studied. For this reason, the measuring cell did not have a reference electrode, which allowed to simplify the experiment to a certain extent. It should be noted that such measurements are more visual, as they allow establishment of the

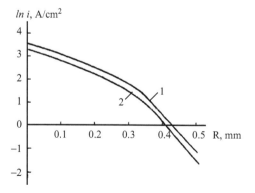

FIGURE 3.42 Influence of the change of electric conductivity of the arc column on the distribution of current on the boundary of slag–weld-pool metal. The cell dimensions (mm): $R_1 = 2.5$; $R_2 = 3$, $H_1 = 3$; $H_2 = 10$. χ_1 (S/mm): [1] 20; [2] 30.

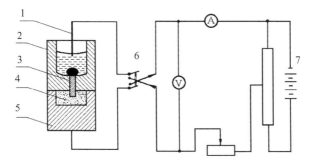

FIGURE 3.43 Schematic of electrolytic cell for investigation of electrocapillary phenomena in steel Sv-08–slag system: [1] auxilary electrode; [2] crucible of fused magnesium; [3] electrode (zirconium diboride); [4] graphite powder; [5] graphite support; [6] pole switch; [7] current source.

magnitude of interphase tension exactly in this section, knowing the current density, flowing through various sections of metal–slag interphase.

In the first experiments, the influence of a constant electric field on the value of interphase tension was studied. Change of interphase tension in the systems of melt of low-carbon steel (Sv-08)–slag, depending on polarity, current density, temperature, and composition of slag, is shown in Figure 3.44. Even though these curves are not the classical electrocapillary curves, reflecting $\sigma_{m-s} = f(\epsilon)$ dependence, the obtained results are indicative of the fact that with superposition of an external constant electric field on the metal–slag system, the change of the magnitude of interphase tension depends on the slag composition and system temperature.

From Lippman equation (Equation (2.19)) for the cathode branch of electrocapillary curve, in which σ_{m-s} value increases with the increase of positive values of φ, $\varepsilon < 0$. Therefore, for systems containing slags consisting of SiO_2–CaO, SiO_2–K_2O and SiO_2–Na_2O, the metal surface has a negative charge.[285] Formation of an excess negative charge in this case is related to transition of iron ions from the metal into slag[286] and of oxygen ions from the slag into the metal. As the metal acquires a negative charge, the cations, present in the slag, adsorb to the interphase from the slag side. Depending on the slag composition, these may be Ca^{2+}, Na^+, or K^+ ions and, to a smaller degree, Fe^{2+} ions (due to its small concentration in the slag). The studied slags, due to their high content of SiO_2, will contain Si^{4+} cations, as well as Si^{2+} cations, which become stable at more than 30 wt% SiO_2 in the slag.[287]

3.6.5 INFLUENCE OF AN EXTERNAL ELECTRIC FIELD

Superposition of an external electric field on the metal–slag system leads to a change of both the magnitude of the charge on the metal surface and of component adsorption into the near-electrode layers, and this will induce a change of the magnitude of interphase tension. For the considered metal–slag systems, the change of the energy of the electric double layer will, apparently, be of great importance, as was reported earlier.[98]

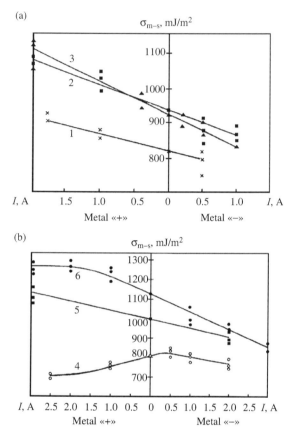

FIGURE 3.44 Dependence of interphase tension on the value and direction of current passing through the Sv-08 steel–slag interface (wt%): (1) 83 SiO_2, 17 K_2O; (2) 83 SiO_2, 17 K_2O; (3) 83 SiO_2, 17 Na_2O; (4) 60 SiO_2, 40 MnO; (5) 50 SiO_2, 50 CaO; (6) 50 SiO_2, 50 CaO. The temperature is 1903 K for 1 and 1803 K for the rest.

It is known[288] that the energy of the electric double layer depends on the metal surface charge ε and layer capacitance C as follows: $W = \varepsilon^2/2C$. Decrease of carbon content in the metal and increase of SiO_2 content in the slag in the absence of FeO in the latter, increase the magnitude of the charge on the metal surface.[241] In addition, presence of a large amount of SiO_2 in the slag should lead to lowering of the capacity of the electric double layer.[241] Increase of the charge on the metal surface and lowering of the capacitance will lead to an increase of the electric energy of the double layer. Consequently, this leads to an enhancement of its role in the change of interphase tension at the superposition of an external electric field on the considered systems. However, the change of the composition of near-electrode phase layers must have the principal role in the change of σ_{m-s} also for these systems.

TABLE 3.8
Values of Interphase Tension in the System of Steel Sv-08–Binary Slag at Interphase Polarization by AC

Slag Composition, wt%	T, K	$i \times 10^2$, A/mm²	σ_{m-s}, mJ/m² With Electric Field	Without Electric Field
SiO₂—83	1803	0.121	948	940
Na₂O—17		0.283	949	
		0.564	951	
SiO₂—83	1803	0.108	942	935
K₂O—17		0.315	940	
		0.606	940	
SiO₂—83	1903	0.138	820	810
K₂O—17		0.307	826	
		0.586	813	
SiO₂—50	1803	0.128	1134	1129
CaO—50		0.294	1132	
		0.585	1132	
SiO₂—50	1853	0.133	998	996
CaO—50		0.361	990	
		0.624	996	
SiO₂—60	1803	0.126	800	805
MnO—40		0.306	815	
		0.578	807	

As A.C. is often used in welding, we have conducted investigations to study the influence of A.C., flowing though the metal–slag interface on the value of σ_{m-s}. The obtained results are given in Table 3.8.

As is seen, an alternating external electric field practically does not affect σ_{m-s} value, which has the following explanation. At low currents, the amplitudes of alternating concentration polarization are equal to[289]:

$$\Delta\varphi_- = I_{m-s}RT/n_v^2F^2c_0\sqrt{D\omega},$$

and at the polarization of the boundary by D.C.

$$\Delta\varphi_d = I_{m-s}\frac{RT}{n_vFI_1}.$$

Here I_{m-s} is the current, flowing through metal–slag boundary; I_1 is the limiting current; n_v is the valence; c_0 is the solution concentration; D is the coefficient of diffusion; F is the Faraday number; R is the absolute gas constant; T is the temperature; ω_c is the current frequency.

Then $\Delta\varphi_{\sim}/\Delta\varphi_{d}$ ratio is equal to $I_1/n_vFc_0\sqrt{D\omega_c}$, if the current value is the same. I_1 value can be determined from the following expression[290]:

$$I_l = \left(\frac{n_vFDc_0}{\delta}\right)\left[1 + 2\sum_{k=1}^{\infty}\exp\left(\frac{-\pi^2k^2D\tau}{\delta^2}\right)\right].$$

At large enough values of time τ the value $I_1 = n_vFDc_0/\delta$, where δ is the thickness of the diffusion layer.

In this case

$$\frac{\Delta\varphi_{\sim}}{\Delta\varphi_d} = \frac{\sqrt{D\omega_\tau}}{\delta}.$$

For a stationary liquid, $\delta = 1\cdot10^{-3}$ mm[291] and D values are of an order of $10^{-8}-10^{-9}$ mm^2/sec.[292]

As the molten metal and slag mix continuously during welding, it may be assumed that $\delta = 1\cdot10^{-3}$ mm. If these values are substituted into the above equation, we will see that $\Delta\varphi$- is much smaller than $\Delta\varphi_d$. This is indicative of the fact that at polarization by A.C. of the metal–slag boundary, the concentration changes in its vicinity will be negligible. Thus, no significant change of the magnitude of interphase tension will occur. Therefore, only in D.C. welding, the magnitude of interphase tension on the metal–slag boundary can be changed by changing the current polarity.

4 Electrode-Metal Transfer and Surface Phenomena

Melting and transfer of electrode or filler metal to the part proceed in the processes of fusion welding, and, primarily, arc welding processes, conducted with consumable electrodes or filler materials. The mode of electrode-metal melting and transfer has a considerable influence on many processes, occurring in welding, and this influence is largely dependent on the kind of welding process. In electroslag welding, the process of electrode-metal melting and transfer is connected with the presence of non-metallic inclusions and impurities in the weld metal. In gas welding, it is related to the composition of the deposited metal and its formation. However, the role of electrode-metal melting and transfer is particularly important in arc welding processes. In this case, they are associated with the efficiency of the welding process, arcing stability, formation and chemical homogeneity of the weld metal, electrode-metal losses for burn-out and spatter, metal interaction with the slag and the gases, etc. Therefore, in this chapter we will focus on electrode-metal transfer in arc welding processes and investigation of the role of surface tension forces in this process. This exercise will provide a knowledge of the facts and factors for influencing and controlling the processes more efficiently.

4.1 FORCES INFLUENCING ELECTRODE-METAL TRANSFER

Electrode-metal transfer is understood to be a totality of events, determining the physical conditions of electrode material, transferring to the part, relative proportions of the different material, and the different physical conditions of their transferring, mode of drop-transfer of the molten electrode material, and dimensions and shape of electrode drops transferring to the part and frequency of their transfer.

4.1.1 The Nature of Electrode-Metal Transfer

It is universally recognized that in arc welding processes, the transfer of the main portion of electrode material to the part is in the drop form. The proportion of vapors and oxides in this case is usually 4–10% and only in some cases, the value is 20–25%. Most of the electrode metal penetrates into the weld pool and together with the molten base metal, it forms a weld, a deposited bead, or a layer after solidification. The rest is lost as burn-out and spatter. The magnitude of the losses largely depends upon the process applied and welding modes.

In electroslag and gas welding, metal losses for burn-out and spatter are very small, and practically all the electrode metal (and in gas welding all the filler metal), transfers to the weld pool in the form of drops.

A drop of the metal melt, formed at the electrode tip, is simultaneously exposed to several forces, the presence and ratio of which are responsible for the drop weight and frequency of drop transfer into the weld pool. The main forces, determining the processes of drop formation and transfer in arc welding are the gravitational force, surface tension force, electromagnetic force, and force of reactive pressure of the vapors. It should be noted that the magnitude and direction of these forces may vary with the change of the conditions and modes of welding. Depending on the used welding process, all of these forces or just some of them may influence the transfer of electrode (filler) metal. Therefore, the most diverse forms of drop transfer of the metal are found in practice, which are inherent in the particular welding process. In order to understand better, the role of each of these forces, let us consider the action of these forces separately.

Numerous studies showed, that in arc welding processes, the gravitational force has a significant influence on electrode-metal transfer only at low currents. In electroslag and gas welding, these forces have a more prominent role, and in this case, they are among the major ones. The gravitational force, depending on the weld position in space, may promote electrode-metal transfer into the weld pool or prevent it. The first case is characteristic for downhand welding, and the second for overhead welding.

4.1.2 The Process of Drop Formation

Surface tension forces are the major forces keep that the drop at the electrode tip. The dimensions of the electrode drops always increase with the increase of these forces. Most of the published reports hold that the surface tension forces contain the electrode metal drop at the electrode tip during the entire process of its growth and detachment. However, by the data of some researchers,[293] in arc welding processes, the surface tension forces change the direction of their action at the drop-detachment stage, thus promoting the breaking up of a liquid-metal bridge, and even impart[294] additional velocity to the drop, transferring into the weld pool.

In order to arrive at a quantitative estimate of surface tension, let us consider the schematic of the drop formation at the electrode tip. During the melting of the electrode or filler wire, a drop of the molten metal forms and grows at the rod tip (Figure 4.1). Drop growth often leads to the formation of a neck, the radius of which (r_n) is smaller, than the electrode radius (r_{el}). Magnitude of the surface tension force (P_σ) in the case of a drop contact with a gaseous medium, i.e. in arc welding in shielding gases or in gas welding, is equal to

$$P_\sigma = 2\pi r_n \sigma_{m-g},\qquad(4.1)$$

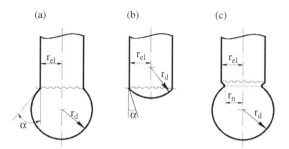

FIGURE 4.1 Appearance of a drop formed at the electrode tip.

and in contact with slag, that is in submerged-arc welding, covered-electrode welding, and electroslag welding,

$$P_\sigma = 2\pi r_n \sigma_{m-s}. \tag{4.2}$$

However, because of the complexity of the determination of r_n value, the value of r_{el} is usually substituted into Equation (4.1) and Equation (4.2). As r_n decreases with the growth of the drop, P_σ value also decreases. The Value of P_σ decreases also with lowering of the values of surface tension (σ_{m-g}) and interphase (σ_{m-s}) tension.

4.1.3 THE ELECTROMAGNETIC FORCE

In arc welding processes, the electromagnetic force has a considerable influence on the electrode-metal transfer. Appearance of this force is due to the interaction of the conductor (through which electric current passes), and the magnetic field, induced by this current. With the passage of current through the molten metal drop and arc column, forces are generated, which try to deform the conductor in the radial direction with a force, proportional to the square of welding current. As the cross-section of the conductor (electrode–drop–active spot–arc column) is variable, an axial component of the electromagnetic force (P_{em}) arises, which is directed from the smaller to the larger cross-section. Therefore, if the dimensions of the active spot are smaller than the diameter of the electrode or the neck, the axial component of the electromagnetic force keeps the drop at the electrode tip and vice versa.

By the data of V.I. Dyatlov,[305] P_{em} is equal to:

$$P_{em} = AI_w^2 [1 + 4.6 lg(r_c/r_{el})], \tag{4.3}$$

where A is the coefficient of proportionality, equal to 0.005 dyn/A^2; I_w is the welding current, A; r_c and r_{el} are the radii of the arc column and electrode, respectively, cm.

4.1.4 The Reactive Forces

Electrode-metal transfer is greatly dependent on the magnitude of the reactive forces. These forces arise as a result of evaporation of a part of the metal from the drop surface and evolution of gases as a result of chemical interaction of the metal melt with the slag or the gas. Metal evaporation in arc welding processes proceeds, mainly, in the active-spot area. It is believed that the resultant of reactive forces (P_r) is applied to the active-spot center and is equal to

$$P_r = (DU_e i_{as}/g_e)^2 \, v_v I_w, \qquad (4.4)$$

where D is the coefficient of proportionality, equal to $3.02 \cdot 10^{-3} \, A^{-1/2}$; U_e is the effective voltage drop at the electrode; g_e is the latent heat of evaporation of the drop material; i_{as} is the current density in the active spot; v_v is the specific volume of the metal vapor at the boiling temperature.

As the current density in the active spot on the cathode is much higher than that on the anode spot, the influence of the reactive forces on electrode-metal transfer is more pronounced in straight-polarity welding. The value, P_r grows at arc constriction, as in this case, the current density in the active spots increases.

4.1.5 Other Forces

In addition to the above forces, the electrode-metal transfer may be further influenced by aerodynamic forces. Yet this will only occur in those cases, wherein powerful plasma (gas) flows arise.

Thus, the magnitude and action of forces influencing the transfer of electrode (filler) metal, depend on the chosen fusion welding process, welding modes, properties of the base and electrode materials, and composition of the slag and the gaseous medium. Therefore, the role of individual forces in the process of electrode-metal transfer can be different, and it will be specific in each concrete case.

4.2 TRANSFER AND SPATTER OF ELECTRODE METAL IN WELDING IN A GASEOUS MEDIUM

As was noted above, in arc welding, most of the electrode metal transfers to the weld pool in the form of drops.

4.2.1 Stages in Drop Transfer

Herein, a distinction is made between the two main forms of drop transfer of the electrode material, namely, without short-circuits and with short-circuiting of the arc gap. Transfer mode is determined by the welding process and the mode, and both the modes often co-exist. Some authors consider the spray transfer as a separate mode, which differs by small dimensions of the drops ($d_d = 0.7 \, d_{el}$) and high frequency of their formation.

In gas-shielded welding, a metal drop formed at the electrode tip, is exposed mainly to the forces of gravity, surface tension, electrodynamic forces, and reactive forces, arising as a result of metal evaporation from the drop surface. The magnitude and ratio of these forces are largely dependent on the shielding gas composition. Argon (inert gas) and CO_2 (active gas) are most often used as shielding gases in the present-day arc welding processes. Mixtures of inert and active gases, for instance, $Ar + CO_2$, $Ar + O_2$, as well as mixtures of active gases, for instance, $CO_2 + O_2$, are quite often used. A mixture of inert gases $Ar + He$ is sometimes applied as the shielding gas. Let us consider the features of electrode-metal transfer for the two most important cases of arc welding, namely consumable-electrode arc welding in argon and in CO_2, as well as in gas welding, as in the latter case, drop formation and transfer proceed in contact of the metal melt with the gaseous medium.

In the case of welding in Ar, globular transfer of electrode metal is observed at low values of the welding current (I_w). Here, drop formation at the electrode tip proceeds for a long time and the drops are of comparatively large dimensions. In this case, the main forces determining the formation and transfer of a large metal drop, will be the forces of gravity and surface tension. The force impact of the plasma flow on the processes of drop formation and detachment for this welding process is just 4–7% of action of aerodynamic forces.[295]

In welding without short-circuiting, the process of drop transfer includes three stages, namely, stable growth, drop detachment, and its flight. The stage of stable growth, occurring under the impact of the forces of gravity and surface tension, is described by the capillarity equation

$$\sigma_{m-g}(1/R_1 + 1/R_2) = \rho g z, \qquad (4.5)$$

where σ_{m-g} is the surface tension of liquid; R_1 and R_2 are the main radii of curvature; ρ is the density; g is the gravity; z is the drop height.

Ivashchenko, and Brodsky[296] considers the process of drop formation and detachment on models, where the liquids used are water and glycerine, having different degrees of viscosity. A feature of the procedure applied in these investigations is the calculation of the capillary pressure, proportional to the average curvature of surface, H, depending on the height of the pendant drop, z

$$P_\sigma = 2H\sigma_{m-g}, \qquad (4.6)$$

where P_σ is the capillary pressure.

For an axially symmetrical surface

$$2H = \{1/[x(1 + x'^2)^{1/2}]\} - \{x''/[(1 + x''^2)^{3/2}]\}, \qquad (4.7)$$

where x is the horizontal coordinate of the drop; x' and x'', respectively, are the first and second derivatives with respect to z.

4.2.2 Drop Detachment

From Equation (4.6) it follows, that the magnitude of the capillary pressure increases with the increase of average surface curvature of a drop, formed at the electrode tip, and of the magnitude of surface tension of the electrode (filler) material.

Experimental determination of a pendant drop contours and calculated values of the capillary pressure are given in Figure 4.2. As is seen, dependence of capillary pressure on drop height for the first contour of the drop, when the neck just starts forming, is linear and hence, this contour corresponds to the stage of a stable growth of the drop. With time, the dependence of capillary pressure on drop height deviates from the linear dependence. In the region of the formed neck, the capillary pressure increases greatly, which facilitates the drop detachment. Therefore, the forces of surface tension will promote detachment of the metal drop from the electrode tip only if the drop growth is accompanied by necking. It should be noted that the effect of this impact should be more noticeable in welding in argon, wherein the magnitude of surface tension of the metal is great.

It has been reported[293] that the forces of surface tension promote breaking up of the bridge between the drop and electrode directly before the drop detachment. However, Beckert et al.[294] provide a solitary report that the forces of surface tension not only promote the drop detachment, but also impart additional velocity to it. Increase of the velocity of the drop may also be related to an increase of capillary pressure in the zone of the formed neck. According to Ivashchenko and Brodsky,[296] presence of capillary pressure leads to appearance of liquid flows in the zones adjacent to the neck. One flow is directed toward the growing drop, and the other toward the part of the drop remaining on the electrode. The fact that not the whole of the drop, but just some part of it, separates from the electrode, has been proved.[297,298] The ratio of the weight of part of the drop, remaining on

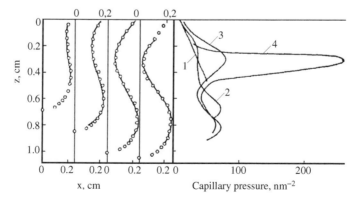

FIGURE 4.2 Parameters of the contour of pendant water drops and the corresponding capillary pressures (points indicate experimental data, lines show the data of, approximating the drop-contour seetrous).

the electrode, to the weight of the entire drop may vary from 0.1 to 0.9, depending on the welding process and modes.

Velocities of the flows, arising in the metal melt as a result of appearance of the capillary pressure may reach a considerable magnitude, as the metal melts are low-viscous liquids. Appearance of such flows will provide additional impetus to the detaching metal drop, which will promote its acceleration.

4.2.3 FORMATION OF METAL SPATTER

Presence of metal flows in the zones, close to the neck, provided their velocity is high enough, will lead, in keeping with Bernoulli law, to the appearance of secondary necking and hence, to the formation of fine drops-satellites. Appearance of fine drops may cause spattering of the electrode metal in consumable electrode argon-arc welding. However, rather few such drops are formed, and, therefore, electrode–metal losses in this welding process are not high. Appearance of metal spatter on the part surface in this case may be related to the removal of gas bubbles from the weld pool.[299]

Labor consumption in the operation of spatter removal largely depends on the strength of the contact between the drop and part surface. Let us analyze, which factors influence and in what manner they influence the strength of the contact between the part surface and the metal drops landing on it.

Temperature of the electrode-metal drops is known to reach 2600 K. When a metal drop, having such a temperature hits the part surface, it starts spreading. Melt spreading in this case proceeds under non-isothermal conditions, as the drop temperature (T_d) is much higher, than the temperature of part surface (T_s). The process of metal-drop spreading lasts until the drop metal has solidified. Duration of drop-metal solidification depends primarily on the initial values of the temperature of the drop and part surface, and may be given by the following expression[94]

$$t_d = (1/\alpha)\left[\ln\left(T_d - T_s\right)/(T_m - T_s)\right], \tag{4.8}$$

where T_m is the melting temperature of the drop material; α is the constant.

According to the data of Fedko and Sapozhnikov,[300] the temperature of metal drops, landing on the base-metal surface, is already just 523–773 K after 1 sec, depending on the drop diameter and thickness of metal being welded. That is the time of drop spreading over the solid-metal surface is very short.

The ratio of the temperatures T_d and T_s is also a major determining factor in the strength of the drop adhesion to the part surface. If at the moment of physical contact of the drop with the part surface, T_s value is high enough, a good chemical interaction will occur between the contacting surfaces, providing a strong adhesion of the drop to the base metal. At lower temperatures of the part surface, the chemical interaction will not proceed over the entire surface of contact, but will just cover a part of it, or will not occur at all.

According to Equation (1.6), the magnitude of adhesion, which characterizes the strength of the metal drop adhesion to the part surface, depends on surface tension of the drop metal and the wetting angle, θ; the better is the part surface wetted by the drop metal, the higher is the magnitude of adhesion. For this specific case, that is, for a given electrode wire and a given shielding gas, the magnitude of surface tension of the molten drop metal will be a constant value. Therefore, the magnitude of adhesion of the metal drop to the solid metal may be influenced only by changing the value of the angle θ.

4.2.4 THE ROLE OF PART PREHEATING

The value of the wetting angle is influenced by various factors. Solid metal wettability by the molten metal may be improved by preheating the solid metal. Therefore, application of part preheating prior to welding should promote the drop spreading over the surface of base metal, higher adhesion, and thus making it more difficult to remove the metal drops from the part surface. This is confirmed by the results of Fedko,[301] according to which, the area of contact of the drop–solid metal and strength of adhesion on this interphase increase with the rise of temperature T_s (Figure 4.3 and Figure 4.4). The strength of adhesion and spreading of the metal drop also depend on the condition of the base-metal surface (Figure 4.3 and Figure 4.4). Greater roughness of the part surface, which promotes wetting of the solid metal by the metal melt (see Section 3.3), enhances the adhesion of the drop to the solid metal. This is confirmed, in particular, by experimental data,[302] according to which the quantity of difficult-to-remove spatter on the part surface is as follows: as-delivered metal, 47%; after sand-blasting of the part surface, 67%; after treatment of the part surface by an emery wheel, 73%.

Strength of adhesion can be lowered by applying onto the part surface prior to welding, a coating, which is poorly wetted by the molten metal. As is seen from Table 3.3, such coatings may be oxide films of MgO or Al_2O_3, on the surface of

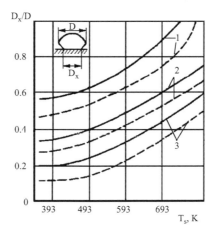

FIGURE 4.3 Change of diameter of the spot at the drop–base metal contact, depending on the initial temperature of the part: (1) part grinding with an abrasive wheel; (2) sand-blasting of the part surface; (3) as-delivered metal.

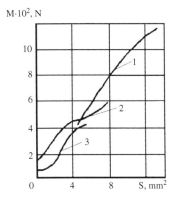

FIGURE 4.4 Change of the strength of drop adhesion to part surface, depending on the area of part–drop contact: (1) part grinding with an abrasive wheel; (2) sand-blasting of the part surface; (3) as-delivered metal.

which the values of wetting angle are greater than $100°$ for the most diverse metal melts in the atmosphere of Ar, H_2, N_2, and CO. In the oxidizing gas medium (air, CO_2), substrate wetting is improved, but value of θ still remains to be quite high, being in the range of $30–50°$ (Figure 4.5).

It should be noted that, when a drop hits the surface of a solid metal, the latter is heated, and this may lead to a greater time required for the drop to spread over the solid-metal surface, lower wetting angle, and greater adhesion of the molten metal to solid metal. Therefore, it is important to know, what and how of the influences that operate in the level of parts preheating, which results from metal drops landing on the base-metal surface.

As the convective flows in the metal drops which land on the solid metal surface are small, it may be assumed that heat transfer from the drops to the part occurs due to heat conductivity, and this process is described by the Fourier equation:

$$dQ_x = \lambda (dT/dx) F dt, \tag{4.9}$$

where dQ is the quantity of heat propagating during time dt due to heat conductivity; λ is the coefficient of heat conductivity; dT/dx is the temperature gradient.

If a drop has the shape of a circle, and spreads under the impact of the forces of gravity and surface tension, the equilibrium value of drop thickness is equal to[274]:

$$\delta_b = \left[\frac{2\sigma_{m-g}(1 - \cos\theta)}{\rho_m g} \right]^{1/2}. \tag{4.10}$$

In contrast, by going over to area of contact F, Equation (4.10) becomes:

$$F = \frac{mg^{1/2}}{[2\rho_m \sigma_{m-g}(1 - \cos\theta)]^{1/2}}, \tag{4.11}$$

where m is the drop weight.

(a)

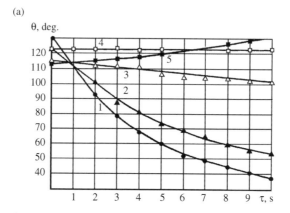

(b)

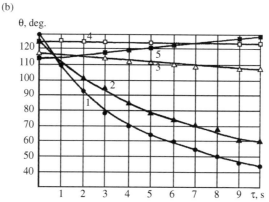

FIGURE 4.5 Change in time of the wetting angle of the surface of corundum by a metal melt in CO_2 (a) and air (b) $T = 1803$ K: (1) armco-iron; (2) St3 steel; (3) 08Kh20N10G6 steel; (4) U8A; (5) alloy Fe–Si (Si = 1.95 wt%).

From Equation (4.11), it follows that the area of contact between a part surface and a metal drop, landing on its surface, will be larger, the greater the drop weight and the lower the density of the drop material, magnitude of the melt surface tension, and the value of wetting angle.

In view of Equation (4.11), the Fourier equation may be written as:

$$dQ_x = \frac{\lambda(-dT/dx)mg^{1/2}}{[2\rho_m\sigma_{m-g}(1 - \cos\theta)]^{1/2}}\,dt. \qquad (4.12)$$

Therefore, local heating of a part, when molten metal drops land on a solid-metal surface, depends on thermo-physical properties of the part material, temperature of the drop and part, drop weight, surface properties of the melt, and wettability of the solid metal by the molten metal.

Now, the latter factor has the most important role, as the value $(1 - \cos\theta)^{1/2}$ changes to the greatest extent among those values, which are included into this equation, as will be shown later (see Chapter 5).

Adhesion of a spatter to the solid metal may be reduced, if appropriate conditions are created for the formation of a slag crust on the surface of drops, hitting the solid metal.

In welding in oxidizing media, the spatter drop surface is always covered by a layer of oxides, the composition of which depends primarily on the composition of the drop metal. In surfacing in air, (the oxidizing potential of oxygen is approximately the same as that of CO_2), using Sv-08G2S wire, the oxide film, formed on the surface of the electrode-metal drop, consists mainly of MnO and SiO_2.[152] Presence of such an oxide film impairs the wettability of the solid metal by the molten drop, and lowers the heat transfer from the melt to the solid metal, eventually reducing the adhesion of the drops to the part surface. Thus their removal from the part surface is simplified.

4.2.5 WETTABILITY AND WELDING CURRENT

Thus, wettability of the solid metal by the molten metal drops influences the strength of the drop adhesion to the part surface.

At low values of welding current, when large-sized drops form, their transfer to the weld pool may also result from short-circuiting. This process may be conditionally subdivided into three stages: (1) formation of a drop at the electrode tip; (2) occurrence of a short-circuit and drop spreading over the pool surface; (3) reduction of the diameter of the liquid bridge between the pool and part of the drop remaining on the electrode and breaking up of the bridge. In this case, the drop forms similar to the case considered above. Here, the second stage has an important role in the process of drop transfer to the weld pool. Experiments show that usually a series of brief short-circuits occur at first between a large drop and the pool, which are accompanied by a transfer of a small amount of the metal into the weld pool. A long-term contact with transfer of the main part of liquid metal forms only after several such short-circuits. It is natural that the processes proceeding at the second stage, depend on the wettability of the pool metal by the drop and the intensity of drop spreading over the weld-pool surface. Under the conditions of argon-arc welding, when the molten metal protection is good, the surfaces of weld pool and drop are free from oxide films. Therefore, wettability and spreading should be good and the duration of the second stage be short.

Reduction of the diameter of the liquid bridge in the third stage of transfer is related to reduction of the free surface of the metal melt, and, therefore, increase of surface tension of the melt should promote breaking up of the bridges, formed during transfer.

With the increase of welding current, the influence of the gravitational force on electrode-metal transfer becomes weaker, and the role of the electromagnetic forces, promoting the drop detachment from the electrode, is considerably enhanced. This leads to the electrode-metal drops becoming smaller. Furthermore, with the increase

of the welding current, the nature of the transfer of the metal changes from the globular to the fine-drop mode. After a certain value of the welding current has been
reached, (the critical current), spray transfer of electrode metal is observed. According to the data of Petrov,[303] the value of critical current is related to surface tension
of the metal melt, and is given by:

$$I_{cr} = k\sqrt{\sigma_{m-g}r_{el}}, \qquad (4.13)$$

where k is the coefficient of proportionality, determined experimentally; r_{el} is the
radius of the electrode wire.

As is seen from Equation (4.13), the smaller the magnitude of surface tension,
the lower is the value of the critical welding current.

4.2.6 THE ROLE OF OXIDIZING GASES

Decrease in the value of critical current due to lowering of surface tension of electrode metal may be achieved by the addition of the oxidizing gas (oxygen or CO_2) to
argon, which is often used in practice. However only at small concentrations of O_2
and CO_2 in argon (5–7%), the effect of their influence on electrode-metal transfer
will be similar (Figure 4.6).[304] With further increase of oxygen content in argon
the fine-drop transfer is preserved, and with increase of CO_2 content, the drop
dimensions increase.

This is attributable to the fact, that unlike welding in argon or in a mixture of
$Ar + O_2$, the reactive forces have a strong influence on electrode-metal transfer,
in a mixture of $Ar + CO_2$. This is similar to the behavior in pure CO_2. According

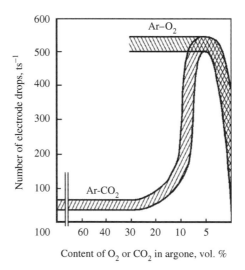

FIGURE 4.6 Influence of the composition of gas mixtures of $Ar-O_2$ and $Ar-CO_2$ on the
number of electrode metal drops ($d_{el} = 1$ mm; $I_w = 250$ A).

TABLE 4.1
Surface, Volume Concentration, and Adsorption
of Silicon in Iron-Based Alloys (T = 1773 K)

Si_v, %	Γ_{Si}, mol/cm^2	Si_s, %
0.80	$3.44 \cdot 10^{10}$	3.11
0.20	$1.07 \cdot 10^{10}$	0.57
0.05	$0.30 \cdot 10^{10}$	0.07

to Dyatlov,[305] this effect is related to a higher degree of constriction of the arc, running in CO_2, which is due to strong overcooling resulting from the high heat conductivity of the shielding gas, and dissociation of CO_2 molecules. In addition, as shown by our investigations, the electrode wires used in welding in CO_2 which usually contain about 1% silicon, have a rather large σ_{sol-g} value. This effect is, to a certain extent, related to an increased concentration of silicon on the melt surface, due to its surface activity in iron-based alloys.

Calculated data, produced by V.I. Yavojsky[306] and given in Table 4.1, are indicative of the fact that surface concentration of silicon is much higher than its volume content.

For these reasons, globular transfer is usually observed in CO_2 welding, particularly in straight-polarity welding, which, similar to argon-arc welding, may proceed with short-circuiting of the arc gap and without it.

Several references report that fine-drop and even spray transfer of the metal in straight-polarity CO_2 welding may be achieved by activation of the electrode wire with alkali and alkaline-earth elements.[307,308] However, when regular wires are used, globular transfer of the electrode metal is found in CO_2 welding. A characteristic feature of CO_2 welding is the more intensive spatter of electrode metal. This is a substantial drawback of CO_2 welding, as the spatter clogs the torch nozzle and detracts it from its shielding properties. Further, as the spatter also lands on the base metal, its removal from the part surface increases the labor consumption in structure fabrication. In addition, the electrode-metal spatter may lead to an increased consumption of electrode wire.

Spatter of electrode metal in CO_2 welding greatly depends on welding current and diameter of the wire used. To reduce spatter in welding with small diameter wire, the welding is performed with short-circuiting.[1]

According to Bukarov and Ermakov,[309] the contact between the drop and weld pool is established immediately, and the nature of the diameter of contact between the drop and the changes in the weld-pool surface with time is the same both for argon-arc and for CO_2 welding (Figure 4.7). However, in welding in Ar, the time for the drop to spread over the weld-pool surface is shorter than in the case of CO_2 welding. This is probably related to the features of wetting and spreading processes. In CO_2 welding, unlike welding in argon, the surface of the drop and weld

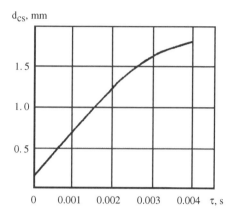

FIGURE 4.7 Change in time of the diameter d_{cs} of the spot of drop–weld pool contact.

pool is covered by a layer of oxide films, which have a higher viscosity than the molten metal. Therefore, spreading of such drops over the weld-pool surface proceeds at a lower velocity.

Presence of oxide films on the weld-pool surface in CO_2 welding lowers also the probability of the spatter of the weld-pool metal. This is owing to removal of gas bubble. This effect is demonstrated by the results of our studies (see Chapter 7). Therefore, mostly electrode-metal spatter occurs in CO_2 welding.

Influence of the composition of the shielding gas on electrode-metal transfer is further confirmed by other data. Surface tension of chromium–nickel steels on the boundary with nitrogen is higher than in argon (Table 3.2). Therefore, its addition to argon increases the critical welding current,[303] when welding wires of such a composition are used.

4.2.7 INFLUENCE OF ELECTRODE WIRE COMPOSITION

Influence of the electrode wire composition on electrode-metal transfer in welding in CO_2 and mixture of Ar + 25% CO_2 was studied by Protsenko and Privalov.[310] Their results are given in Figure 4.8. As is seen, silicon has the strongest influence on the characteristics of the electrode-metal transfer in welding in pure CO_2 and the gas mixture. In the opinion of Protsenko and Privalov,[310] this is attributable to the fact that silicon has the highest surface activity in iron-based alloys, compared to other studied alloying components (Mn, Cr, Ni, Mo, Al, Ti). Therefore, silicon concentration on the drop surface is higher than that of other alloying components, so that it will bind the surface-active oxygen and prevent its penetration into the metal melt better than other alloying elements.

Thus, in arc welding with solid wires, both in argon and CO_2, the process of metal transfer from the electrode into the weld pool is largely determined by the surface properties of electrode metal and surface phenomena (wettability and

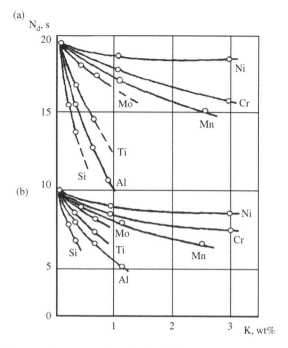

FIGURE 4.8 Influence of concentration K of alloying elements in the welding wire on the frequency of drop transfer in CO_2 welding ($d_{el} = 2$ mm): (a) $I_w = 200-220$ A; $U_a = 27-28$ V; (b) $I_w = 400-430$ A; $U_a = 33-34$ V.

spreading) on the boundary of the weld pool–electrode-metal drop. As flux-cored wires are used in gas-shielded welding along with the solid electrodes, let us consider the features of electrode-metal transfer also for this case.

4.2.8 FLUX-CORED WIRES AND FILLER METAL

General features of metal transfer in gas-shielded arc welding are characteristic also for the case, when flux-cored wires are used. Here also, despite some special features,[311] the melting process is of a periodic (drop) nature. Therefore, similar to solid-wire welding, lowering of surface tension of electrode-drop metal leads to a reduction of its weight.[298,312] Influence of alloying components on the weight of electrode-metal drops corresponds to surface activity of these components.

The gas welding process is also accompanied by melting of the filler metal, drop formation, and its transfer to the weld pool at the contact of the molten metal with the gaseous medium. However, in gas welding, no electric current flows through the filler wire, and, hence, electrodynamic forces are absent. Moreover, in this case, the drop temperature is lower than with arc welding processes, that is why no noticeable evaporation of the metal occurs, so that the reactive forces do not have any significant influence on the transfer of molten metal drops. Therefore, in gas welding, mostly the forces of gravity and surface tension influence drop transfer into the weld pool. Thus, the process of formation and transfer of the molten metal drops may be controlled also in gas welding by varying the magnitude of the surface tension of the filler metal.

4.3 TRANSFER OF THE ELECTRODE METAL CONTACTING THE SLAG

The contact of the molten electrode metal with the slag occurs in covered-electrode welding, submerged-arc welding, and electroslag welding.

4.3.1 THE UNDERLYING FACTORS

In covered-electrode welding, performed in regular modes, two kinds of electrode-metal transfer are observed, namely globular and fine-drop. As shown by numerous investigations, the kind of transfer depends on welding modes, as well as the composition and thickness of the coating, and kind and polarity of the current applied. As in covered-electrode welding, the drop of metal hanging on the electrode tip comes into contact with the slag, formed by the melting of the coating; when the drop separates, it has to overcome the forces of interphase tension on the metal–slag boundary. The magnitude of interphase tension may vary (see Chapter 3) with the change of the composition of the metal and the slag and the temperature of the metal–slag system at the passage of D.C. through the metal–slag boundary.

4.3.2 THE IMPACT OF INTERPHASE TENSION

Experimental data on the interconnection of the frequency of drop formation and the magnitude of interphase tension on the metal–slag boundary are given by Mazel.[313] In those experiments, a thin layer of the single-component coating was applied onto a metal rod. Welding was performed in the atmosphere of nitrogen or CO_2, and the number of the drops formed was estimated by the number of short-circuits. Welding current and arc length were maintained approximately constant in all the tests.

Performed studies showed (Figure 4.9), that drop detachment from the electrode tip is usually facilitated at lowering of the magnitude of interphase tension on metal–slag boundary. Gaseous medium had little effect on electrode-metal transfer.

4.3.2.1 Basic-Type Coating

Influence of interphase tension on electrode-metal transfer in covered-electrode welding is further confirmed by the results of other investigations. It is known[1,313] that in welding with basic-type electrodes, globular transfer is observed in a wide range of welding modes. Enlargement of electrode-metal drops in this case is attributable to two causes.[1] With electrodes with a basic coating, the atmosphere of the arc contains about 30 vol.% CO_2. Dissociation of CO_2 causes constriction of the arc column and active spots, leading to an increase of the reactive forces, which in the covered-electrode welding usually are smaller than in welding with active

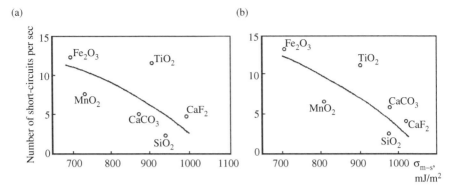

FIGURE 4.9 Dependence of the number of short-circuits in covered-electrode welding on the magnitude of interphase tension on the boundary of electrode metal–slag: (a) welding in the atmosphere of nitrogen; (b) CO_2 welding.

gases. The consequent effect on the axial component of the electromagnetic force prevents the drop detachment. This is further related to a high value of interphase tension on the boundary of the electrode metal–slag, which in this case contains mainly CaO and CaF_2, the presence of which in the slag increases[90] the σ_{m-s} value. In the published work,[298,314] it was noted that in the presence of CaF_2 and $CaCO_3$ in the coating, increase of $CaF_2/CaCO_3$ ratio leads to a greater diameter of the drop and short-circuit duration (Figure 4.10). Increase of $CaCO_3$ content in the coating through the dissociation of the latter in heating, even though it causes

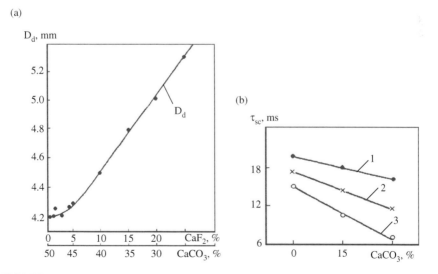

FIGURE 4.10 Dependence of dimensions (a) and duration of existence (b) of electrode-metal drops on the content of CaF_2 and $CaCO_3$ in electrode coating: [1] CaF_2 in the inner layer; [2] CaF_2 uniformly distributed in the coating; [3] CaF_2 in the outer layer.

an increase of reactive forces, simultaneously promotes enhanced oxidation of the metal drops, and, hence reduction of the σ_{sol-s} value and drop size. Processing properties of electrodes with a basic type of coating can be improved,[314,315] using two-layer coatings, wherein the layer, adjacent to the electrode rod, is made of a material on the boundary with which the interphase tension of the metal melt decreases.

For instance, use of electrodes[315] with an inner layer of $CaCO_3$, and an outer layer of CaF_2, allowed reducing the dimensions of the formed drops. This was achieved at a high total content of CaF_2 in the coating (Figure 4.10).

4.3.2.2 Acidic Coating

In welding with electrodes with an acidic and rutile coatings, fine-drop transfer of electrode metal is observed. Small dimensions of the drops are due to comparatively low σ_{m-s} values, as the slags, formed at melting of the electrode coatings in this case, contain components (MnO, FeO), that lower the value of interphase tension on electrode metal–slag boundary. For electrodes with acidic and rutile coatings, dependence of drop dimensions on welding current is noted (Figure 4.11). Increase of current density leads to a noticeable reduction of the weight of the drops, detached from the electrode.

In covered-electrode welding, an electric current flows through the electrode-metal drop–slag interphase. This may change the magnitude of the interphase tension on the metal–slag boundary. The extent of this influence, as was shown earlier, depends on slag composition, density of current flowing through the metal–slag boundary, and the temperature of the system. In this connection, change of the kind and polarity of

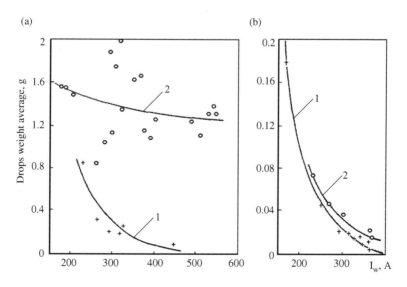

FIGURE 4.11 (a) Dependence of the average weight of the drops on welding current in welding with UONI-13/45 electrodes of 8 mm diameter and (b) OMM-5 electrodes of 5 mm diameter [1] straight polarity; [2] reverse polarity.

electric current used in welding should lead to a change of the mode of electrode-metal transfer. This is confirmed by experimental data on welding with electrodes with different types of coatings.

4.3.3 THE INFLUENCE OF WELDING POLARITY

In welding with electrodes with a basic type of coating of UONI-13/45 grade, change of welding current polarity has a marked influence on the weight of electrode metal drops (Figure 4.11). In welding with electrodes with an acidic coating (OMM-5 electrode), the weight of electrode metal drops remained practically the same both at straight- and reverse-polarity welding (Figure 4.11). Change of polarity and kind of the applied current has little influence on drop weight in welding with rutile-type electrodes (ANO-4 electrode). In this case, the drop weight in welding with A.C and D.C of straight and reverse polarity is approximately the same (Figure 4.12). Increase of current density leads to lowering of the weight of the molten metal, which remains on the electrode tip. This is indicative of the drop detachment boundary shifting toward the electrode, as a result of increase of the melt temperature.

4.3.3.1 Polarity and Single-Component Coatings

A.A. Erokhin studied the influence of single-component coatings on electrode-metal transfer[297] at different polarity of the welding current. Liquid glass was used to apply coatings on 4 mm diameter rods of Sv-08 wire, coating weight being 3–5% of that of the rod. Results of the performed investigations are given in Table 4.2.

As is seen, the weight of drops at straight polarity is smaller than that of drops at reverse polarity in all the cases. In manual arc welding, as the welding current densities are comparatively smaller, electrodynamic forces, acting on the drop should hinder its detachment.[305,313] These forces are greater at straight polarity than at reverse polarity. Judging by the values of the melting coefficient, derived by Erokhin,[297] it can be stated, that at least in three cases, the reactive forces, acting on the drop, are also greater, than at reverse polarity. Location of the heated spot at straight and reverse polarity in welding with electrodes with coatings of marble and quartz sand was the same, at the bottom of the drop. Therefore, the observed change of the drop weight cannot be

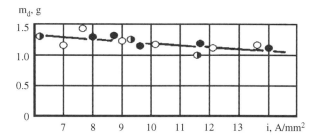

FIGURE 4.12 Influence of density, polarity, and the kind of current on the weight of drops in welding with a rutile-type electrode (ANO-4): (○) reverse polarity; (●) straight polarity; (◑) A.C.

TABLE 4.2

Influence of Coating Composition and Welding Current Polarity on Electrode Drop Weight

Coating	Average Weight of the Drop, at Polarity, mg	
	Straight	Reverse
Marble	46.0	100.0
Quartz sand	71.5	74.0
Fluorspar	31.1	124.0
Titania	18.2	46.6

attributed to the impact of electromagnetic or reactive forces nor to change of the heated spot location on the drop. However, the obtained results are in good agreement with the data on the change of interphase tension in metal–slag system at superposition of a constant external electric field on it.

In submerged-arc welding, drop transfer of the electrode metal is observed. Investigations conducted using high-speed x-ray photography showed that in submerged-arc welding, the drop dimensions depend on welding current polarity. For instance, in welding with AN-20 flux, which consists mainly of SiO_2, CaF_2, MgO, and Al_2O_3 and practically no FeO or MnO, a reduction of the number of formed drops is observed at reverse polarity, when σ_{m-s} value on the electrode metal–slag boundary increases at superposition of an external electric field (Figure 4.13).

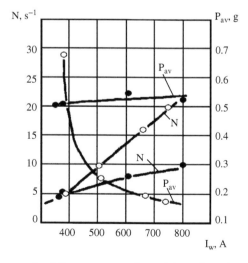

FIGURE 4.13 Influence of welding current on the weight and number of transferring drops ($U_a = 40$ V): (○) reverse polarity; (●) straight polarity.

4.3.3.2 Polarity Effect with Flux

We conducted experiments on bead deposition on plates of St3 steel with Sv-08 wire, using fluxes consisting of $60\%SiO_2-40\%MnO$ and $17\%Na_2O-83\%SiO_2$, at D.C. of straight and reverse polarity. In this case, it is established that when a flux, containing SiO_2 and MnO is used, no noticeable disturbance of deposited metal formation occurs in welding either at direct or reverse polarity at high welding speeds (0.0228 m/s), which is ensured by a high frequency of formation of electrode-metal drops. However, in this case, σ_{m-s} value is small (Table 3.4) and changes only slightly with the change of the polarity of the current, flowing through the metal–slag boundary (Figure 3.44).

4.3.3.2.1 Submerged-Arc Welding with a Flux

The situation is different in submerged-arc welding, using a flux, consisting of Na_2O and SiO_2. In this case, in reverse polarity welding, when σ_{m-s} value reaches 1100 mJ/m^2 on the electrode metal–slag boundary (Figure 3.44), no weld formation is found and individual metal drops solidify on the plate already at the welding speed of 0.011 m/s and higher. This is indicative of a low frequency of drop formation. In straight-polarity welding, when σ_{m-s} value is comparatively small (equal to about $800-850$ mJ/m^2), the deposited bead formed even at the welding speed of 0.0228 m/s. This value is close to the value of interphase tension in the system of steel Sv-08–slag, consisting of SiO_2-MnO.

4.3.3.2.2 Electroslag Welding

In electroslag welding, the electrode metal also transfers into the metal pool as drops. However, in electroslag welding, the temperature of electrode drops is much lower than with the arc processes, and the reactive forces in this case do not have any noticeable influence on electrode-metal transfer. Drop size and frequency of their transfer into the weld pool in electroslag welding are determined primarily by the gravitational force, interphase tension force, and electromagnetic force. With decrease of current density, the role of electromagnetic forces becomes smaller and the influence of surface forces becomes stronger. Therefore, all the factors, which influence σ_{m-s} value (composition of metal and slag, slag-pool temperature, kind and polarity of current) will influence the dimensions of the formed drops. This is also confirmed by the experimental data.

According to Latash and Medovar,[316] increase of slag-pool temperature, which should lead to lowering of interphase tension in the metal–slag system, results in the reduction of drop dimensions. Experimental data[316] on the influence of an external electric field on drop dimensions, derived using the mercury model are indicative of the important role of electrocapillary phenomena in electroslag welding processes.

Thus, the forces of surface tension are important in the processes of electrode-metal transfer with all fusion welding processes. Therefore, using the techniques, which allow changing the magnitude of these forces, it is possible to control the electrode-metal transfer, thus influencing the performance of welded joints and the deposited metal.

5 Formation of Weld and Deposited Metal

Dimensions and shape of the weld have a significant influence on the quality of the welded joint. Weld-metal resistance to solidification cracking, strength characteristics of the welded joint, composition of the weld and deposited metal, electrode-metal consumption, etc. are other associated factors. In this connection, much attention is given to studying the processes of formation of the weld and the deposited metal.

Influence of the modes of welding and surfacing, and of various techniques (workpiece preheating, superposition of an external electromagnetic field, inclination of workpiece and electrode) on the processes of formation of the weld and the deposited metal has now been investigated in a rather detailed manner. The influence of some physical properties of welding consumables (metal density, slag viscosity) on weld formation has also been studied to a certain extent.

However, the role of surface properties of the metal and the slag, as well as of the surface phenomena, proceeding at the boundaries of the contacting phases, in the processes of formation of the weld and deposited metal is absolutely insufficiently studied. Therefore, in this chapter, the main attention is given to consideration of the influence of surface properties and phenomena on the processes of formation of the weld (deposited metal) and development of weld-shape defects (undercuts, lacks-of-penetration, lacks-of-fusion, burns-through, craters, rolls).

5.1 WELD AND SERVICEABILITY OF A WELDED STRUCTURE

Achieving the service reliability of a welded structure has always been one of the major tasks of welding fabrication. These issues are currently given ever greater attention. This is related to more complex conditions of service of the welded structure arising due to higher working stresses, wider range of service temperatures, operation in aggressive media, etc. In addition, welding is very often applied for fabrication of items, where the configuration and dimensions promote initiation of hazardous stress concentrations. Welders have to deal more often with high-strength materials and materials with high reactivity and high melting temperature, as well as novel materials, for instance, composite materials.

5.1.1 CAUSES OF STRENGTH DETERIORATION

It is known that the theoretical strength of a metal (which is characterized by the strength of adhesion between the atoms), is hundreds of times higher than its practically exhibited strength. Lowering of strength of metals is attributable to many reasons, namely, imperfections of the metal crystalline lattice, change of its structure, and for welded joints also to the presence of defects.

Practically, in material processing, various defects always form, irrespective of the kind of the technological process. The kind of defects and mechanism of their formation depend on the features of the chosen technological process.

The welding process is no exception to the generalization that manufacturing leads to development of defects. Similar to other manufactured products, presence of defects in a welded joint lowers the performance of the welded structure to varying degrees and promotes its failure.

Brittle fractures are the most hazardous for welded items. They occur suddenly and most often at stresses much lower than the design stresses. In this case, the fracture sites may be located both on the main (load-carrying) and secondary structural elements and their appearance is related to imperfections of a welded item or to the presence of defects in a welded joint.

Presence of stress-raisers is a necessary condition for brittle fracture initiation. Therefore, the most hazardous defects in a welded joint are cracks or crack-like defects (undercuts, lacks-of-penetration, lacks-of-fusion), which (Figure 5.1)

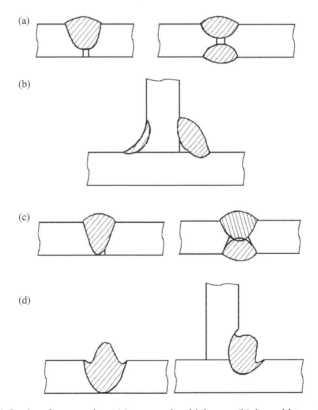

FIGURE 5.1 Lacks-of-penetration: (a) across the thickness; (b) in weld top; (c) along the edge; and (d) undercuts in the weld.

develop in welding. These defects, being plane, are rather difficult to detect by x-ray photography, which makes them even more dangerous. Therefore, it is necessary to avoid these defects in a welded joint during welding. This cannot be achieved without a fair knowledge of the mechanism of their initiation and giving allowance for all the factors influencing their development.

The process of failure of any material consists of several stages: (1) crack initiation; (2) its stable growth up to a certain critical value; and (3) spontaneous crack propagation.

As the fracture process starts generally with the appearance of an embryo crack, presence of cracks or crack-like defects in the metal is a factor, which makes it prone to fracture.

However, not any cracks and hence, crack-like defects can be the cause for metal fracture. For low-carbon steels, for an item under tension in service, cracks of $(2-2.5) \cdot 10^{-3}$ m are not hazardous.[317] Therefore, not all the cracks, present in the metal, may lead to structure failure, and the higher the ability of the metal to undergo plastic deformation, the lower is the probability of brittle fracture of this metal, and the admissible crack dimensions become greater.

Analysis of the cases of failure of welded structures, operating in the most diverse conditions, indicates, that their fracture most often starts from defects, present in the welded joint or the base metal. Defects, located in the welded joint, are more hazardous[318] (Figure 5.2). This is related to the presence of residual stresses in the welded structure, which result from non-uniform high-temperature heating of the metal in welding. The connection between breaking stresses and the presence

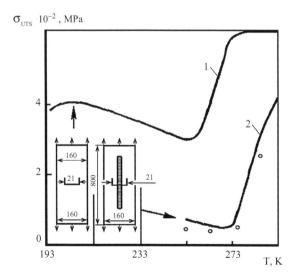

FIGURE 5.2 Temperature influence on the value of breaking stresses in steel St3sp (thickness of 25 mm) in the presence of a defect in the initial material (1) and the weld zone (2) dots indicate the calculated data.

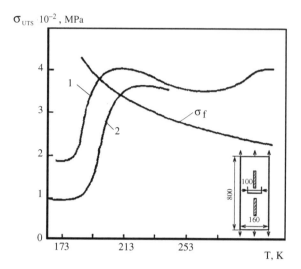

FIGURE 5.3 Dependence of nominal breaking stresses on the temperature for samples without residual (1) and with residual (2) stresses for St3sp steel, 20 mm thickness.

of residual stresses in an item is particularly pronounced at sub-zero temperatures (Figure 5.3). Therefore, the influence of defects present in the welded joint on the strength of the welded structure depends largely on the conditions of operation, distribution, and magnitude of residual, as well as working stresses arising during operation, and the totality of these stresses.

5.1.1.1 Cracks and Crack-Like Defects

As was noted earlier, cracks are the most hazardous of all the defects, found in welds. Presence of cracks in a welded joint often is the sole cause for catastrophic failure of welded structures. Cracks weaken the section of welds or elements being welded, leading to lower static strength of the joint. Moreover, being stress-raisers, cracks significantly lower also the dynamic strength of welded joints.

Lacks-of-penetration constitute the most hazardous of crack-like defects that develop during weld formation. Results of fatigue testing of butt joints with similar lacks-of-penetration, indicate that the influence of lacks-of-penetration on the joint fatigue strength (σ_f) depends on the presence and nature of residual stresses (Figure 5.4). The greatest endurance is found for samples with lacks-of-penetration, located in the zone of compressive residual stresses, although the weld area in these samples was two times greater than that in other samples.[318] Therefore, the influence of lacks-of-penetration on the welded joint's performance depends on the magnitude and kind of residual stresses, as well as on the depth of lack-of-penetration and to a smaller extent on its length (Figure 5.5 and Figure 5.6).[1,318]

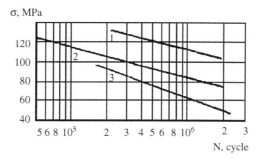

FIGURE 5.4 Influence of residual stresses on the endurance limit of butt-welded joints with lacks-of-penetration located in the zone. Compressive residual stresses (MPa): (1) 140; (2 and 3) 80 and 200.

Influence of defect size becomes insignificant, if lacks-of-penetration are located in the zone of high tensile residual stresses.[1] In this case, not very large lacks-of-penetration are just as hazardous as large ones, which is confirmed by the data, given in Figure 5.7. Adverse influence of lacks-of-penetration on the strength properties of a welded joint becomes stronger[1] with increase of hydrogen content in the weld metal.

5.1.1.2 Undercuts

Undercuts have a substantial influence on welded joint's strength. Being planar defects, they induce a considerable stress concentration and noticeably lower the static and, particularly, dynamic strength of a welded joint for the most diverse materials. Fatigue strength of butt-welded joints with undercuts depends on the depth of the undercut, magnitude of tensile residual stresses, and type of the joint.[1] Influence of the undercut depth and level of tensile residual stresses for butt-welded joints's is shown in Figure 5.8. As is seen, the fatigue life of a welded joint decreases with increase of undercut depth. However, the greater the

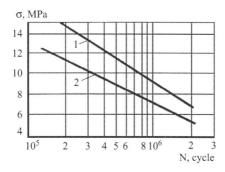

FIGURE 5.5 Change of fatigue resistance of butt-welded joints with lacks-of-penetration of different depth, located in the zone of high tensile residual stresses. The size of lake-of-penetration (mm); (1) $2 \cdot 160$; (2) $8 \cdot 160$.

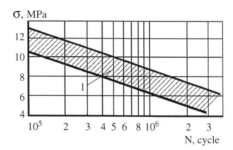

FIGURE 5.6 Change of fatigue resistance of butt-welded joints with lack-of-penetration of different lengths: (1) area of scatter of testing results located in the zone of high tensile stresses.

thickness of the metal being welded, the greater is the admissible depth of the undercuts. Therefore, it is more correct to evaluate the influence of undercut depth on the fatigue strength of the welded joint by the ratio of admissible size of undercut d to sheet thickness δ, as is proposed.[319] According to the data of Iida, Sato, and Nagai's, welded joint fatigue strength decreases by 5% at $d/\delta = 0.01$, by 10% at $d/\delta = 0.01$ and by 15% at $d/\delta = 0.02$.[319] In this case, similar to lacks-of-penetration, the adverse influence of undercuts is enhanced, if they are in the zone of high tensile residual stresses. It is established[318] that the fatigue strength of welded joints with undercuts depends little on the mechanical properties of the steel used as the base metal (Figure 5.9).

Lacks-of-fusion, which also are planar crack-like defects, lower the static and to a greater extent, the dynamic strength of the weld by reducing its cross-section.

An inadmissible defect of a weld is the presence of burns-through, which are cavities in the weld, resulting from flowing out of the weld-pool metal.

5.1.1.3 Craters

Craters are also defects of weld shape which are depressions formed during the arc extinction. Formation of a crater reduces the weld section. The crater, as a rule, has shrinkage looseness, which often develops into a crack.[1] Therefore, these defects also impair the quality of welded joint.

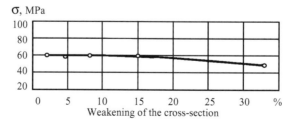

FIGURE 5.7 Dependence of endurance limit of butt-welded joint on weakening of the cross-section of a weld with lacks-of-penetration.

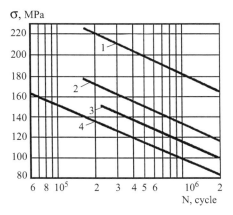

FIGURE 5.8 Results of fatigue testing of samples without undercuts; (1) and with undercuts (2 to 4). Undercut depth (mm): (2) 1.0–1.5; (3 and 4) 2.0–3.5. Residual stress (MPa): (2 and 3) 100; (4) 200.

5.1.1.4 Rolls

Rolls are one of the defects, arising during the formation of a weld or deposited layer. These defects form when the molten metal flows over the base-metal surface without fusing with it. Most often, they develop in arc welding of single-pass square butt welds, as well as in fillet welds by inclined electrode, and in surfacing particularly by an indirect arc. Comparatively, such defects seldom form less frequently in electroslag welding. Although these defects are external, they are rather difficult to reveal at a visual inspection of a weld or deposited bead. Presence of rolls in a weld or deposited metal impairs their performance, therefore, when detected, these defects are usually removed by machining.

In addition to defects, the welded joint's performance is greatly influenced by the shape and dimensions of the penetration zone, bead and weld root, and the

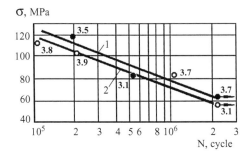

FIGURE 5.9 Results of fatigue testing of butt-welded joints with undercuts, depending on the mechanical properties of the base metal: (1) steel St3; (2) steel IZ-138; numbers at the points correspond to the depth of the undercut. (●) steel St3; (○) IZ-138.

smooth transition from the base metal to the deposited metal. Therefore, let us consider the features of formation of a weld or a deposited bead, as well as the formation of weld-shape defects (lacks-of-penetration, undercuts, burns-through, lacks-of-fusion, craters, rolls), and we will start with the process of the penetration zone formation.

5.2 FORMATION OF THE PENETRATION ZONE

A welded joint's performance depends to a certain extent on the dimensions and shape of the penetration zone. In particular, these parameters are largely responsible for weld-metal susceptibility to hot cracking, removal of gas bubbles, and non-metallic inclusions from the weld pool, etc. In consumable-electrode arc welding additional factors such as the amount of base metal participating in weld formation, and, hence its composition, structure, and properties also depend on the penetration area.

5.2.1 SHAPE AND DIMENSIONS OF THE PENETRATION ZONE

Dimensions and shape of the penetration zone depend on many factors. Irrespective of the fusion welding process, they are always related to thermo-physical properties of the base metal and welding modes. For instance, in arc welding, the shape and dimensions of the penetration zone are influenced by the value of welding current, arc voltage, welding speed, electrode extension, relative position of the electrode and the part in space, etc. In addition to the above factors, for the most diverse types of welded joints and fusion welding processes, the configuration and shape of the penetration zone, in particular, penetration depth, should depend also on the surface properties of the metal.[320,321] To some extent, this is indicated by the data,[322,323] obtained on the decrease of the depth and increase of the width of the penetration zone in tungsten electrode argon-arc welding of steels, subjected to refining remelting. Of course, this is a case of the presence of less surface-active impurities.

However, experimental investigations of the influence of the metal's surface properties on the shape and dimensions of the penetration zone were conducted only in some studies.[274,324–327] In this case, it is established that increase of surfactants content in the base metal leads to a noticeable increase of the penetration depth for different fusion welding processes, namely, nonconsumable electrode argon-arc welding, laser welding, and electron beam welding.

In welding processes, formation of the penetration zone depends to a certain extent on the kind of the used electrode. Let us therefore consider the connection between the penetration zone shape and the surface properties of the metal separately for consumable and nonconsumable electrode welding.

In nonconsumable electrode welding without the filler wire, the shape of the penetration zone at incomplete penetration of the metal (Figure 5.10) is characterized by dimensions H_p and B_w, and at compete penetration by B_w, B_w', h, and h', respectively.

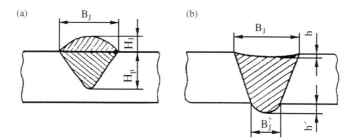

FIGURE 5.10 Shape of a weld: (a) with incomplete penetration; (b) with full penetration.

5.2.1.1 Penetration Depth

Penetration depth, H_p is one of the most important characteristics of the shape of the penetration zone. In keeping with the modern concepts on the penetration mechanism,[328] in arc welding, the molten metal is driven from the front part of the weld pool into its tail part under the impact of the arc pressure. This result in uncovering of the base metal under the arc, which is melted by the arc heat and continues to be driven into the weld-pool tail part. As a result, the level of liquid metal in the weld-pool tail part rises higher than that of the metal in its front part. This induces a hydrostatic pressure in the weld pool. Displacement of the liquid metal surface, which affects the penetration process depends on the curvature of the weld-pool surface and magnitude of surface tension of the metal melt.

According to the data of Russo, Kudoyarov and Suzdalev,[321] under the condition of dynamic equilibrium between arc pressure P_a, hydrostatic pressure $P_h = H_p\rho_m$, and surface tension of the melt P_s, which is a mandatory condition for normal running of the welding process, the following dependence is in place

$$H_p = \frac{P_a - P_p}{\rho_m}, \tag{5.1}$$

where ρ_m is the metal density.

From Equation (5.1) it follows that penetration depth in the welding of a certain material, can be increased by lowering the surface tension of the metal or increasing the arc pressure. The latter can be achieved by increasing the welding current or energy concentration in the heated spot. It is not always possible to increase the penetration depth by increasing the welding current, as in this case only the penetration depth increases, and the weld width changes very little. This results in an unfavorable shape of the penetration zone. Moreover, heat input into the part is increased, and this may lead to development of undesirable structures in the weld and heat-affected zone (HAZ), as well as higher deformations of the structure. When a consumable electrode is used, increase of welding current also leads to a larger amount of the molten electrode metal. This results in a greater height of the deposited bead, formation of an abrupt transition from the deposited metal to the base metal, that is, stress-raisers form on the boundary between the base and deposited metal.

The above and some other drawbacks may be avoided, if the base-metal penetration depth is increased not due to increase of the welding current but due to decreasing of the magnitude of the metal's surface tension.

5.2.2 THE INFLUENCE OF BASE-METAL SURFACE PROPERTIES

Investigations of the influence of base-metal surface properties on the shape and dimensions of the penetration zone in nonconsumable (tungsten) electrode welding were conducted in a ADSV-5 automatic machine, using argon as the shielding gas; VDM-1601 rectifier was the current source.

Plates of steel St3, 08Yu and transformer steel E3A 4-mm thick were used as the base metals. In order to change the surface properties of the metal, FeO and FeS powders were applied to the plate surface, using an adhesive. This way, sulfur and oxygen were added to the weld pool, which are characterized by a high surface activity in iron and iron-based alloys.

Performed experiments showed that the change of the base-metal's surface tension has little influence on the weld width in the nonconsumable electrode welding. Depth of metal penetration changes in a much more noticeable manner with the addition of surfactants into the weld pool. Change of H_p value depends on the content of these components in the metal, initial composition of the metal, and welding current (Figure 5.11 through Figure 5.13).

The most marked increase of penetration depth with addition of sulfur and oxygen was observed for St3. This effect is enhanced with the increase of welding current (Figure 5.12). Presence of deoxidizer elements (Si, Al) and elements, binding sulfur in the metal (08Yu, E3A steels) lowers the influence of sulfur and oxygen on the penetration depth.

The effect of the influence of surfactants on the change of penetration depth is also manifested, although to a smaller degree, when twin-arc welding is used

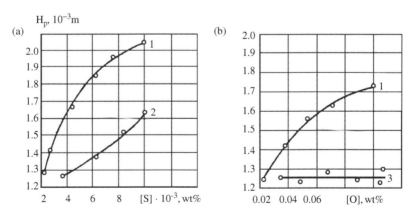

FIGURE 5.11 Dependence of penetration depth on the content of sulfur (a) and oxygen (b) in base metal: ($I_w = 160$ A; $U_a = 12$ V; $v_w = 0.007$ m/s): (1) steel St3; (2) steel 08Yu; (3) steel E3A.

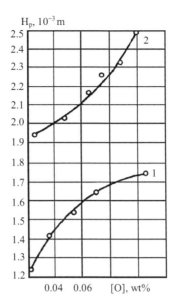

FIGURE 5.12 Change of the depth of metal penetration (St3 steel), depending on the content of oxygen in it. I_w (A): (1) 160; (2) 220.

(Figure 5.14). This is probably, related to the change of hydrostatic pressure in the weld pool, which is to some extent confirmed by the influence of electrode spacing, l_e, on the H_p value. Analyzing the obtained results on the influence of oxygen and sulfur on the change of penetration depth, it is easy to see, that upon addition of sulfur to steel St3 the penetration depth is somewhat greater, than that in the presence of oxygen in the metal.

This does not correspond to their influence on the surface tension of iron-based melts. However, this contradiction is illusory, and it has the following explanation. It is known that in the presence of oxygen in the metal the viscosity of the metal melt is

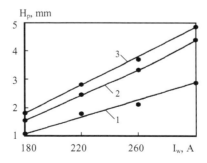

FIGURE 5.13 Dependence of the metal penetration depth on welding current (T = 1803 K; U_a = 16 V; v_w = 0.0044 m/s). σ_{m-g} (mJ/m^2)(T = 1803 K): (1) 1264; (2) 918; (3) 926. The surfactant is O_2 in (2) and S in (3).

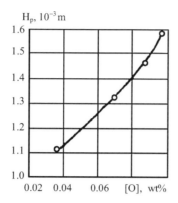

FIGURE 5.14 Change of the depth of base-metal penetration, depending on the content of oxygen in it twin-arc welding: $l_{el} = 6$ mm; $I_w = 220$ A; $U_a = 12$ V; $v_w = 0.007$ m/s.

much higher than at the same content of sulfur in the metal.[329] Therefore, with the addition of oxygen into the metal, melt displacement into the weld-pool tail part will become difficult, and this means that the melt will remain under the arc, which will result in H_p lowering.

5.2.3 THE ACTION OF SURFACTANTS

Addition of surfactants, will not, obviously, change the characteristics of the arc discharge, as their content in the metal is very low. This is confirmed to some extent by the results of experiments, conducted by us on welding plates of St3 steel 10-mm thick by a defocused electron beam. Application of such a beam provided a penetration zone, close in its shape to that in the nonconsumable electrode arc welding. As is seen from Figure 5.15, in this case also, addition of surface-active sulfur into the metal increased the depth of penetration.

The connection between the content of oxygen and sulfur in the metal and change of its penetration depth is also noted in some investigations.[322,330,331] Increase of the depth of penetration of the base metal is also observed in welding in oxidizing gas media,[332] when transition of oxygen from the gas phase into the electrode metal is possible. This will lead to lowering of its surface tension and change of the nature of electrode-metal transfer.

Change of the depth of metal penetration with the addition of surfactants into it is most probably related to the change of the hydrodynamic processes in the weld pool.

Opinions differ as to the causes of appearance of metal flows in the weld pool. Some investigators associate metal flows to the mechanical impact of the arc, others to the electromagnetic forces that arise in welding, still others associate the metal flows to the presence of a surface tension gradient of the metal in the weld pool. Without denying the influence of various factors on the hydrodynamic processes in the weld pool, let us consider the role of surface phenomena in greater detail.

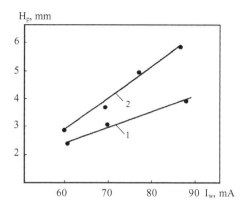

FIGURE 5.15 Change of the metal penetration depth under the impact of a defocused beam, depending on the current value ($U_{ac} = 30$ kV; $v_w = 0.011$ m/s) $T = 1803$ k: $\sigma_{m-g} = $ (mJ/m^2): (1) 1264; (2) 861.

Using various mathematical models, different researchers produce differing results. According to the data a study on the movement of a metal in the weldpool,[333] the influence of the melt surface tension, σ on the stirring of the weld-pool metal can be neglected, if the following inequality is satisfied:

$$\sigma \ll \rho_m \lambda^2 g, \tag{5.2}$$

where λ is the scale of motion, which may be taken to be equal to the length of weld pool; ρ_m is the metal density; g is the acceleration of gravity.

From Equation (5.2) it follows that at $\lambda \gg \sqrt{\sigma/\rho_m g}$, the surface forces have no influence on the motion of the liquid.

In the single-pass welding of 10-mm thick steel, λ value is close to 10^{-2} m, and $\sqrt{\sigma/\rho_m g}$ to $4.5 \cdot 10^{-3}$ m. Therefore, for this case, the influence of surface tension on molten metal motion can be neglected,[332] as λ is by an order of magnitude greater than $\sqrt{\sigma/\rho_m g}$. The influence of surface tension on the motion of the metal melt will become more pronounced only in the welding of sheet metal, when $\lambda \approx 10^{-3}$ m.[333]

However, the magnitude of surface tension of the weld-pool metal is not a constant value, as was assumed in the above study.[333] Under the actual conditions of welding, individual sections of the weld-pool have different temperatures and also differ by the content of the surface-active components. That is, they will also have different values of surface tension. Presence of a surface tension gradient leads to displacement of the surface layers of the metal melt. At a positive curvature of weld-pool surface, the molten metal will move from the sections with a high surface tension to those with a lower surface tension.

5.2.4 THERMOCAPILLARY PHENOMENA

If the change of surface tension in the weld pool is related only to different temperatures of the metal surface, the so-called thermocapillary convection arises, as was predicted by Marangoni. The intensity of such a convection is characterized by a non-dimensional parameter, namely Marangoni number

$$Mn = \left(\frac{L\Delta T}{\rho v a}\right)\frac{\partial \sigma}{\partial T},$$

where a is the diffusivity; L is the characteristic linear dimension; v is the kinematic viscosity.

Let us analyze, how temperature distribution over the weld-pool surface may influence the molten metal motion.

According to Frolov, the temperature in the zone of the arc active spot at low currents (4–17 A) reaches 2600 K on the anode, and 2400 K on the cathode. At currents characteristic for the welding processes (100 A and more),[147] the temperature in the active spots reaches the metal boiling temperature.[1] On the other hand, in the vicinity of the fusion zone, the temperature is close to that of metal melting, that is, temperature gradient in the weld pool reaches 200–300 (and more) degrees per millimeter.

It is known,[333] that the surface tension of the molten metal becomes zero at the critical temperature T_{cr}, determined from the expression $T_{cr} = 1.7\ T_{boil}$ (T_{boil} is the boiling temperature of a given metal). Iron boiling temperature, by the data of various investigators, varies between 3043 and 3473 K. With a linear dependence of surface tension of the metal melt on temperature, its surface tension at any temperature is given by the following formula:

$$\sigma_T = \sigma_o\left(\frac{T_{cr} - T}{T_{cr} - T_o}\right),\tag{5.3}$$

where σ_o is the metal surface tension at temperature T_o.

At temperature $T_o = T_{mel} = 1812\,K$, the surface tension of molten iron, σ_o, is close to $1800\ mJ/m^2$,[190] then in the active spot zone, $\sigma_{3473K} = 1069\ mJ/m^2$. In reality, the metal's surface tension in the active spot is even smaller, and by our data, it is approximately $300–350\ mJ/m^2$.

As the dimensions of the weld pool are small, the gradient of surface tension, arising as a result of the temperature gradient is quite observable, and it is greater across the weld-pool width. For a weld pool 10-mm wide and 30-mm long, this gradient will be about $300\ mJ/m^2$ mm across the weld-pool width, and just $50\ mJ/m^2$ mm along its length. Therefore, the surface tension gradient will have the greatest influence on molten metal motion across the weld-pool width, and its influence will be smaller along the pool length.

Therefore it is clear that metal convection in the weld pool, associated with the surface tension forces, restricted to the arc welding alone.

According to Olshansky, Gutkin, and Girimadzhi, in electron beam welding by a moving beam, the velocity of thermocapillary flows in the weld pool is described by the following equation:[324]

$$V = \left(\frac{y^2}{2\eta}\right)\frac{\partial}{\partial x}\left(\frac{\delta}{r}\right) + \left(\frac{y}{\eta}\right)\frac{\partial \sigma}{\partial x}, \tag{5.4}$$

where r is the radius of metal film curvature in a section, normal to the crater axis; η is the metal viscosity; x and y are the current coordinates along the length and across the thickness of the molten metal film.

Calculations, conducted in a study,[334] using Equation (5.4), demonstrated, that the main mass of the metal is transferred from the high-temperature zone into the solidification zone due to thermocapillary convection.

5.2.4.1 Evidence from Laser Welding and Arc Welding

The connection between the process of metal melt transfer and the thermocapillary effect is also noted in laser welding.[2,335,336]

Similar to the earlier work,[334] analysis of the laser beam impact on a plane layer of incompressible liquid of thickness h for a deep weld pool ($h \gg d$, where d is the size of the heat source) revealed[2,335] that at the initial moment of time, the function of the metal flow current has the form:

$$\psi = \left\{\frac{2tT_o\rho^{-1}xd(d-z)}{[(d-z)^2 + x^2]^2}\right\}\left(\frac{d\sigma}{dT}\right), \tag{5.5}$$

and the stationary function of current at the same temperature distribution on the melt surface is as follows:

$$\psi = \left\{\frac{T_o dxz}{2\eta[x^2 + (d-z)^2]}\right\}\left(\frac{d\sigma}{dT}\right). \tag{5.6}$$

Thus, in nonconsumable electrode arc welding, and particularly in electron beam and laser welding, the molten metal transfer in the weld pool proceeds due to thermocapillary convection. Probably only in gas and electroslag welding, the thermocapillary effect will not have such a prominent role, due to a more uniform heating of the weld-pool surface and a much smaller temperature gradient.

In the case of a full penetration of the metal, values of h and h' at the same values of B_J and B'_J depend on the weld-pool weight P_G, arc pressure P_a, and metal surface properties P_s. Weld-pool equilibrium under the impact of the above forces, provided the system is isothermal, is written as:

$$P_G + P_a - P_s^o - P_s^r = 0, \tag{5.7}$$

where P_s^o, P_s^r are the surface tension forces on the outer and root parts of the weld pool, respectively.

The surface tension force depends on the radii of curvature of the metal-pool surface

$$P_s = \sigma_{m-g} \left(\frac{1}{r_1} + \frac{1}{r_2} \right), \qquad (5.8)$$

where r_1, r_2 are the main radii of curvature.

With a certain error it may be assumed that the outer and inner surfaces of the penetration zone are part of a cylinder. Then, $r_2 = 0$, and Equation (5.8) becomes:

$$P_s = \frac{\sigma_{m-g}}{r_1}. \qquad (5.9)$$

As $B_J > B'_J$, then $r_1^o > r_1^r$, and, hence, at the same values of σ_{m-g}, $P_s^o < P_s^r$. However, it is, probably, not quite correct to neglect the P_r^0 value due to the high temperature of weld pool, as is usually assumed. This is attributable to the following causes. First of all, in the heating of metal melts even when σ_{m-g} value decreases with temperature rise, overheating of the metal above the melting temperature by 300–700 degrees does not lead to a great decrease of surface tension. Second, in the presence in the metal of elements, having a high surface activity, the value σ_{m-g} does not drop as dramatically with temperature rise, as for a pure metal melt. This is related to a decrease in the adsorption of surfactants with the rise of the melt temperature. A similar phenomenon was observed, for instance, when studying the surface tension of armco-iron at different temperatures in oxidizing gaseous media (Section 3.3).

As follows from Equation (5.7), the smaller is $P_s^o + P_s^r$ (which is possible at decrease of σ_{m-g} at unchanged values of curvature radii), the greater will be the difference $P_G + P_A - (P_s^o + P_s^r)$ and, therefore, h and h' values will increase.

Experiments on welding the samples of steel St3, the surface tension of which was varied by adding sulfur to the weld-pool metal confirm this postulate (Figure 5.16). As is seen, lowering of the magnitude of surface tension of weld-pool metal results in an increase of values h and h'. It should be noted, that the shape and dimensions of the penetration zone in nonconsumable electrode welding can be changed also by adding surfactants to the shielding gas composition. For instance, according to Heiple and Burgard, the optimal shape of the penetration zone at a greater depth of penetration than in welding in pure argon, can be achieved by adding just 0.05–0.14% SO_2 to argon.[337] Moreover, this does not lead to deterioration of the mechanical properties of the welded joint, as sulfur content in the weld metal in this case is not higher than 0.003%.

5.2.4.2 Consumable Electrode Welding

Influence of surface properties of the metal on the formation of the penetration zone was studied in consumable electrode welding in shielding gases and using fluxes. The first series of experiments were conducted in ADPG-500 automatic machine. In this case, Sv-08 and Sv-08G2S electrode wires of 1-mm diameter were used and the shielding gases were Ar, CO_2, and $CO_2 + N_2$. Plates of steels

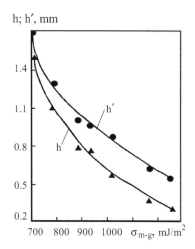

h; h′, mm

FIGURE 5.16 Dependence of h and h' values on the surface tension of steel St3. $I_w = 220$ A; $U_a = 15$ V; $v_w = 0.0053$ m/s.

St3 and E3A were used as the base metals. Prior to welding, the plates were ground and their thickness was 5.5-mm. Shielding gas flow rate in all the experiments remained constant at $900 \cdot 2.8 \cdot 10^{-7}$ m^3/s.

Selection of these welding consumables is attributable to the following causes. For steels St3 and Sv-08, when argon is replaced by CO_2, a decrease of σ_{m-g} value respectively by 510 and 543 mJ/m^2 is observed (Section 3.3). For Sv-08G2S steel, such a change of the gaseous medium composition leads to a decrease of σ_{m-g} by just 98 mJ/m^2, and for E3A steel the effect is to a certain increase of surface tension value. Application of $CO_2 + N_2$ gas mixture is due to the fact that at CO_2 replacement by N_2, the oxidizing potential of the gaseous medium decreases, but, in view of the close thermo-physical properties and ionization potentials of nitrogen and carbon dioxide, the properties of the arc discharge should not change significantly.

All the experiments on gas-shielded welding were conducted at reverse polarity in the following welding modes: $I_w = 320 - 340$ A; $U_a = 26 - 28$ V.

5.2.5 EFFECT OF CURRENT POLARITY

In the above sections, we dealt with experiments conducted at reverse polority mode. Now let us discuss the effect of charging the polarity of the current applied. In a second series of tests on submerged-arc welding, experiments were conducted in an ABS automatic machine using a Sv-08 wire of 4 mm diameter. Plates of St3 steel 8-mm thick were used as the base metal. In these experiments, fluxes of the following composition (wt%) were applied: SiO_2 — 75, Na_2O — 25, and SiO_2 — 60, MnO — 40. Fluxes were produced by fusing SiO_2 with Na_2CO_3 and SiO_2 with MnO in graphite crucibles in argon atmosphere. Selection of fluxes, base, and electrode metals was due to different influence of a constant external electric field on the interphase tension, σ_{m-g} of the metal–slag boundary for the given systems. According to our data, change of polarity for the system of low-carbon steel–SiO_2–Na_2O slag leads to a marked change of interphase tension (see Section 3.4). At negative potential on the metal, the σ_{m-g} value decreases, and at positive potential it increases. For a system with SiO_2 and MnO slag, the change of current polarity has little influence on σ_{m-g} value.

In submerged-arc welding, the welding current was 470–490 A, and the arc voltage was 34–36 V.

First of all, it should be noted that replacement of argon by CO_2 and the increase of carbon dioxide content in $N_2 + CO_2$ mixture led to a widening of the penetration zone, B_s by just 4–16%, with a greater increase being observed in bead deposition with Sv-08 wire on plates of St3 steel. The smallest increase of B_s was found in the case of bead deposition on E3A steel with Sv-08G2S wire. Width of the penetration zone changed slightly also in submerged-arc welding, using a flux, consisting of SiO_2 and MnO. In this case, with the change of current polarity, B_s value changed by just 2–9%. In welding using flux of SiO_2 and Na_2O, the width of the penetration zone in straight-polarity welding is higher than in reverse-polarity welding by approximately 40–43%.

In welding in argon, the depth of base-metal penetration differs comparatively little with different combinations of the tested wires and plates. Figure 5.17 shows the change of metal penetration depth, depending on the welding speed for the following systems: Sv-08 wire–St3 steel (curve 1) and Sv-08G2S wire–E3A steel (curve 2). In the welding of St3 steel plates with Sv-08G2S wire, and of E3A steel plates with Sv-08 wire, H_p had intermediate values. It is also seen that the penetration depth is higher than in the case of welding E3A steel plates with Sv-08G2S wire.

The value of H_p changes to a much greater extent for the studied systems in CO_2 welding. The greatest depth of penetration is found in the deposition of beads on the plates of St3 steel with Sv-08 wire (Figure 5.17), and the smallest in the case of bead deposition on E3A steel with Sv-08G2S wire.

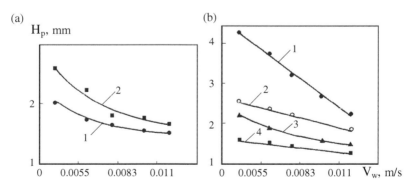

FIGURE 5.17 Dependence of the metal penetration zone depth on welding speed: (a) argon-arc welding; (b) CO_2 welding. [1] Sv-08 wire–St3 steel; [2] Sv-08G2S wire–St3 steel; [3] Sv-08 wire–E3A steel; [4] Sv-08G2S wire–E3A steel.

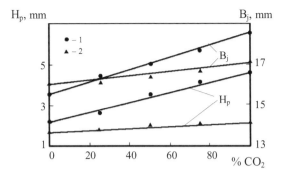

FIGURE 5.18 Dependence of penetration zone width and depth on CO_2 volume content in $N_2 + CO_2$ mixture in surfacing with Sv-08 wire of steels St3 (●) and E3A (▲).

Data on welding in CO_2-N_2 mixture are indicative of the change of the dimensions of the penetration zone being related to the oxidizing potential of the shielding gaseous medium. The higher the oxidizing potential of the gas phase, the greater is the depth and width of the penetration zone (Figure 5.18), this being to a higher degree characteristic of the case of bead deposition with Sv-08 wire on the plates of St3 steel, and to a smaller degree in the bead deposition on E3A steel.

Results of experiments on submerged-arc welding indicate that in welding with SiO_2–Na_2O flux, the penetration depth is greater in straight polarity and smaller in with reverse polarity (Figure 5.19). In the latter case, at welding speeds >0.011 m/sec, no weld was formed, but individual metal drops were deposited. In submerged-arc welding with a flux of SiO_2 and MnO, the penetration depth was practically the same both in straight and at reverse-polarity modes.

Analysis of the derived data indicates that in consumable electrode welding, the dimensions of the penetration zone are related to surface properties of the base and electrode metals. Similar to nonconsumable electrode welding, change of surface tension of the metal melt has a more marked influence on the penetration depth and a smaller influence on weld width. In welding in the atmosphere of argon,

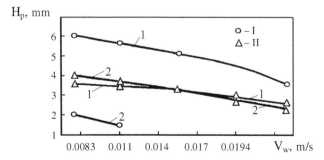

FIGURE 5.19 Dependence of metal penetration depth on welding speed at straight (1) and reverse (2) polarity in surfacing with SiO_2–Na_2O (○) SiO_2–MnO (△) fluxes.

the penetration depth is greater, when the deposition is performed with Sv-08G2S wire on E3A steel, that is, for those materials, where the surface tension is smaller.

In welding in CO_2 (in contact with which steels St3 and Sv-08 have the smallest σ_{m-g} values), a penetration zone with the highest H_p and B_w values forms in bead deposition with Sv-08 wire on the plates of St3 steel. Increase of CO_2 content in $CO_2 + N_2$ mixture also leads to a noticeable increase of H_p and B_w at bead deposition with Sv-08 wire on St3 steel.

A clear relationship between the surface properties of the metal and dimensions of the penetration zone is also observed in submerged-arc welding. With the use of a flux consisting of SiO_2 and MnO, when the change of current polarity does not have any noticeable influence on the magnitude of interphase tension on metal–slag boundary, both H_p and B_w remain practically the same in welding at straight and at reverse polarity.

The situation is different in submerged-arc welding with a flux of SiO_2 and Na_2O. In this case, in straight-polarity welding, the depth and width of the penetration zone are much greater than in reverse-polarity welding. In other words, the dimensions of the penetration zone are increased, when the value σ_{m-s} on electrode metal–slag boundary is smaller, and, therefore, the frequency of formation of electrode metal drops is higher.

As was already noted, change of the dimensions of the penetration zone and, first of all, of its depth, is largely dependent on the hydrodynamic processes in the weld pool, although the opinions on the causes of development of metal flows in the weld pool are different.

The results of our experiments indicate that value H_p can change significantly due to just the change of surface tension of the metal melt at the same welding speeds, with practically unchanged arc pressure and the same electrodynamic forces. This effect, however, is dependent on the welding speed and is particularly noticeable at small values of v_w. Moreover, this effect is further enhanced at increase of welding current. All this is indicative of the fact that the surface properties of the metal and arc pressure as well as electrodynamic forces and welding speed have an influence on the shape and dimensions of the penetration zone.

Influence of surface properties of the electrode and base metals on the dimensions of the penetration zone is, probably, determined by two causes. First of all, change of surface tension of the electrode metal leads to a change of the dimensions of the drops and the mode of electrode-metal transfer. This proposition accordance with the published data,[338,339] has a significant influence on the dimensions of the penetration zone. Second, as was already mentioned, the dimensions of the penetration zone depend on the velocity and direction of the metal flow in the weld pool. The establishment of the flow is due to the presence on the weld-pool metal surface of sections, having different surface tension. In submerged-arc welding, however, the contact of the metal melt with the highly viscous slag will result in a lower velocity of motion of the weld-pool metal than in the case of metal contact with the gaseous medium. This is in particular indicated by the data of Kuzmenko according to whom the velocities of metal flows in the weld pool in submerged-arc welding are approximately 110-125 mm/s.[340] For this reason, in submerged-arc welding, the change of the depth of metal penetration will depend more on the surface tension of electrode metal and the associated mode of transfer.

In the welding of dissimilar metals, for instance, of copper to steel, wires of copper alloys with the melting temperature lower than that of the steel part are often used as the electrode material. In this case, the processes of flowing and propagation of the metal melt in the gap between the parts being welded are important in the formation of the penetration zone, primarily in butt welding. It is particularly important to know, how the electrode metal spreads over the surface of the steel part, as steels are wetted by copper melts worse than solid copper.

5.2.6 Experimental Findings

We conducted experiments to study the ability of copper alloys to flow into a slot gap between the steel samples under the impact of an arc discharge on the melt. The experimental procedure was as follows.

Flat polished steel plates (steels St3 and 12Kh18N10T) of $20 \times 20 \times 10$ mm^3 size were assembled with a fixed gap between the edges, which was varied from 0.1 to 1.0 mm during the experiments. Samples of the studied copper alloy (MNZhKT5-1-0.2-0.2, M3r, BrKMts3-1) were placed on the assembled plates, located symmetrically relative to the gap (Figure 5.20). The thus prepared samples were mounted on the work-table of the vacuum chamber of the installation, which is described in Section 2.5. The chamber was first sealed, pumped down, and filled with argon, and the copper alloy was melted by the electric arc struck between the tungsten electrode and the sample the of studied material. In this case, the copper alloy was melted under the impact of the arc discharge and flowed into the gap between the plates. The welding current, arc voltage, arcing time, and the gap between the plates were controlled during the experiment. After the arcing was interrupted and the chamber has cooled down, the samples were taken out and broken up at the soldered joint. The value of the area of flowing (F) was selected as the criterion to evaluate the ability of the metal melt to flow into the gap between the parts, as the most objective value.

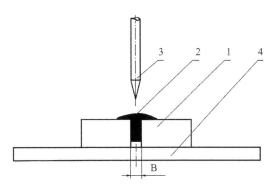

FIGURE 5.20 Schematic of arrangement of the studied samples in the vacuum chamber: (1) steel plates; (2) copper alloy; (3) tungsten electrode; (4) table with preheating element.

5.2.6.1 Influence of Welding Current

In the first experiments, the influence of the kind and strength of the welding current and its polarity on the copper melt flowing into the gap between the steel plates was investigated. As is seen from Figure 5.21, which gives the data for the copper alloy (MNZhKT5-1-0.2-0.2), the copper melt penetration into the gap between the plates is markedly enhanced with the increase of welding current. The A.C. arc has a more pronounced influence on this process than the straight-polarity D.C. arc. In this case, the difference in the area of the melt flowing into the gap for steels St3 and 12Kh18N10T is very small at a short duration of the time of the arc action on the melt (10 s). Increase of arcing time from 10 to 20 s markedly enhances the area of the copper melt flowing into the gap between the steel samples. Similar influence of the value and kind of welding current on copper melt penetration into the slot gap between the steel plates is also observed for all the studied materials. The weight of the copper alloy sample has a relatively small influence (Figure 5.22) on the size of the area of the copper melt flowing into the gap between the steel parts.

5.2.6.2 Duration of Arc Discharge Action

The time of the arc discharge action on the copper alloy has a much greater influence on the process of the melt flowing into the gap between the plates. This is manifested in the application of both A.C. and D.C. arcs. The results, shown in Figure 5.23, for the use of A.C. arcs reveal that longer time of the arc action on the copper melt, leads to a significant increase the value of F. With the extension of the arcing time from 10

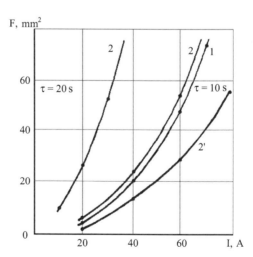

FIGURE 5.21 Influence of the value and kind of welding current on the area of flowing of MNZhKT5-1-0,2-0,2 melt into the gap between the steel plates ($U_a = 12$ V; $B = 0.1$ mm): (1 and 2) A.C. arc; (2′) straight-polarity D.C. arc.

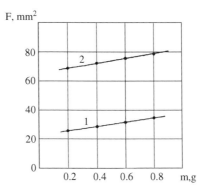

FIGURE 5.22 Influence of the sample mass on the area of flowing of MNZhKT5-1-0,2-0,2 melt into the gap between the steel plates under the impact of on arc discharge on the copper sample ($I_w = 30$ A; $U_a = 12$ V; $B = 0.1$ mm; $\tau = 30$ s): (1) St3 steel; (2) 12Kh18N10T steel.

to 30 sec at the current of 30 A, a progressing difference in the areas of melt flowing is observed for steels St3 and 12Kh18N10T. This is, apparently, related to different thermo-physical properties of these steels. In addition, at interaction of the copper melt with 12Kh18N10T steel, components of hard steel (Fe, Ni, Cr) transfer into the melt, such a transfer should result in the increase of σ_{m-g} value of the melt, increase of its melting temperature and viscosity. Hence it results in worse spreading of the melt over the surface of solid metal.

5.2.6.3 Dimensions and Spacing of the Steel Plates

Influence of the size of the gap between the steel plates on the area of flowing of the copper melts was also studied. As follows from the experimental results

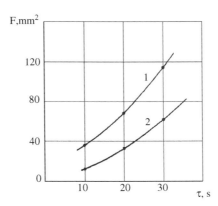

FIGURE 5.23 Influence of the arc action time on the area of the melt flowing (MNZhKT5-1-0.2-0.2) into the gap between steel samples ($I_w = 30$ A; $U_a = 12$ V; $B = 0.1$ mm): (1) St3; (2) 12Kh18N10T.

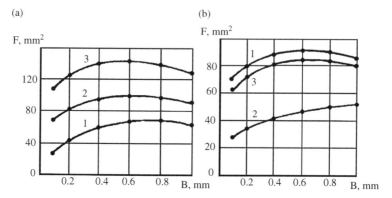

FIGURE 5.24 Change of flowing area of copper melts depending on the value of the gap between the parts of St3 (a) and 12Kh18N10T (b) steels: (1) M3r copper; (2) BrKMts3-1 bronze; (3) MNZhKT5-1-0, 2-0, 2 alloy. $I_w = 30$ A; $U_a = 12$ V; $\tau = 30$ s.

(Figure 5.24), maximum sizes of the area of copper melts flowing correspond to the gap values, which are in the range of 0.6–0.8 mm.

For the studied systems the values of the area of melt flowing into the slot gap are given in Figure 5.25 at the following constant parameters: $I_w = 30$ A;

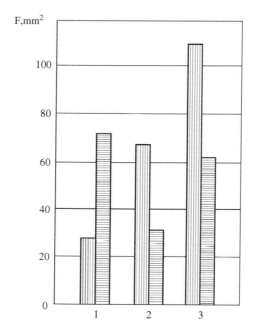

FIGURE 5.25 Values of the area of melt flowing into the gap between the steel plates under the impact of an arc discharge ($I_w = 30$ A; $U_a = 12$ V, $B = 0.1$ mm; $\tau = 30$ s): (▦) St3 steel, (▤) 12Kh18N10T steel. (1) M3r copper; (2) BrKMts3-1 bronze; (3) MNZhK5-1-0,2-0,2 alloy.

$U_a = 12$ V; $B = 0.1$ mm. As is seen, the value of the flowing area essentially depends on the composition of the contacting phases.

5.2.6.4 Calculations

Mathematical processing of the derived results allowed obtaining empirical dependencies of the size of the flowing area on welding current, arcing time, and the gap between the parts. The expression, describing the combined influence of the above parameters on the size of the flowing area, has the following form:

$$F = (2.7 \cdot 10^{-2} I_w + 1.47 \cdot 10^{-4} I_w^2)(0.65 + 7.44 \cdot 10^{-3} I_w + 8.33 \cdot 10^{-5} I_w^2)$$
$$\times (71.0 - 5.53\tau + 0.19\tau^2)(0.9 + 0.78B - 0.6B^2). \tag{5.10}$$

As is seen from Equation (5.10), the welding current value has the strongest influence on F value.

The process of formation of the penetration zone depends on molten metal flowing into the gap between the parts being welded not only in the welding of butt joints, but also in the welding of tee and fillet joints. In this case (Figure 5.26), the influence of gravitational forces will be weak and the velocity of molten metal flowing into the gap between the parts will be given by the following equation[68]:

$$V = \frac{\delta \sigma_{m-g} \cos \theta}{6\eta_m x}, \tag{5.11}$$

where x is the distance, filled by the melt and defined by the time of the process of metal flowing τ, $x = \sqrt{\tau \delta \sigma_{m-g} \cos \theta / 3\eta_m}$.

From Equation (5.11) it follows that the velocity of the melt flowing into the gap between the parts in the welding of tee and fillet joints will be significantly affected by the wettability of the solid metal by the metal melt. Therefore, use of techniques, allowing improvement of wettability on the solid metal–metal melt interface (removal of oxide film, preheating of parts, etc.) should lower the probability of formation of lacks-of-penetration in the welding of tee and fillet joints, which is confirmed by the data from practical work on part welding.

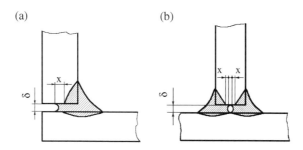

FIGURE 5.26 Metal flowing into the gaps in welding fillet (a) and tee (b) joints.

Thus, the conducted investigations indicate that when developing technologies of welding and surfacing it is necessary to allow for the surface properties of the electrode and base metals, and to control them, in order to ensure the required size and shape of the penetration zone in nonconsumable and consumable electrode welding.

5.3 FORMATION OF THE WELD BEAD AND ROOT

Shape and dimensions of weld reinforcement, and smooth transition from the base metal to the deposited metal have a marked influence on welded joint is performance. The importance of these factors is particularly high for critical structures operating at alternating loads.

5.3.1 ISSUES IN THE FORMATION OF BUTT WELDS

Issues of formation of butt welds and development of the methods to calculate their dimensions depending on welding modes, are considered in the works of many researchers. These studies showed that the shape and dimensions of weld convexity are largely dependent on welding modes and thermo-physical properties of the metal being welded. Alongside these factors, however, the process of weld formation is further affected by the surface tension forces. One of the first researchers, who paid attention to the connection between the surface properties of the metal and formation of the weld bead was G.Z. Voloshkevich,[341] who showed that the shape and dimensions of weld reinforcement are dependent on the magnitude of the capillary constant of the interface

$$a_c = \sqrt{\sigma/g\Delta\rho}, \tag{5.12}$$

where $\Delta\rho$ is the difference of densities of the contacting phases.

According to Voloshkevich,[341] the profile of bead surface is described by an equation, which relates the curvature of weld profile to the surface tension of weld metal and hydrostatic pressure $\sigma/r = g\Delta z + C_o$ (r is the radius of curvature of bead surface in the point being considered; z is the coordinate of the point being considered; C_o is the constant, dependent on the selection of the origin of coordinates). For the case, when the origin of coordinates coincides with the vertex of surface curvature, $C_o = \sigma/R_o$, where R_o is the radius of curvature of the surface in this point.

5.3.2 THE GRAPHICAL METHOD

Solution of this equation by the methods of graphic integration[280] showed a satisfactory coincidence of the calculated and actual shape of the deposited bead. Under the actual conditions, however, weld formation differs from the conditions of a liquid drop spreading over the surface of a solid, due to the base-metal melting by the arc. This makes it more difficult to apply the method of graphic integration to study the influence of different factors on the shape and dimensions of the weld bead.

It should be noted that this and subsequent equations were obtained for the case of thermodynamic equilibrium at a constant temperature. In reality, the weld pool, including its tail part, is characterized by a pronounced non-uniformity of temperature distribution, which may affect the shape and dimensions of weld reinforcement.

Representing the surface of the molten metal of the weld pool in the area of the solidification front as a cylindrical surface and assuming the bead to form under the impact of the forces of gravity and surface tension, an equation[264] was obtained, describing the cross-sectional profile of weld reinforcement:

$$B = \int_0^h \frac{(x^2/2a_c^2 + x/R_o - 1)}{\sqrt{1 - (x/2a_c^2 + x/R_o - 1)^2}} \, dx + C, \tag{5.13}$$

where B is the bead width at a given h; h is the distance from the upper point of the profile to the horizontal plane; x is the coordinate of the height of any point of the bead profile; R_o is the radius of curvature of bead profile in the vertex.

I.L. Emelyanov also suggested a procedure to determine the shape and dimensions of the deposited bead, taking into account both the metal surface properties, and the modes of welding: values of welding current, arc voltage, welding speed, etc.[81] However, other studies[264,341] used approximate methods of graphic integration of the capillarity equation, which makes studying the process of weld reinforcement formation much more complicated, particularly in welding in positions other than the downhand position. Nonetheless, this work performed taking into account the surface properties of the weld pool metal, demonstrated the good agreement of the calculated and actual shape of weld reinforcement.

Calculated equations, which allow determination of the main geometrical parameters of weld reinforcement, were derived in by Berezovsky and Stikhin.[263] For butt welding in the downhand position, these equations become:

$$C = \sqrt{z_o^2 + 2(1 - \cos \varphi)} - z_o;$$

$$b_o = 2 \int_0^\varphi \frac{\cos \varphi \, d\varphi}{\sqrt{z_o^2 + 2(1 - \cos \varphi)}}; \tag{5.14}$$

$$f_o = b_o(C + z_o) - 2 \sin \varphi.$$

Here, C is the dimensionless reinforcement of the weld; $C = c/a_c$; z_o is the dimensionless distance from the reinforcement vertex to the origin of coordinates, $z_o = a_c/R_o$; b_o is the dimensionless width of reinforcement; $b_o = B/a_c$; f_o is the dimensionless area of reinforcement, $f_o = F/a_c^2$.

Solving the system of Equation (5.14) allows calculating the reinforcement height C and the angle of base-metal transition to the deposited metal, φ with known values f_o

and b_o. The latter values are calculated by the specified welding mode. It should be noted that angle φ is not equal to the angle of wetting, as shown in Berezovsky and Stikhin.[263]

Calculated dependencies [Equation (5.14)] were obtained on the assumption, that the weld convexity is formed in the weld-pool tail part and, therefore, the arc pressure does not have any noticeable influence on the formation of weld reinforcement. In a number of cases, however, in welding by a low-power arc of metals, characterized by a high heat conductivity, the weld pool is short. In this case, the influence of the arc on the formation of weld reinforcement cannot be ignored.

The advantage of the method, proposed in the work of Berezovsky and Stikhin,[263] is the ability to apply it in calculation of the shape and dimensions of the bead for different types of welded joints. So, in narrow-gap welding, the shape of the meniscus in the gap is given[342] by the same system of equations, as in the case of formation of reinforcement in butt welding, only f_o is found from the following expression:

$$f_o = \pm \frac{(F_d - BL)}{a_c^2}, \tag{5.15}$$

where F_d is the cross-sectional area of the deposited metal; $L = \sqrt{q/\pi e c \gamma T_f - B^2/4}$ (q is the heat input; $c\gamma$ is the heat capacity of the metal). The plus sign is taken for a convex meniscus and the minus sign for the concave one.

5.3.3 THE ROLE OF TEMPERATURE FIELDS AND INCLINATION ANGLES

Influence of surface properties of the metal on the formation of weld reinforcement in the downhand position is further noted in different studies.[343–345] In submerged-arc welding, the formation of weld reinforcement is strongly influenced[346] also by the surface tension of the slag.

Features of weld formation are related to surface phenomena, in particular, to wettability also in dissimilar metal welding. This is very significant in those cases when the metals being welded differ significantly by their thermo-physical properties. For instance, in arc welding of steel to copper, the temperature fields in the parts will have the following form, provided the arc moves precisely along the butt (Figure 5.27). As is seen, in this case, the symmetry of the temperature fields relative to the weld axis is disturbed. The isotherms of a copper part run ahead of those of steel, and the width of the zone of a copper part preheating is several times greater than that of the zone of steel preheating. However, in this narrow zone, the temperature of a steel plate is much higher than that of a copper plate. This will lead to different wettability of the parts being joined by the molten electrode metal and hence, the deposited bead will have different angles of transition from the base to the deposited metal and different width on the copper and the steel plates. In order for the deposited bead to be symmetrical, it is necessary for the steel and copper plates wettability by the electrode metal to be the same. This may be achieved either by preheating the copper part, or by applying coatings on the copper part surface, which are readily wetted by molten electrode metal, or by the shifting the electrode to the copper part in order to preheat it better.

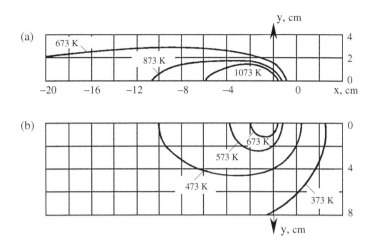

FIGURE 5.27 Shape of temperature fields in the steel (a) and copper (b) plates of 1 cm thickness in arc welding; $q = 4200$ W; $V_w = 0.002$ m/sec

5.3.3.1 Angle of Inclination

The role of surface forces in the formation of the weld bead is enhanced in welding or surfacing in the vertical, horizontal, and particularly, overhead positions. This is associated with the change of the relation of forces, acting on the weld-pool metal. In welding of horizontal welds on a vertical plane, primarily the forces of gravity and surface forces act on the weld-pool metal. The forces of mechanical impact of the arc on the surface of the weld-pool tail part, where the bead forms, are relatively small, and, therefore, their influence may be ignored.

The possibility of the bead forming on inclined surfaces under the impact of the forces of gravity and surface tension, may be evaluated, using Ya.I. Frenkel's formula.[347] According to Frenkel',[347] rolling down of a liquid of a certain weight starts at a certain critical angle of inclination of the surface, α, which depends on the value of adhesion of the liquid to the part surface:

$$2rW_A = mg \sin \alpha, \tag{5.16}$$

where r is the radius of the drop, which is on the surface of a solid; m is the drop weight; g is the acceleration of gravity.

From Equation (5.16) it follows that the value of the critical angle of inclination of the surface of the part, at which the weld-pool metal starts rolling down, depends on the value of W_A, and hence, on the surface properties of the solid and molten metals, as well as interphase energy on the melt–solid metal boundary.

Numerous investigations indicate that for the most diverse liquid–solid systems the dependence between $(m/2r)$ and $(1/\sin \alpha)$ and the angle of inclination of this straight line is equal to W_A/g. Deviation of this dependence is observed only with

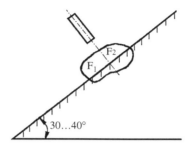

FIGURE 5.28 Schematic of measuring the weld pool fluidity in bead deposition on an inclined plane.

good wettability of the solid by the liquid, when the value of adhesion is close to that of cohesion strength of the liquid.

The connection between the surface phenomena and bead formation on inclined and vertical surfaces follows from the results of investigations, derived in the study of Ryabtsev, Kushkov, and Chernyak[348] who used a procedure, proposed by A.A. Erokhin,[24] which essentially consisted in the following. A horizontal bead is deposited on an inclined sample (Figure 5.28), the bead cross-section being asymmetrical due to the weld-pool fluidity. Ratio of the areas of the bead upper and lower parts $\gamma = F_1/F_2$ was chosen as the criterion for evaluation of the fluidity of weld-pool metal. According to the derived results, with increase of welding current (which leads to increased weld-pool weight), γ ratio also increases, which is in accordance with Equation (5.16). Investigation of the influence of alloying components (C, Cr, P, Cu) demonstrated, that increase of the content of surfactants (C, P) in the metal leads to a higher fluidity of the weld-pool metal. When the influence of fluxes (AN-348A, AN-15M, AN-20, AN-26) on weld-pool fluidity was studied, it was also noted that the pool fluidity was the highest in the deposition of beads, using AN-348A flux, on the boundary with which the metal interphase tension is small.

The role of surface tension forces in weld formation in overhead welding is considered in sorce studies.[349,350] With the same assumptions [349] as in the waste of Berezovsky and Stikhin,[263] equations were derived, which allow the main dimensions of weld reinforcement to be determined:

$$C = z_0 - \sqrt{z_0^2 - 2(1 - \cos\varphi)},$$

$$B_0 = 2\int_0^\varphi \cos\varphi \Big/ \sqrt{z_0^2 - 2(1 - \cos\varphi)}\ \mathrm{d}\varphi, \qquad (5.17)$$

$$F_0 = b_0(c - z_0) + 2\sin\varphi,$$

where the designations are the same as in Equations (5.14).

5.3.4 SURFACE TENSION VS. GRAVITATIONAL FORCE

A feature of welds, produced in overhead welding, is the fact that when the weld pool has reached certain dimensions, the gravitational forces cannot be compensated by the surface tension forces. In this case, the metal flows out of the weld pool. As demonstrated by experiments,[350] metal flowing out of the weld pool occurs in its tail part. This is related to the fact that the metal is contained in the weld-pool head part in addition to the surface tension forces, also by the forces of arc pressure. Calculations show,[350] that at welding current of 200–230 A, approximately 50% of the weld-pool molten metal is contained by the surface tension forces. In this connection, in overhead welding, various techniques are often used to contain the weld-pool metal, which make the welding process more complicated and increases its cost.

According to Chudinov and Taran,[350] a certain limit width of convexity B_{max} exists. Above the limit free formation of the weld in the overhead position becomes impossible at any area of the deposited metal. Similarly, there also exists a limit for the area of convexity, F_{max}. Both B_{max} and F_{max} depend only on the capillary properties of the liquid metal. Disturbance of weld formation may occur when the condition of interface stability holds, if the hydrostatic forces are not balanced by those of surface tension. This is because of too deep a penetration.

Normal weld formation in overhead welding is found, if the following inequality holds[349]

$$H_a + C < a_c^2/R_o,   \quad (5.18)$$

where H_a is the average depth of the weld pool in the tail part; C is the height of reinforcement, R_o is the radius of reinforcement profile in the vertex.

As follows from Equation (5.18), the capillary forces have the strongest influence on the process of weld formation in the overhead position, as the capillary constant, included into this inequality is squared.

At one-sided welding with complete penetration, the welded joint's performance is largely dependent on the shape and dimensions of the weld root. In gravity welding, weld-root formation is determined by the relation of forces, acting on the weld-pool metal. They are the gravity force, welding arc pressure, as well as surface tension of the liquid metal. The magnitude and direction of these forces depend on the welding mode, geometry of edge preparation, and weld position in space. According to Tyulkov and Turbin,[351] the thickness of the molten metal layer, which may form in the gravity position is equal to:

$$H = 2\sigma \cos\theta/b_y.   \quad (5.19)$$

Here, b_y is the width of the weld, found from expression $b_y = b + \varepsilon$, where b is the weld width after the weld-pool solidification, and ε is the magnitude of linear shrinkage of the weld-pool metal.

Application of Equation (5.19) is particularly effective in the case of welding structures, which require a guaranteed penetration of the metal, but no backing can be used to form the weld root, for instance, when are produced butt joints on thick-walled pipelines.

If melting backing can be used to form the weld root, then the balance of forces, acting on the molten metal is given by the following equality[352]:

$$P_a + P_{hs} + P_{cf} = P_s + P_m + P_{ad}, \tag{5.20}$$

where P_{hs} is the hydrostatic pressure of the weld-pool mass, referred to as the area of full penetration; P_{cf} is the force arising at liquid-metal motion from the weld-pool head part to its tail part; P_m is the force of surface tension of weld-pool surface; P_s is the force of surface tension of the slag formed in the melting of the backing; P_{ad} is the work of adhesion between the slag and the solid surface of the metal and the backing.

A condition of the absence of defects (macro rolls, slag inclusions, dents), characteristic of the back bead, is satisfying the following inequality

$$P_a + P_{cf} + P_{hs} < P_s + P_m + P_{ad}. \tag{5.21}$$

Analysis of the forces, acting on the molten metal showed,[352] that mostly the force of surface tension of the slag can counteract the sum of forces $P_{ad} + P_{hs} + P_{cf}$. According to the performed calculations, it is necessary for the surface tension of the slag to be not less than 500 mJ/m^2.

Sound formation of the back bead of the weld is possible, if the following inequality holds[353]:

$$P_{m-s} + P_m > P_a + P_{cf} + P_{hs}, \tag{5.22}$$

where P_{m-s} is the force of interphase tension on the boundary of molten metal and the slag of the backing.

Thus, selection of a backing, which ensures formation of the slag, on the boundary with which σ_{m-s} value for this metal is not less than 500 mJ/m^2, allows improvement of formation of the weld-back bead.

5.3.4.1 Angle of Transition

It is known,[354,355] that the welded joint's performance is often dependent not only on the weld configuration, but also on the smoothness of base-metal transition to the deposited metal. Recently, argon-arc treatment of welded joints is becoming accepted which, along with the change of the HAZ structure, metal degassing, and change of the shape and dimensions of non-metallic inclusions, also allows increasing the angle of transition from 160 to 180°.

According to Belchuk and Naletov,[356] the angle of transition from the base metal to the deposited metal, φ, is related to the radius of transition R by the dependence, $R = A/\varphi$. Studying the features of formation of the zone of transition from the base metal to the deposited metal allowed deriving[356] another expression, defining the connection between R and φ

$$R = \Delta L ctg(\varphi/2), \tag{5.23}$$

where $\Delta L = (B_S - B_L)/2$, and B_S, B_L are the width of the isotherm with temperature of solidus and liquidus for the base metal, respectively.

For many steels and metals, the value of ΔL changes in the range from 0.1 to 0.4 mm.[357] As shown by Berezovsky and Stikhin,[263] the value of the angle, φ_{k-} depends on the width of the reinforcement, area of the deposited metal, and capillary constant, that is related to the surface properties of the molten metal.

Influence of surface phenomena on weld formation is noted[327,358,359] for various processes of fusion welding.

5.4 FORMATION OF THE DEPOSITED METAL

Unlike welding, in surfacing the depth of penetration and thickness of the deposited metal (layer) often should be small, and the maximum weld width is sought, as this improves the process efficiency.

5.4.1 WETTING ANGLE

It is an estabilished fact that wetting angle, θ is the main characteristic of the interphase of the three phases (in this case solid metal–molten metal–gas or slag). Under the condition of a quite shallow depth of penetration $\cos \theta$ value is equal to:

1. In surfacing in a gaseous medium

$$\cos \theta = (\sigma_{s-g} - \sigma_{s-l})/\sigma_{m-g}, \qquad (5.24)$$

2. In submerged-arc surfacing

$$\cos \theta = (\sigma_{s-s} - \sigma_{s-l})/\sigma_{m-s} \qquad (5.25)$$

Taking into account Dupre equation (2.6), Equation (5.24) and Equation (5.25) become:

$$\cos \theta = (W_A/\sigma_{m-g}) - 1, \qquad (5.26)$$
$$\cos \theta = (W_A/\sigma_{m-s}) - 1. \qquad (5.27)$$

From Equation (5.26) and Equation (5.27) it follows that the shape of the deposited bead, which is related to the value of θ, depends on the value of adhesion between the molten metal and the solid metal and on the magnitude of surface or interphase tension, depending on the kind of the shielding used for the metal melt. As long as $W_A > \sigma_{m-g}$ (σ_{m-s}), $\cos\theta > 0$, that is, wetting of the solid metal by the molten metal will be good, and angle $\theta < \pi/2$. Not only the shape of the deposited bead, but also the minimum possible thickness of the deposited metal is related to the value of the wetting angle.

In order to establish this dependence, let us consider the process of spreading of the melt drop over the surface of the solid metal. As the formation of the deposited bead proceeds outside the zone of action of the arc or other heat source, we will not take into account the influence of the pressure of the heat source on the molten metal or the reactive forces. In this case, formation of the deposited bead is determined only by the action of the forces of gravity and surface tension.

In spreading of a drop of the melt over the surface of solid metal, the total change of surface energy of the system per unit surface are will be equal to $\Delta\sigma = \sigma_{m-g} + \sigma_{sol-1} - \sigma_{sol-g}$.

If a drop has the shape of a circle of radius r, then the surface energy increases by a value of $2\pi r\Delta\sigma dr$, and the potential gravitational energy on the condition that the center of gravity is located at the mid-height of the drop, will be reduced by the following value:

$$\frac{1}{2}V_d\rho_m gd\delta_d,$$

where V_d is the drop volume; ρ_m is the melt density; g is the acceleration of gravity; δ_d is the drop height.

For a large drop of a small height, it may be assumed, that $V_d \approx \pi r^2 \delta_d$. Then, $d\delta_d/\delta_d = -2dr/r$, and the condition of equilibrium of the spreading process will be written as:

$$\pi r^2 \Delta\sigma \left(\frac{d\delta_d}{\delta_{eq}}\right) = \frac{1}{2}\pi r^2 \delta_{eq}\rho_m gd\delta_d, \tag{5.28}$$

where δ_{eq} is the equilibrium value of drop thickness.

Hence, the minimum possible thickness of the deposited layer $\delta_{eq} = (2\Delta\sigma/\rho_m g)^{1/2}$ or, in view of Equation (5.24):

$$\delta_{eq} = \left[2\sigma_{m-g}\frac{(1 - \cos\theta)}{\rho_m g}\right]^{1/2}. \tag{5.29}$$

From Equation (5.29), it follows that the value of δ_{eq} depends comparatively little on σ_{m-g} and ρ_m, as they appear in the equation in the power of $1/2$. It is obvious that the deposited layer thickness will be largely determined by the value of the angle of wetting of the processed surface by the metal melt, as the difference $(1 - \cos\theta)^{1/2}$ depends almost linearly on the value of θ up to values of $\theta = 120°$ (Figure 5.29).

Since the formation of the deposited bead is largely dependent on the value of the wetting angle, all the factors influencing the value of θ will influence the shape and dimensions of the deposited bead. However, two factors, probably, influence the process of the deposited metal formation to the greatest extent. These are: (1) the base-metal temperature and (2) the value of surface tension of the melt spreading over the solid metal surface.

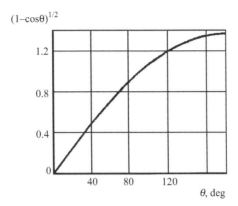

FIGURE 5.29 Dependence of the value of $(1-\cos\theta)^{1/2}$ on the value of θ.

5.4.2 EXPERIMENTS WITH ELECTRON BEAM SURFACING

The connection between metal-melt spreading and base-metal temperature is confirmed, in particular, by the results of investigations on electron beam surfacing, conducted by us. Surfacing of plates of steel (St3) and copper (M1) was performed in a ELV-4 unit by a beam of electrons, taken out into the atmosphere with an energy, $E = 1.5\,\text{MeV}$ and a power $P = 30\text{--}50\,\text{kW}$. Copper plates of thickness, δ_{pl} of 4.5 and 12 mm were surfaced using the following surfacing materials: copper powder, copper plates ($\delta_{pl} = 2.5\,\text{mm}$), and powders of the following grades: PR-N80Kh13S2R, PR-N70Kh17S4R4, PG-KhN80SR2. The last category of powders are expected to increase the wear resistance of copper products, exposed to high temperatures in service. Wear-resistant powders PR-N80Kh13S2R, PR-N70Kh17S4R4, and M6F3 were used in surfacing of steel plates of 5, 8, and 12 mm thickness.

In surfacing of all the materials, it is noted that the geometry of the deposited bead is disturbed with the increase of deposition rate (at $P = \text{const}$). The deposited bead is first narrowed, then bridges form and, finally, formation of individual drops of the deposited powder is observed (Figure 5.30). The same effect of drop formation was also noted by A.G. Grigoryants in surfacing with powder materials by the laser beam.

5.4.2.1 The Critical Rate of Deposition

Analyzing the experimental results (Figure 5.31), it is readily seen that in bead deposition on copper plates, the process of the deposited bead formation becomes disturbed after a certain deposition rate has been achieved, which is critical for this specific case. The value of this rate depends on the thickness of the deposited metal layer and to a smaller degree on powder composition. In deposition of copper materials (both powder and plate), the values of critical rates are somewhat higher than in deposition of chromium–nickel powders. The observed disturbance

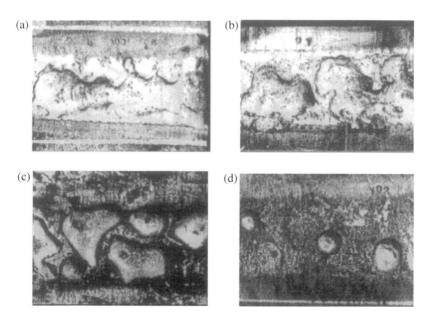

FIGURE 5.30 Formation of the deposited bead (PR-N80Kh13S2R) on copper plates ($\delta_{pl}=$ 12 mm; $I_w = 30$ mA; $v = 8-9$ Hz). V_s (m/s): (a) 0.001; (b) 0.0012; (c) 0.0014; (d) 0.0016.

of the deposited bead formation with increase of the deposition rate is related to decrease of the heat input, which becomes insufficient for substrate preheating up to such a temperature, which ensures good wetting of the solid copper by the molten metal or its spreading. Optimal values of energies, which provide a good formation of the deposited layer, probably, depend on the volume of the powder being melted and the thickness and composition of plates being surfaced.

Thus, formation of the deposited metal in surfacing of copper plates with powdered materials and plate electrodes depends on the energy of the electron beam, and, hence on base metal heating and the associated wettability of the solid metal by the metal melt. At insufficient heating of the part, the value of θ in the peripheral sections of the deposited bead will be much greater than the equilibrium value of the wetting angle. In this case, as was shown in Section 3.3, the melt gets contracted in the direction from greater θ values to its smaller values. This is exactly the cause for the formation of bridges in the deposited layer and then also the formation of individual metal drops. Now, the wettability can be provided by a correct selection of the surfacing modes.

Investigations on surfacing of steel parts (St3) with powdered materials demonstrated (Figure 5.31), that in this case also formation of the deposited metal depends on the magnitude of the heat input, and, hence, on the solid metal heating.

Thus, formation of good metal deposit is related to surface phenomena, and is guaranteed only if the part surface has the temperature necessary for a good wettability of the solid metal by the metal melt.

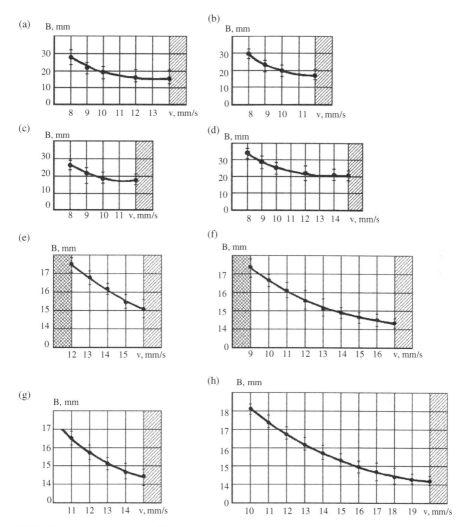

FIGURE 5.31 Dependence of the deposited layer width, B in surfacing of copper (a, b, c, d) and steel (e, f, g, h) plates on the surfacing speed: (▒ Area of drop formation; ▓ Area of through penetration. δ_{pl} (mm), δ_{pow} (mm), I_w (mA), v(Hz) are. (a) 12, 30, 8–9; (b) 12, 2, 30, 8–9; (c) 2.5, 10–20, 30, 8–9; (d) 2.5, 10–20, 30, 8–9; (e) 5, 1.5, 12, 8–9; (f) 8, 1.5, 20, 6; (g) 5, 3.5, 12, 6; (h) 12, 1.5, 30, 6. Material deposited: Cr–Ni powder for all cases excepting (d) wherein it was M1 copper.

In order to achieve greater width of the deposited layer experiments on surfacing were performed, using transverse oscillations of the electron beam. Mixing of the base and the deposited metal (Figure 5.32) is observed in the peripheral sections, which may lead to the change in the properties of the deposited metal. Therefore, let us consider this process in greater detail.

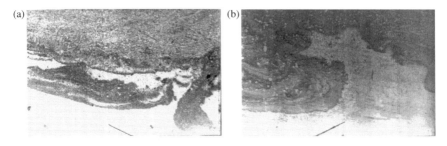

FIGURE 5.32 Structure of fusion boundaries ($\times 200$) of a copper sample, deposited by a scanning beam, when N80Kh13S2R (a) and N70Kh17S4R4 (b) powders are used for surfacing.

5.4.2.2 Electron Beam Oscillations

With at electron beam oscillations, the temperature in the opposite points of the deposited metal layer changes from boiling temperature, T_{boil} at the moment of the beam being in this point, to a temperature, close to the melting temperature of the deposited metal, T_{mel} in the opposite point. Therefore, thermocapillary phenomena can be one of the possible causes for the observed mixing of the base and the deposited metal.[360]

If we assume that as found in the study on concentrated capillary flows[361] the resistance force, dF_1 acting on each element of surface, dS for a liquid flowing around a flat plate is equal to:

$$dF_1 = \alpha(\eta\rho U^3/x)^{1/2}dS, \tag{5.30}$$

where α is a coefficient, equal to 0.332; η is the dynamic viscosity of liquid; ρ is the density of the liquid; U is the velocity of motion of the liquid layer; x is the distance from dS section to the point of beam action.

This force is balanced by those of surface tension dF_2, induced by the gradient of surface tension on the metal melt surface, $dF_2 = (d\sigma/dx)\,dS$.

Equating these forces and integrating with respect to x between zero and B, we obtain:

$$U = \{4(\Delta\sigma/\alpha)^2\eta\rho B\}^{1/3}, \tag{5.31}$$

where $\Delta\sigma = \sigma_{Tmel} - \sigma_{Td}$ is the gradient of surface tension across the width of deposited layer, B.

5.4.3 Temperature Gradient and the Speed of Displacement

Presence of a temperature gradient leads to a change in the viscosity (η), the density (ρ), and the surface tension of the melt (σ) in the path of movement of the metal flow, which are described by the following dependencies[362]:

$$\eta = \eta_0 \exp(E/RT), \tag{5.32}$$

where R is the absolute gas constant; η_0 and E are the constants, the magnitude of which depends on the composition of the deposited material. For powders with a high content of nickel, they may be assumed to be equal to the corresponding values for pure nickel. Then, $\eta_0 = 0.1663$ mN s/m^2 and $E = 50.2$ mJ/mole[362]

$$\rho = \rho_0 + (T - T_0)\frac{d\rho}{dT}, \tag{5.33}$$

where $d\rho/dT$ is a constant, equal to 1.160 mg/(cm$^3 \cdot$ K)

$$\sigma_T = \sigma_0 \frac{T_c - T}{T_c - T_0}, \tag{5.34}$$

where σ_T, σ_0 are the surface tension of the melt at temperatures T and T_0, respectively; T_c is the temperature, at which the magnitude of surface tension becomes equal to zero; $T_c = 1.7\, T_{boil}$.[2]

Calculated values of U, obtained in view of dependencies of Equation (5.32), Equation (5.33), and Equation (5.34) are shown in Figure 5.33 and Figure 5.34. As is seen, the velocity of the molten metal flow depends on the width of the surfaced region, and the flow velocity rises with reduction of the deposit width.

It should be noted that U values, derived in view of temperature dependencies of η, ρ, and σ, are close to the experimental data, given by Najdich, Zabuga, and Perevertajlo,[361] and they are lower, when these dependencies are not taken into account.

In order to evaluate the values of velocities, at which the vortex-type flows can form causing mixing of the base and the deposited metals, let us introduce the dimensionless surface velocity[363]

$$u = UH/v, \tag{5.35}$$

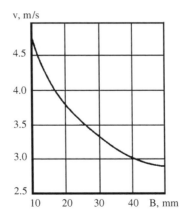

FIGURE 5.33 Change of the speed of metal displacement depending on the width of deposited bead.

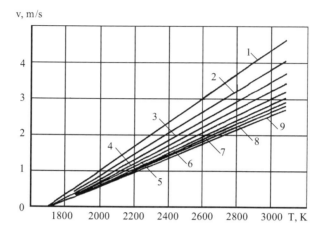

FIGURE 5.34 Change of the speed of metal displacement depending on the temperature for different values of the deposited bead width, mm: (1–9) 10, 15, 20, 25, 30, 35, 40, 45, 50.

which may be represented as two components:

$$u_m = -\mathrm{Re}_m/k_m \quad \text{and} \quad u_G = -G_r/k_G,$$

where k_m and k_G are the coefficients, which determine the nature of the liquid metal flowing; Re_m and G_r are the dimensionless parameters, calculated by the following formulas:

$$\mathrm{Re}_m = |\sigma| G H^2/\rho v^2 \quad \text{and} \quad G_r = \beta g G H^4/v^2,$$

where $|\sigma|$ is the temperature coefficient of surface tension of the metal melt; $G = (T_{boil} - T_{mel})/B$; H is the depth; ρ is the melt density; v is the kinematic viscosity; β is the temperature coefficient of volume expansion; g is the acceleration of gravity.

Under the conditions of transition of the laminar flow into a vortex-type flow

$$G_r = (k_G/k_m)\mathrm{Re}_m, \quad u = \mathrm{Re}_m/k_m + G_r/k_G,$$

then, using U values, calculated by the above formula [Equation (5.31)], k_M and k_G can be determined for different width B of the deposited bead. For a change of B in the range, 10–50 mm, the coefficients k_M takes the values from 6 to 45; and $k_G = (244–1390) \cdot 10^{-3}$. Calculated values of dimensionless parameters Re_m and G_r vary in the range of $(14–1150) \cdot 10^3$ and 530–35250, respectively. Influence of various parameters on the value of coefficients k_M and k_G is shown in Table 5.1.

According to the data of Camel, Tison, and Favier,[363] the flow of the molten metal may be characterized as viscous M_1 or surface M_2 Marangoni flow, laminar viscous G_1 or laminar G_2 flow of surface layers.

TABLE 5.1
Values of Coefficients k_M and k_G

Width of Deposited Bead, m	T, K	$Re_m \times 10^3$	G_r	k_m	$k_G \times 10^{-3}$
0.01	2128	74	2640	18	656
	2628	415	13759	32	1064
	3073	1149	32350	45	1389
0.02	2128	37	1320	12	413
	2628	208	6880	20	670
	3073	575	17625	29	875
0.04	2128	18	660	7	260
	2628	104	3440	13	430
	3073	292	8812	18	551
0.05	2128	14	528	6	244
	2628	83	2752	11	354
	3073	234	7050	16	475

Conditions of existence of each of these flows are described by the following inequalities:

$$M_1 - Re_m < k_m^3, \qquad \left(\frac{k_m}{k_G}\right) G_r < Re_m; \tag{5.36}$$

$$G_1 - G_r < k_G^2, \qquad Re_m < \max\left(\frac{k_m G_k}{k_G}, \frac{G_r^{3/2}}{k_G^{3/2}}\right); \tag{5.37}$$

$$M_2 - k_m^3 < Re_m, \qquad \min\left(G_r^{3/4}, \frac{G_r^{3/2}}{k_G^{3/2}}\right) < Re_m; \tag{5.38}$$

$$G_2 - k_G^2 < G_r, \qquad Re_m < G_r^{3/4}. \tag{5.39}$$

Proceeding from the obtained data, it is established that conditions Equation (5.38) are satisfied in surfacing by an oscillating electron beam. Therefore, Marangoni-type flow of surface layers is the most probable kind of the metal melt flow in electron beam surfacing.

Dependence of the shape of the weld and deposited bead on the surface properties of the metal and slag, as well as the wetting angle, opens up wider possibilities for changing their configuration. It is obvious that the geometrical dimensions of the weld and the deposited bead may be changed not only by varying the modes of welding and surfacing, but also by adding surfactants to the metal, creating a certain

gas atmosphere, applying special fluxes, changing the potential of the metal surface
due to external emf, using preheating of the parts by changing θ, etc.

5.5 DEFECTS OF WELD SHAPE AND SURFACE PHENOMENA

The main defects of weld shape, which have a significant influence on welded joint's
performance are: undercuts, lacks-of-penetration, lacks-of-fusion, burns-through,
and craters. Let us consider the role of surface properties and phenomena in the
process of formation of these defects, as well as rolls.

From practical experience of making weldments, it is known that undercuts are
most often found in automatic welding processes, especially in fillet welding. They,
however, quite often form also in butt welding, as a rule at high welding speeds.

The cause for their formation can be deviation of the electrode from the weld
axis, or too high arc voltage. In both the cases, a deeper penetration of one of the
edges is observed, leading to the formation of a groove, which remains after
solidification of the weld-pool metal.

5.5.1 GROOVE FILLING AND UNFILLING

Groove unfilling by the metal melt and formation of undercuts is largely determined
by the ratio of the rates of metal solidification and groove filling by the metal melt.
Therefore, undercuts can be removed by two methods: (1) lowering of the rate of
solidification of weld-pool metal (V_{cr}) (in other words, extension of the time of
weld-pool existence) and (2) as increase of the speed of groove filling by liquid
metal (V_{spr}). There exists an optimal ratio of V_{cr} and V_{spr}, at which no undercuts
form. It is obvious that no undercuts will form, if the time required for groove
filling with the metal, τ_{spr}, is smaller than the time of existence of weld pool, τ_{ex}.
Now, if $\tau_{spr} > \tau_{ex}$, the undercuts will form.

As follows from Equation (1.1), duration of weld-pool existence, τ_{ex}, becomes
longer with the increase of weld-pool length. On the other hand, the weld-pool
length depends primarily on thermo-physical properties of the metal being welded
and the power of the used heat source. Therefore, in welding of a certain metal,
the duration of weld-pool existence may be increased by using multi-electrode
welding, preheating the part, or lowering the welding speed. The first of these techniques is used, mainly, in welding of very long rectilinear welds, for instance in the
case of large-diameter pipes, the second, for comparatively small items, and the
third technique is not optimal, as it lowers the process efficiency.

However, the time of weld-pool existence can be extended also by other techniques, for instance, in submerged-arc welding, as will be shown further on (see
Section 9.5), by increasing the shunting current flowing through the slag. This
may be achieved by using an additional nonconsumable electrode in the slag pool.

As was already noted, undercuts can be eliminated not only by extending the
time of weld-pool existence, but also by increasing the rate of the molten metal
spreading. As this technique is related to surface properties of the materials being
welded and wettability, let us consider it in greater detail.

5.5.1.1 Extending the Time of Weld-Pool Existence

In order to clarify the factors, influencing this process, let us consider the sequence of undercut development (Figure 5.35). In welding, depending on the kind of the applied shielding the system will consist of three phases: molten and solid metals, and molten flux, or gas. In both the cases, various chemical reactions proceed at the interphases, which will change the magnitudes of interphase tension on these boundaries. Therefore, the considered systems will be non-equilibrium ones and hence, the value of the wetting angle will change with time. In all the cases, however, the value of wetting angle is determined by the ratio of the values of specific free energies.

In spreading of a melt over the solid surface, the free energy of the system will be reduced, and, therefore, the spreading process will proceed spontaneously, only if the following inequalities hold:

$$\sigma_{s-g} - \sigma_{s-l} > \sigma_{m-g}, \tag{5.40}$$

$$\sigma_{s-g} - \sigma_{s-l} > \sigma_{m-s} \tag{5.41}$$

Here, Equation (5.40) is valid for the case of welding in a gas atmosphere or vacuum and Equation (5.41) is valid for submerged-arc welding.

The driving force (P) of metal spreading over the surface of the solid, in view of Young's equation, may be written as follows:

$$P_{m-g} = \sigma_{m-g}(\cos \theta_{\tau m-g} - 1); \tag{5.42}$$

$$P_{m-s} = \sigma_{m-s}(\cos \theta_{\tau m-s} - 1). \tag{5.43}$$

Here, θ_τ is the instant value of the wetting angle. Thus, the magnitude of the driving forces of spreading depends on the value of the wetting angle.

5.5.1.2 Surface Roughness

The value θ_τ depends not only on the ratio of the specific free energies on the interphase, but also on the roughness of the solid surface, angle of surface inclination, etc.

Allowing for the surface roughness, Equation (5.42) and Equation (5.45) become:

$$P_{m-g} = \sigma_{m-g}(k \cos \theta_{\tau m-g} - 1); \tag{5.44}$$

$$P_{m-s} = \sigma_{m-s}(k \cos \theta_{\tau m-s} - 1), \tag{5.45}$$

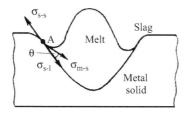

FIGURE 5.35 Relation of surface energies in formation of undercuts.

where k is the coefficient of roughness, equal to the ratio of the true area of the surface to the area of projection of this section onto a plane.

As the true area is always greater than its projection on a plane, the coefficient k is always greater than unity and therefore, roughness of the solid surface promotes liquid spreading over it.

Displacement of point A (Figure 5.35) is only possible in the case, if the driving force of the metal spreading is greater than the forces hindering the spreading. From Equation (5.44) and Equation (5.45) it follows, that this may be achieved by reducing the wetting angle. This is because in systems with a variable wetting angle, melt displacement proceeds from greater values of wetting angles to smaller ones. With this purpose, welding by an inclined electrode (backward-inclined) can be used. With such an arrangement, the part edges are preheated. This is also true for the welding of parts having coated edges and which are readily wetted by molten metal drops.

Thus, the formation of undercuts depends to a certain extent on the wettability of the solid metal by the metal melt, and the process of undercut formation can be controlled by changing the wettability.

5.5.2 MULTILAYER WELDING

Lacks-of-penetration have a marked influence on the welded joint's performance. It is known that lacks-of-penetration are most often observed across the base-metal thickness, in the apex of the angle and along the edges of the parts being welded. In multilayer welding, lacks-of-penetration between the individual layers are sometimes found.

Lacks-of-penetration across the base-metal thickness often result from incorrect selection of the welding mode and, primarily, insufficient welding current, which does not permit achieving the required depth of penetration. However, as the depth of penetration depends also on the surface properties of the base and electrode metals (Section 5.2), lacks-of-penetration can be eliminated by reducing σ_{m-g} and σ_{m-s} values, depending on the applied shielding medium.

Presence of lacks-of-penetration between the layers in multilayer welding or surfacing is often related to the presence of remnants of the slag crust on the surface of the underlying layer. On the other hand, removal of the slag crust is associated (Section 9.5) with the magnitude of adhesion between the metal and slag, and becomes more difficult with increase of W_A.

In view of the fact, that the process of formation of the weld proper proceeds in a short time interval, formation of lacks-of-penetration will be influenced to a certain extent also by the speed of the melt filling the gap between the parts being welded. The influence of various factors on this process, including the surface properties and phenomena has been discussed in Section 5.2.

When arc welding is used, burns-through form mainly in the welding of the sheet metal. From Equation (5.7) it follows, that in downhand welding of a butt joint, the probability of formation of the burns-through depends on the mass of the weld-pool metal, arc pressure, and surface properties of the metal.

For a thin metal, the mass of the weld-pool metal, and hence, the value of P, increases with the increase of heat input, chiefly, due to widening of the weld. This, in its turn leads to increase of the main radii of curvature r_1 and r_2. In addition, if increase of the heat input was due to increase of welding current, this will also lead to higher arc pressure.

5.5.2.1 Formation of Burns-Through

From the points presented above, it is clear that by increasing the heat input, it is possible that the arc pressure and the hydrostatic pressure will exceed the surface tension forces, which will lead to flowing out of the weld-pool metal and formation of burns-through. It is obvious that in gravity welding, burns-through may be avoided either by reducing the volume of the weld pool or the arc pressure, or increasing the surface tension of the metal.

Influence of surface properties of weld-pool metal on the formation of burns-through was studied in ADPG-500 automatic machine. Welding wires Sv-08 and Sv-08G2S of 1.2 mm diameter and 3-mm thick plates of St3 and E3A steels were used. Welding was conducted at reverse-polarity D.C. in CO_2 atmosphere at a constant arc voltage (24–26 V) and a shielding gas flow rate of $900 \cdot 2.8 \cdot 10^{-7} \, m^3/sec$. As is seen in Figure 5.36, when such materials are used, (for which the surface tension drops markedly in contact with an oxidizing medium), formation of burns-through occurs at smaller values of the welding current, that is, at lower arc pressure and smaller mass of the molten metal.

5.5.2.2 Formation of Craters

Surface properties of the metal also have a certain influence on the formation of craters in the weld. In arc welding, these defects develop as a result of the impact of the following forces. Arc pressure P_a and reactive forces P_r arising as a result of evaporation of the metal in the active spot zone. These forces try to depress

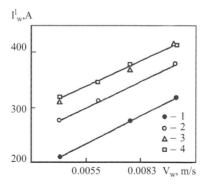

FIGURE 5.36 Dependence of burn-through current value on the welding speed (CO_2 welding): (1) Sv-08 wire, St3 steel; (2) Sv-08G2S wire, St3 steel; (3) Sv-08 wire, E3A steel; (4) Sv-08G2S wire, E3A steel.

the metal-pool surface. Contrarily forces of surface tension, P_s and hydrostatic pressure, P_{hs} try to level the pool surface. Therefore, the crater forms, if the following inequality holds:

$$P_a + P_r > P_s + P_{hs}. \tag{5.46}$$

In arc welding, particularly, at low values of welding current, the reactive forces are comparatively small; this value thatefore may be neglected, and then, after transformation, Equation (5.46) becomes:

$$P_a > \left(\frac{2\sigma_{m-g}}{r_c}\right) + \left(\frac{\rho_m}{h}\right), \tag{5.47}$$

where r_c is the radius of the crater-bottom curvature; h is the maximum depth of the crater. From Equation (5.7), it follows that the crater depth increases with decrease of metal density and its surface tension, and also with an increase of the arc pressure. Such an influence of metal density and arc pressure is confirmed by experiments, the results of which are given by Lin and Eager.[364] A more marked influence of the value of welding current was further noted,[364,365] which is attributable both to an increase of arc pressure, and the appearance of vortex-type flows in the weld pool with increase of the welding current.

In keeping with the mathematical model, accepted by Potekhin,[365] the influence of the forces of surface tension of the metal on the dimensions of the crater is not strong. However, the experiments conducted by us are indicative of this influence being quite pronounced. In submerged-arc welding using AN-348A flux with Sv-08 wire on plates of St3 steel (the surface tension of which was varied by adding surface-active sulfur into the weld pool), the crater depth increased from 0.7 to 1.5 mm with increase of sulfur content in the metal from 0.052% to 0.43%. In CO_2 welding with Sv-08G2S wire on plates of steel St3 and E3A, the crater depth is greater when the base metal is steel St3, which has a smaller σ_{m-g} value on the boundary with CO_2 (Figure 5.37). In all the cases, the crater depth increased with increase of welding current.

In welding by the electron or laser beam, the role of reactive forces in the process of crater formation becomes greater and therefore, P_r should be substituted instead of P_a in Equation (5.47).

5.5.2.3 Formation of Rolls

An important feature of the formation of the horizontal welds are (1) the undercuts, which often develop in the upper part of the weld, and (2) rolls which develop in the lower part (Figure 5.38).

Considering the weld pool in equilibrium under the impact of the forces of gravity and surface tension, and neglecting the force of arc pressure and displacement of

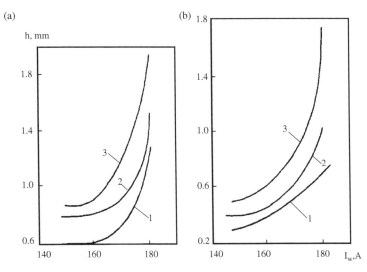

FIGURE 5.37 Dependence of the weld crater depth on the welding current value ($v = 0.0042$ m/sec): (a) Sv-08G2S wire, St3 steel; (b) Sv-08G2S wire, E3A steel. U_a (V): (1) 40; (2) 35; (3) 30.

metal in the weld pool, Suzdalev et al.[366] obtained the following dependence:

$$Z = -\frac{\rho B^3}{18\sigma\sqrt{3}}.$$ (5.48)

As is seen from this equation, in order to lower the height of the roll, it is necessary to reduce the weld width, B or increase the surface tension of the weld-pool metal. Similar results on the influence of the surface properties of the metal and weld

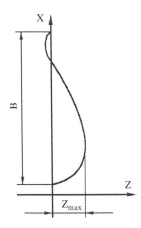

FIGURE 5.38 Schematics of the horizontal weld formation in the vertical plane.

width on roll formation of roll in horizontal welds were derived by variation-energy method in the study of Berezovsky, Stikhin and Bakshi.[367]

It should be noted that rolls often form in surfacing by an indirect arc. In this case, with the use of wire electrodes, the amount of the molten metal is more than two times greater than that in surfacing by a direct arc, and heating of the surfaced part is not high enough. Therefore, the hugh mass of molten electrode metal, that comes into contact with the comparatively cold metal, does not spread over the surface of the part. An effective method to prevent the rolls in this case, is preheating the parts. In surfacing of the steel parts, the rolls do not normally form, if the pre-heating temperature is about 973–1073 K.

Thus, formation of weld-shape defects (undercuts, lacks-of-fusion, lacks-of-penetration, burns-through, craters, rolls) is related to the surface properties of the metal to varying degrees. This opens up wider possibilities for the elimination of these defects by changing the surface properties of the metal and slag, or the wettability of solid metal by the metal melt.

6 Non-Metallic Inclusions

Normally, non-metallic inclusions are always present in welds made by any fusion welding method. The only exceptions are electron beam and electroslag welding which may provide conditions that allow non-metallic inclusions in the weld metal to be markedly decreased. All the rest of the fusion welding methods are usually characterized by an increased content of non-metallic inclusions in the weld or the deposited metal compared with that of the initial welding consumables used. Composition of non-metallic inclusions, their quantity, size, shape, and distribution in the weld may have a marked effect on the mechanical properties of a welded joint. Therefore, despite the fact that welds are, owing to the small volume of the weld pool, an inconvenient object for investigation of non-metallic inclusions, quite considerable number of papers are dedicated to this issue.

Results of the investigations conducted showed that the composition of non-metallic inclusions formed in the welds depends in many respects upon the welding consumables and methods employed. The majority of the studies so far carried out give no more than the final results of the process of formation of non-metallic inclusions, that is they consider composition of inclusions, their shape, location, etc., Issues associated with kinetics of initiation of non-metallic inclusions, their growth, and removal have not been paid attention. However, it is absolutely clear that without investigation of these processes it is difficult to identify the ways of decreasing the content of non-metallic inclusions in the weld or providing such inclusions the right shape and size so that they would exert no substantial effect on the weld quality.

Meanwhile, the processes of formation, growth, and removal of non-metallic inclusions are closely related to surface properties and phenomena. We will consider this relationship by an example of oxide and sulfide inclusions, which are the most common inclusions in the welds and have the greatest effect on the performance of welded joints.

6.1 NON-METALLIC INCLUSIONS IN THE WELDS

6.1.1 Types of Non-Metallic Inclusions

All non-metallic inclusions contained in the weld metal can be subdivided into two groups: (1) inclusions formed in the weld pool during welding as a result of different physical–chemical processes, that is inclusions of an endogenous origin and (2) inclusions introduced into the weld pool from the outside, that is, inclusions of an exogenous origin. The major part of non-metallic inclusions contained in the weld are of an endogenous origin. Formation of such inclusions in the weld pool is favored by enrichment of the molten metal with impurities, which is caused by liquation processes and decrease in the solubility of the impurities by cooling of the metal melt. The exogenous formation of non-metallic inclusions in the weld can take place as a result of transfer of part of the electrode covering to the weld pool (in the form of individual droplets or together with the electrode metal), transfer of oxides present on the surface of parts welded, incomplete removal of the slag crust from the surface of a previous bead in multi-pass welding and surfacing, etc.

Shape and size of inclusions contained in the weld depend in many respects upon their melting temperature which, in turn is determined by a composition of the inclusions. As seen from the data of Table 6.1, most oxide inclusions are based on $SiO_2-MnO-FeO$, $SiO_2-MnO-Al_2O_3$, or $SiO_2-FeO-Al_2O_3$, and only

TABLE 6.1
Composition of Oxide Non-Metallic Inclusions in Covered-Electrode Welding, CO_2, and Submerged-Arc Welding

Electrode/Wire/Flux	Type of Inclusion	Content in Inclusions, wt%				
		SiO_2	MnO	FeO	Al_2O_3	CaO
Uncovered electrode	Coarse	45.81	19.56	18.18	—	—
	Dispersed	78.21	10.22	3.18	—	—
Rhodonite–Perovskite electrode	Coarse	23.3	32.1	14.0	5.1	—
	Dispersed	52.4	9.4	8.1	1.1	
Ts-3	Coarse	46.0	20.3	28.1	4.2	—
	Dispersed	68.7	8.0	4.1	9.1	
OMM-5	Coarse	40.1	23.1	14.1	8.1	—
UONI-13/55	Coarse	27.2	28.2	28.2	7.1	—
VSTs-2[a]	—	36.2	29.5	9.5	—	20.2
TsM-9	—	45.99	31.02	2.76	22.23	—
Sv-08G2SA[b]	—	34.97	24.51	4.26	36.26	—
Sv-08G2S[b]	—	29.68	32.25	5.86	32.21	—
Electrode with chalk covering	—	19.0	11.15	54.0	—	—
Sv-08 + AN-5[c]	—	95.6	0.1	2.6	0.6	1.1
Sv-08 + AN-348A[c]	—	50.6	22.6	0.6	23.1	3.1
Sv-08 + AN-348A	—	53.6	32.5	0.5	12.0	0.8
Sv-08 + AN-20[c]	—	72.4	9.5	1.7	15.6	0.8
Sv-08G + AN-20[c]	—	71.2	1.3	1.1	25.5	0.9
OSTs-45[c]	—	43.7	40.4	1.9	14.0	—

[a]Inclusions contain 4.3% TiO_2 and 20.2% CaO + MgO; [b]CO_2 welding; [c]Submerged-arc welding.

Source: Novozhilov, N.M., *Fundamentals of Metallurgy in Gas-Shielded Welding*, Mashinostroenie, Moscow, 1979, p. 231; Lubavsky, K.V., *Problems of Theory of Welding Process*, Mashgiz, Moscow, 1948, p. 86; Alov, A.A., *Avtogen. Delo.*, 4, 4, 1945; Alov, A.A., *Problems of Theory of Welding Processes*, Mashgiz, Moscow, 1948, p. 5; Mazel, A.G., Tarlinsky, V.D., and Sbarskaya, N.P., *Svar. Proizvod.*, 9, 39, 1967; Podgaetsky, V.V., *Non-Metallic Inclusions in Welds*, Mashgiz, Moscow, 1962, p. 84.

in some cases, for example, in the welding using VSTs-2 electrodes, the basic components of the oxide inclusions are SiO_2, MnO, and CaO + MgO.

Sulfide non-metallic inclusions most often contain FeS and MnS (Table 6.2). In addition, complex oxysulfide inclusions may also be found, but in this case, the inclusions are based on 2–3 components. Therefore, to determine the melting point of non-metallic inclusions contained in the welds, it is possible to use binary and ternary constitutional diagrams of $FeO–SiO_2–Al_2O_3$, $FeO–SiO_2–MnO$, $MnO–SiO_2–Al_2O_3$, and $FeS–MnS$.[372,373]

If we add compositions of non-metallic inclusions given in Table 6.1 and Table 6.2 to the corresponding constitutional diagrams, we may notice that in the majority of cases, the melting point of oxide inclusions is high and often in excess of 1773 K. Only some inclusions characterized by an increased content of FeO and MnO have a melting point lower than 1573–1673 K. The melting point of sulfide inclusions is lower and ranges from 1443 to 1573 K.

In addition to inclusions, the compositions of which are given in Table 6.1 and Table 6.2, the welds often contain also individual inclusions of corundum, quartz, magnesia, and calcium oxide.[368,370,374] These inclusions transfer to the weld from the base metal, welding wire or slag, or they may be formed in the weld pool, if metal contains much aluminum, silicon, etc. These inclusions have a relatively high melting point, which is 2323 K for Al_2O_3, 1998 K for SiO_2, and 3073 K for MgO.

As temperature of the weld pool in submerged-arc welding of low-carbon steel, as mentioned earlier, is within a range of 1943–2143 K, both liquid and solid

TABLE 6.2
Composition of Sulfide Inclusions Contained in the Welds

Electrode/Wire/Flux	Content in Inclusions, wt%	
	FeS	MnS
TsM-7	52.4	46.6
TsM-9	65.4	34.6
Covered electrode:		
Chalk	81.2	18.8
Ore–acid	69.5	30.5
Flux OSTs-45	67.2	32.8
Sv-08GSA[a]	74.0	26.0
Sv-08GS[a]	50.7	49.3
Sv-08G2S[a]	42.8	57.2

[a]CO_2 welding

Source: Novozhilov, N.M., *Fundamentals of Metallurgy in Gas-Shielded Welding*, Mashinostroenie, Moscow, 1979, p. 231; Alov, A.A., *Avtogen, Delo.*, 4, 4, 1945.

non-metallic inclusions are formed in the molten metal under welding conditions. Some of them, and primarily sulfide inclusions, may remain in the liquid state also after the weld-pool metal begins to solidify.

6.1.2 THE INFLUENCE OF NON-METALLIC INCLUSIONS

6.1.2.1 Thermal–Structural Factors

Non-metallic inclusions contained in the weld may have a substantial effect on mechanical properties of a welded joint. As non-metallic inclusions are stress-raisers, this effect should greatly depend upon the size, shape, and distribution of inclusions in the weld metal, as well as upon the binding forces at the inclusion–metal interface and relationship between elastic constants of the inclusions and matrix.

According to the reported data,[374–377] thermal–structural stresses amounting to significant values can be formed near a non-metallic inclusion and inside it, caused by different thermal expansion coefficients of this inclusion and the matrix. So, according to the data given in study,[378] these stresses amount to 588 MPa. Development of such stresses may lead to the formation of pre-fracture zones, where a fracture initiates and further develops even without application of an external load.

Formation of thermal–structural stresses near inclusions is associated with their chemical composition. This adds to the significance of a proper selection of alloying elements, because it determines composition of the inclusions formed in the weld metal. Thus, according to the data reported studies,[375,376] stress fields are induced by cooling around the inclusions of Al_2O_3 and Ca-aluminates having a lower thermal expansion coefficient than the metal. In the metal, which contains inclusions having a higher thermal expansion coefficient than steel, for example, MnS, cavities are formed at the inclusion–metal interface, this also having an adverse effect.

6.1.2.2 Strength of Welded Joints

Most researchers involved in studies of the effect of non-metallic slag inclusions on the static strength of welded joints come to a conclusion[379] that tensile strength of the weld metal on low-carbon steels hardly changes with the content of the inclusions amounting to 10% of the cross-sectional area of the weld. However, shape, distribution, and phase composition of non-metallic inclusions, even if small in size ($d_{in} < 1$ μm), exert a substantial effect[304] on values of impact toughness (KCV) and critical brittle temperature of the weld metal. Decrease in the KCV value is especially pronounced, if the weld metal contains acute-angled non-metallic inclusions. The presence of clusters of non-metallic inclusions in the weld metal also leads to decrease in the KCV.

According to the data given in a study,[304] in submerged-arc welding of low-carbon steel, the weld metal contains silicate-type complex inclusions of a typical size of less than 2 μm. Electron microscopy revealed dispersed oxide particles of $(2-3) \cdot 10^{-9}$ mm present in the ferrite grain of such welds. Increase in quantity of

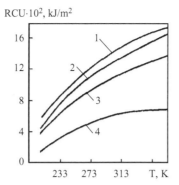

FIGURE 6.1 Effect of the content of silicate non-metallic inclusions in the weld on its impact toughness (percentage of the area occupied by inclusions in the area of the weld): (1) 0.28–0.3%; (2) 0.34–0.52%; (3) 0.104–0.11%; (4) 0.0196%.

the dispersed silicate inclusions in the weld metal was shown to lead to a decrease in impact toughness of the metal (Figure 6.1).

A similar effect on the impact toughness of the weld metal by oxide non-metallic inclusions is also found[304] in the welding of low-alloy steels (Table 6.3).

As seen from Table 6.3, that the presence of refractory inclusions of alumina leads to a greater decrease in the ductility of the weld metal. This may be caused by both angular inclusions of this composition and high values of thermal–elastic stresses induced at the inclusion–metal interface.

It should also be noted that the presence of oxide non-metallic inclusions in the weld metal in the welding low-alloy steels leads (as proved by our investigations), to embrittlement of metal. This is evidenced (Figure 6.2) by the mode of fracture of a weld containing non-metallic inclusions for instance, welding of steel 09G2S using flux-cored wire. The type of fracture of specimens cut from the weld after impact testing is indicative of the fact that embrittlement of the metal occurs with non-metallic inclusions contained in the weld metal. In addition, increase in the size

TABLE 6.3
Dependence of Impact Toughness of Weld Metal upon its Content of Non-Metallic Inclusions (Steel 16GNMA)

Type of Non-Metallic Inclusions Contained in the Weld	Impact Toughness $KCV \cdot 10^2$, kJ/m^3, at a Content of Non-Metallic Inclusions, %				
	0.015–0.02	0.025–0.03	0.035–0.04	0.055–0.06	0.075–0.08
Aluminosilicates (dominance of Al_2O_3)	21.8	18.0	14.1	8.5	7.0
Silicates	24.6	21.5	21.2	12.5	7.8

(a)   (b)

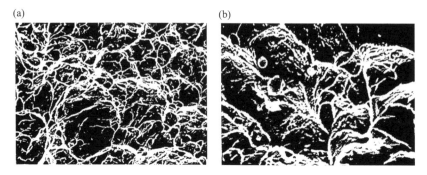

FIGURE 6.2 Fracture of a specimen of the weld after impact tests at 293 K without non-metallic inclusion (a) and with non-metallic inclusion (b) ($\times 1000$).

of a non-metallic inclusion makes this effect more pronounced (Figure 6.3). Embrittlement of the weld metal containing non-metallic inclusions is aggravated also with increase in the test temperature.

The presence of non-metallic inclusions in the weld metal can favor formation of other defects as well. For example, sulfide inclusions, which often have the melting point below the solidification temperature of the metal, cause initiation of solidification cracks, whereas the presence of oxide non-metallic inclusions favors formation of porosity.

Therefore, as seen from this brief review, non-metallic inclusions contained in the weld exert a marked effect on the performance of welded structures, especially when they operate at low temperatures and under alternating loads.

Since non-metallic inclusions which are most often found in the weld are of an endogenous origin, we will consider the process of their formation in more detail, starting from their nucleation.

(a)   (b)

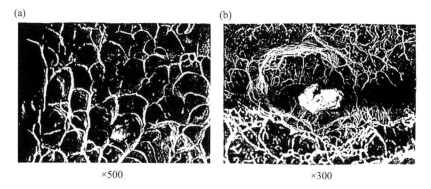

×500   ×300

FIGURE 6.3 Variations in character of fracture of specimens cut from the weld depending upon the size of non-metallic inclusion.

6.2 NUCLEATION OF NON-METALLIC INCLUSIONS IN THE WELD POOL

According to the current concepts, formation of a new phase inside a mother liquor begins from emergence of stable nuclei. A condition, which is required for such nuclei to be formed, is oversaturation of the weld-pool metal with a new phase material. It may be caused by decrease in solubility of the material in cooling of the melt, or may result from the formation of different compounds due to occurrence of chemical reactions.

6.2.1 THE KINETICS OF NUCLEATION

Kinetics of formation of non-metallic inclusions in the molten metal was studied by many researchers. As a rule, to estimate the probability of formation of the new phase nuclei, they used the equation suggested by Ya.I. Frenkel'.[380] In accordance with this equation, the intensity of nucleation of a new phase is equal to

$$I = Ae^{-\Delta G/kT}, \tag{6.1}$$

where I is the intensity of nucleation of a new phase, A is the coefficient determined by calculations, ΔG is the change in the free energy caused by formation of a nucleus having a critical size, k is the Boltzmann's constant, T is the absolute temperature and e is the natural logarithm base.

The coefficient A can be found from the following relationship[42]:

$$A = n'\left(\frac{\sigma_{\text{m−in}}}{kT}\right)^{1/2}\left(\frac{2V}{9\pi}\right)^{1/3}\frac{nkT}{h}, \tag{6.2}$$

where n' is the number of atoms on the surface of the nucleus having a critical size, V is the volume per atom of the initial phase, n is the number of atoms per unit volume of the initial phase, $\sigma_{\text{m−in}}$ is the tension between the phases at the metal–precipitating inclusion interface and h is the Planck's constant. The rest of the designations are the same as in Equation (6.1).

As is seen, the coefficient A is not a constant value, it rather changes with a change in composition of the precipitating inclusion. According to the data of Turpin and Elliot,[381] the value of A for Al_2O_3 is 10^{26}, for $FeO-Al_2O_3$, it is 10^{25}, for SiO_2, 10^{28} and for FeO, 10^{30}.

Free change energy for the formation of a stable nucleus of the new phase is determined from the following equality:

$$\Delta G = \frac{16\pi\sigma_{\text{m−in}}^3 M_{\text{ph}}^2}{3\rho^2 R^2 T^2 \left[\ln(C/C_s)^2\right]} \tag{6.3}$$

where M_{ph} is the molecular mass of the forming phase, ρ is the material density, C/C_s is the oversaturation of the melt with the new phase material, and R is the universal gas constant.

However, the probability of formation of non-metallic inclusions is characterized not only by the intensity of their nucleation, but also by the size of a formed nucleus. The nuclei formed in the melt will be stable and capable of further growing only if their size reaches a certain critical value.

In the case of a liquid inclusion, its shape seems to be close to spherical one, while the critical size of such a nucleus is equal to:

$$r_{cr} = \frac{2\sigma_{m-in}M_{ph}}{\rho RT \ln{(C/C_s)}}. \tag{6.4}$$

If a refractory crystalline inclusion is formed, its critical size characterized by the mean size of a crystal face will be equal to:

$$h_i = \frac{2\sigma_{m-in}^i}{\rho RT \ln(C/C_s)}, \tag{6.5}$$

where σ_{m-in}^i is the interphase tension at the boundary of the ith face of the inclusion with the metal melt.

As a rule, the value of σ_{m-in}^i is unknown. Therefore, in calculations involving some approximation, it is possible to use a mean value of the specific interphase energy at the melt–inclusion interface, σ_{m-in}. In addition, because of the small size of the inclusion nuclei, their shape can be considered close to spherical. Therefore, in the formation of crystalline inclusions, their critical size can also be determined from Equation (6.4).

6.2.2 OVERSATURATION AND ITS EFFECTS

Analysis of Equation (6.1) through Equation (6.4) allows a conclusion that the intensity of nucleation of non-metallic inclusions in the weld pool is associated with certain values for the degree of oversaturation of the melt with a precipitating material, and the interphase tension between the phases at the melt–inclusion interface.

Oxide inclusions are formed in the weld pool as a result of chemical reactions, which in a general form can be written as follows:

$$m[\text{Me}] + n[\text{O}] = (\text{Me}_mO_n).$$

Then the value of oversaturation C/C_s can be found from the following expression:

$$\frac{C}{C_s} = \frac{K_a}{K_e} = [\%\text{Me}]_a^m[\%\text{O}]_a^n$$

where K_e is the equilibrium constant of deoxidation reaction, $[\% \text{Me}]_a$ and $[\% \text{O}]_a$ are the actual concentrations of a de-oxidizing element and oxygen dissolved in the melt.

Therefore, the oversaturation value is determined from the ratio of the product of solubilities of certain components under actual conditions to the product of solubilities of the same components in the equilibrium state.[381] Numerous calculations are indicative of the fact that oversaturation of the melt with the precipitating phase material should be very high for the stable nuclei of oxide non-metallic inclusions to be formed. For example, according to the data of a study,[382] $C/C_s = 80–800$ for SiO_2 and $3 \cdot 10^6 – 3.6 \cdot 10^{14}$ for Al_2O_3. However, as seen from Equation (6.4) and Equation (6.5), the decrease in the value of interphase tension between the phases at the metal melt–inclusion interface must decrease the required oversaturation value.

6.2.2.1 Precipitation Phenomena

Calculations made using Equation (6.4) and Equation (6.5) evidence that inclusions consisting of iron and manganese oxides may spontaneously precipitate from the insufficiently deoxidized steel even at a comparatively low oversaturation. If the precipitating phase contains SiO_2, in addition to FeO and MnO, the intensity of formation of the stable nuclei dramatically decreases, and they grow in size. Since an addition of sulfides of different materials to slag leads to a marked decrease in the value of the interphase tension at the metal–slag boundary (Figure 6.4), one might expect that sulfide oxides in the weld-pool metal would be readily formed. However, because of the fact that sulfur, in contrast to oxygen, possesses an unlimited solubility in liquid iron, formation of sulfide inclusions depends in many respects upon the presence of sulfide-forming components in the metal. The strongest de-sulfurizing agents are alkaline-earth and rare-earth elements, as well as Al and Zr. However, all these elements are at the same time the strong de-oxidizing agents. Therefore, their high activity with respect to sulfur may become pronounced only after deoxidation of steel.

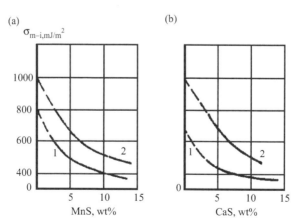

FIGURE 6.4 Variations in the value of interphase tension at the liquid iron–slag interface depending upon the content of MnS (a) and CaS (b) in the slag of an initial composition (wt%): [1] 10 CaO; 60 MnO; 30 SiO$_2$; [2] 10 CaO; 30 MnO; 60 SiO$_2$. (From Mikiashvili, Sh. M. and Samarin, A.M., *Physicochemical Principles of Steel Production*, Nauka, Moscow, 1964, p. 42.)

This is proved by the results of Gulyaev,[383] according to which within the range of solidification rate investigated, $(0.165–83.8) \cdot 10^{-4}$ m/sec, the quantity of silicate inclusions did not change, whereas sizes of the inclusions grew to some extent with increase in the solidification rate from $(0.165$ to $3.3) \cdot 10^{-4}$ m/sec. Further increase in the solidification rate to $83.3 \cdot 10^{-4}$ m/sec had no effect on the sizes of the silicate inclusions. The content of sulfide inclusions decreased from 65 to 5% with increase in the cooling rate within the same range, and the radius of the inclusions decreased from 10 to 1 μm.

Similar results were obtained also in other studies as well,[371,384] where it was established that sulfides located in the weld along the fusion line of the base metal with the deposited one had the smallest sizes, and those located in the center of the upper part of the weld had the largest sizes. This is attributable also to a different rate of cooling in these regions of the weld. Therefore, conditions of transition of the metal from the liquid to the solid state have a substantial effect on the process of formation of sulfide inclusions.

As shown in Chapter 1 (Section 1.4), sulfur is a strong liquation element, which may have a marked effect on the oversaturation of the melt with sulfur and, hence, on the formation of the new-phase nucleus. Besides, because of a high solubility of sulfur in steel, oversaturation required for the formation of sulfide inclusions can be formed only at the end of the solidification period. Besides, as follows from Equation (1.25), it will have the highest value near the solid–liquid metal interface.

The values of interphase tension obtained after some holding of the metal–slag system are usually used for calculations to determine I and r_{cr}. This results in over-values of oversaturation required for the formation of the stable nuclei of non-metallic inclusions. This takes place for several reasons.

6.2.3 ELECTROCHEMICAL CONSIDERATIONS

As is well known,[66] the difference in the electric potentials is formed between two electrically conducting phases (metal–slag) in contact, which is caused by the for-mation of a electric double layer. In this case, the value of the interphase tension at the metal–slag boundary is as follows[347]:

$$\sigma_{m-in} = \sigma_0 - \frac{C_d \varphi^2}{2}, \tag{6.6}$$

where σ_0 is the value of the interphase tension at the zero-charge potential; C_d is the capacitance of the double layer, φ is the electric potential.

Therefore, the higher the capacitance and potential of the electric double layer, the lower the value of σ_{m-in} and the higher the probability of formation of the nucleus of a non-metallic inclusion. According to the data of a study,[284] addition of iron oxides or sulfides to the slag in contact with the molten metal leads to a dramatic increase in the value of potential, φ (Figure 6.5) and, after some time, to its decrease to the initial value.

Formation of a nucleus, because of a short time of the process, is characterized by a drastic increase in the concentration of a precipitating material (FeO, FeS)

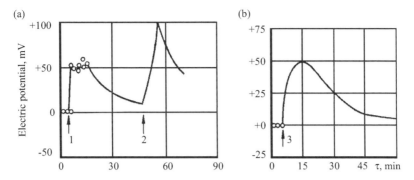

FIGURE 6.5 Change in time of electric potential in the cast iron–slag system ($CaO–Al_2O_3–SiO_2$) with an addition of iron oxides (a) and iron sulfide (b) to slag: [1] addition of Fe_3O_4; [2] addition of Fe_2O_3; [3] addition of FeS.

in some limited volume of the metal. Therefore, formation of an inclusion containing a substantial amount of FeO or FeS may lead to a similar substantial increase in the potential as in the case of addition of iron oxides or FeS to the slag. According to Equation (6.6), this must lead to a decrease in the value of σ_{m-in} and, hence formation of the stable nucleus in a homogeneous metal melt is made much easier.

Formation of non-metallic inclusions may take place at lower oversaturations, if two or more components simultaneously participate in the formation of a nucleus. According to Laptev,[385] in the formation of the nucleus of a non-metallic inclusion from two components, the critical number of particles in the nucleus, n_{cr} (given that the values of σ_{m-in} for pure components are close to each other), is equal to:

$$(n_I + n_{II})_{cr} = \frac{32\pi\sigma_{m-in}^3 V_m^2}{3\left[RT\ln\left(\dfrac{C^I}{C_s^I} + \dfrac{C^{II}}{C_s^{II}}\right)\right]^3} \qquad (6.7)$$

If the nucleus is formed from one component:

$$n_{cr} = \frac{32\pi\sigma_{m-in}^3 V_m^2}{3\left[RT\ln\left(\dfrac{C}{C_s}\right)\right]^3}, \qquad (6.8)$$

where V_m is the molar volume of the precipitating phase material, and I and II are the indices designating the first and second components, respectively.

The relationship between C^I/C^{II} and K was derived[385] by determining the values of n_{cr} and $(n_I + n_{II})_{cr}$ from Equation (6.7) and Equation (6.8) and by considering a relationship of $n_{cr}/(n_I + n_{II})_{cr} = K$ at $C^{II}/C_s^{II} = 1$ and a varying the value of C^I/C_s^I:

C^I/C_s^I	K
1	∞
1.2	80.8
1.4	8.4
1.6	4.0
2.0	2.0
3.0	2.0
10.0	1.13

6.2.4 THERMODYNAMIC CONSIDERATIONS

It follows from the above discussion that formation of a stable two-component nucleus requires a smaller quantity of particles than for the formation of a stable one-component nucleus. This difference is especially pronounced, if the values of C^I/C_s^I and C^{II}/C_s^{II} differ but slightly from each other.

Solution of this problem in a general form[386,387] proved the conclusion that considering the energy involved, formation of a two-component nucleus is always more preferable than formation of a stable nuclei from pure components. As seen from Table 6.4 compiled from the data of a study,[387] the value of the required oversaturation substantially decreases in the case of formation of oxides at a different content of deoxidizers and oxygen in the metal. In addition, the degree of oversaturation, at which the stable nuclei are formed, depends[387] upon the ratio of the concentrations of the reagents and varies with the value of oversaturation of each of the components participating in the formation of a nucleus.

All the statements considered hold true for the formation of stable nuclei in a homogeneous environment. Under actual conditions, the weld pool always comprises some interfaces: between solid and molten metal or between refractory inclusions introduced from the outside and a melt. Therefore, it is highly probable that formation of a new phase will be of a heterogeneous nature. In this case, the process of formation of non-metallic inclusions is determined in many respects by

---

TABLE 6.4

Calculated Values of Required Oversaturations of the Melt with Oxides in the Formation of a Stable Nucleus from One or Two Components

Deoxidizer	(Me_mOn)	(FeO)	$(FeO \cdot Me_mO_n)$
Al	$1.91 \cdot 10^{10}/7.81 \cdot 10^{10}$	0.468/2.741	148.9/203.0
Ti	188/3500	0.162/0.338	15.82/33.03
Si	125/953	0.346/0.5-8	13.35/19.55

Note: Minimum values of C/C_s are followed by the maximum values.

---

the degree of wetting of a substrate on which the new phase is precipitated with the material of this phase.

According to Khollomon and Turnbull,[42] a change in the free Gibbs energy of the system in heterogeneous nucleation of inclusions is equal to:

$$\Delta G_{het} = \Delta G_{hom} \cdot f(\theta).$$ (6.9)

Function $f(\theta)$ is determined from the following expression:

$$f(\theta) = \frac{1}{4}(2 + \cos\theta)(1 - \cos\theta)^2.$$ (6.10)

It follows from Equation (6.9) and Equation (6.10) that the better the wetting of the substrate with the new phase material, the easier the precipitation of the new phase on the substrate. So, for $\theta = 5°$, the relationship, $\Delta G_{het}/\Delta G_{hom}$ is equal to $0.1 \cdot 10^{-4}$; for $\theta = 10°$, it is equal to $0.17 \cdot 10^{-3}$; for $\theta = 30°$ it is $1.012 \cdot 10^{-1}$. Therefore, the presence of readily wettable interfaces in the weld pool makes formation of non-metallic inclusions much easier. As the majority of inclusions found in welds are oxides, we will give them the main consideration.

According to Table 6.5 compiled on the basis of data of,[388–390] the contact angles formed by oxide melts on the surface of fused magnesia, chamotte, and quartz are usually low, and at 1523–1673 K, they vary from 0 to 20°. In the wetting of the surface of a solid metal with oxide melts, the contact angle is equal to 55° at a temperature close to the melting point of the metal.[388]

The measured values of the contact angles in the wetting of different materials with iron sulfide (after holding for 5 sec) at a temperature of 1373 K are given below:

30KhGSA	9.0°	07Kh25N13	47.0°
U10	9.5°	Corundum	31.5°
E3A	24.5°	Quartz	37.5°
1Kh18N9T	35.5°	Iron (II) oxide	0°

Therefore, oxide and especially sulfide inclusions may readily precipitate along the grain boundaries in a metal, as well as at oxide inclusions, much so if the latter contain a substantial amount of FeO.

The probability of formation of oxides and sulfides at the interfaces is proved by numerous facts. Results of metallographic and microchemical examinations evidence[391] that almost all the sulfide-type inclusions contain also oxide inclusions. Besides, oxides covered with sulfides make up the central part of such inclusions. As precipitation of oxide inclusions precedes precipitation of sulfide ones, the oxide inclusions formed in the weld pool are likely to play the role of substrates for the precipitating sulfides. In the welding of pure iron, for example in welds on armco-iron, sulfide films are formed along the grain boundaries. In this case, the low content of oxygen and deoxidizers in the metal prevents formation of the oxide inclusions.[371] All this is indicative of the fact that the shape of the sulfide inclusions and their location in the weld depend to a large degree upon the surface phenomena.

TABLE 6.5
Values of Contact Angles Formed by Oxide Melts on Refractory Materials

CaO	SiO$_2$	FeO	Fe$_2$O$_3$	Al$_2$O$_3$	MgO	MnO	T, K	Contact Angle, (deg)
				Fused Magnesia Substrate				
—	40.0	60.0	—	—	—	—	1503	6
24.0	11.1	41.8	12.8	7.9	2.3	—	1743	4
43.8	20.0	12.6	1.8	16.4	5.3	—	1723	12
33.1	20.0	11.6	2.7	5.7	7.2	17.9	1693	2
				Chamotte Substrate				
28.4	12.4	39.6	4.4	11.6	4.0	—	1683	6
24.4	11.07	43.6	10.6	7.71	2.4	—	1683	0
10.0	32.0	58.0	—	—	—	—	1553	0

Note: According to the data of Levin, A.M.[388] the value of the contact angle in wetting the magnesite substrate with an oxidizing slag is equal to zero ($T = 1873$ K).

This, it is clear that formation of stable nuclei of non-metallic inclusions greatly depends upon the surface properties of a material precipitating from the melt, as well as upon the wettability of the surfaces, on which a new phase is formed, with the material of this phase.

The role of the surface phenomena in the process of growth of non-metallic inclusions is not less important.

6.3 COARSENING OF NON-METALLIC INCLUSIONS IN THE WELD POOL

A study of the process of coarsening of non-metallic inclusions is of a considerable interest, as the size of non-metallic inclusions determines the mechanical properties of a welded joint as well as the removal of the inclusions from the weld pool.

6.3.1 FUNDAMENTALS OF NUCLEI GROWTH

Growth of nuclei of non-metallic inclusions formed in the molten metal may take place either as a result of adsorption of ions from the melt because of oversaturation, or as a result of joining of individual fine inclusions at their collision. In the first case, the rate of growth of the inclusions is determined by the processes of diffusion of reagents to the surface of the inclusions. This mechanism of growth seems to be more characteristic of sulfide inclusions, as they are formed at the moment when the major part of metal has already solidified, and the mobility of the inclusions, as well as the probability of their collision, are very low.

The second process is likely to dominate for oxide inclusions, as they are formed at the moment, when the volume of the molten metal is for the free movement of the inclusions. However, oxide inclusions may grow also due to diffusion processes.

Consideration[274] of the process of diffusion growth of non-metallic inclusions shows, that in this case, the role of the surface phenomena is not high. An equation valid for growth of oxide and sulfide inclusions due to diffusion has the following form[392]:

$$r_{in}^2 = r_0^2 + 2V_mD(C_\infty - C_r)\tau, \tag{6.11}$$

where r_{in} is the final radius of an inclusion; r_0 is the initial radius of the inclusion; V_m is the molecular volume of a precipitating material; D is the diffusion coefficient of an impurity in the melt; C_∞ is the concentration of the material in the bulk of the melt; C_r is the concentration of the material on the surface of a growing particle; τ is the time of growth.

6.3.2 COALESCENCE AND COAGULATION

As seen from Equation (6.11), the final size of an inclusion in its diffusion growth depends primarily upon the difference of the concentrations of a material in the inclusion and the melt, the value of the material diffusion coefficient, and the process time. Therefore, as the time of existence of the weld pool is short, increase in size of non-metallic inclusions, especially the oxide ones, due to diffusion processes, is hardly probable.

Consider the second possible process of coarsening of non-metallic inclusions due to their joining. Joining of non-metallic inclusions may take place either, where the interface, which separates them disappears or where the interface persists. In the first case, the process is called coalescence and in the second it is called coagulation. Coalescence takes place in the merging of liquid inclusions, which may be of either an identical or different composition. Coagulation most often occurs in the collision of solid inclusions or solid and liquid inclusions, as well as in the collision of liquid inclusions having a limited mutual solubility. Coagulation is perikinetic, if it takes place when the effect of the forces is similar in all directions. If the probability of collision of particles in one direction is higher than in the rest of the directions, coagulation is orthokinetic.

Because the weld pool contains particles of different size, shape, and density, there is intensive stirring of the molten metal, the orthokinetic coagulation is most important in the welding processes.

Presence of a large number of inclusions in the weld pool results in the establishment of a developed interface between the non-metallic inclusions and the melt. Such a system will possess a substantial free energy, and will tend to decrease it. This can be achieved only, if many fine inclusions are replaced by a smaller number of the coarse ones. The magnitude of the free energy decrease in the merging of the particles depends upon their aggregate condition.

In coalescence of liquid inclusions,[393] the decrease in the free energy is directly proportional to the interphase tension and reduction in the metal–inclusion interface: $-dG = \sigma_{m-in} \, dS$. It is well known, that stable colloidal systems can exist only, if values of the interphase tension are much lower than 1 mJ/m^2. Under actual conditions, the values of σ_{m-in} are much higher. Therefore, the inclusions dispersed in the weld-pool metal have the form of a very unstable emulsion. In this connection, the process of coagulation of liquid non-metallic inclusions may occur spontaneously, and the higher the value of σ_{m-in}, the higher the completeness of this process.

Let us consider that two liquid non-metallic inclusions having the same size ($r_1 = r_2$) and the same composition join together. If the volume of a newly formed inclusion is equal to the sum of volumes of the inclusions, that have already joined, the surface energy of the new non-metallic inclusion will decrease, compared with the surface energy of the two initial inclusions. The decrease is given by the following equation:

$$-dG = 4\pi r_1 \sigma_{m-in}(2 - 2^{2/3}) \qquad (6.12)$$

When a liquid inclusion merges with a solid one, decrease in the free energy is as follows:

$$-dG = \sigma_{m-in} dS_{m-in} + \sigma_{m-s} \, dS_{m-s}$$

However, formation of the interface between liquid and solid inclusions leads to a growth of the free energy by a value of $dG = \sigma_{s-1} \, dS_{s-1}$. Then, the total change in the free energy of the system in coagulation of solid and liquid particles will be as follows:

$$-dG = [(\sigma_{m-in} + \sigma_{m-s}) - \sigma_{s-1}]dS$$

where σ_{m-in}, σ_{m-s}, and σ_{s-1} are the interphase tensions between metal and liquid inclusion, metal and solid inclusion, and between solid and liquid inclusions, respectively; dS is the change in the interface between liquid and solid inclusions and metal melt, and between the liquid and solid inclusions.

As interphase tensions at the inclusion–metal boundaries are much higher than those at the boundaries of the inclusions, decrease in the free energy in the merging of the molten and solid inclusions is substantial. Therefore, this process may occur spontaneously.

6.3.3 THE EFFECT OF OXIDES

It is a known fact, that solid oxides can be readily wetted with oxide melts (see Table 6.4). That is why, when liquid oxide inclusions collide with solid oxides, especially if the latter are of very small sizes, the solid particles may be fully covered with a liquid shell. It is likely that the time of the shell formation process is short, as the rates of spread of the oxide melt droplets over the solid laminae of Al_2O_3, SiO_2, and MgO amount to significant values.[394] The presence of solid inclusions covered with silicate endogenous inclusions in the weld is noted in the study of Mazel'.[370]

The free energy of the system in the coagulation of solid particles will change by the following value:

$$-\mathrm{d}G = (\sigma_{m-s} - \sigma_{s-s})\,\mathrm{d}S, \tag{6.13}$$

where σ_{m-s} and σ_{s-s} are the interphase tensions at the boundaries of the metal melt and solid inclusion, and of the solid inclusions, respectively.

It follows from Equation (6.13), that the process of coagulation of solid particles occurs spontaneously in the case, where $\sigma_{m-s} > \sigma_{s-s}$.

According to Baptizmansky, Bakhman, and Dmitriev,[395] collision of two solid non-metallic inclusions poorly wetted with the metal melt is accompanied by the capillary phenomena, resulting in a pulling of the melt from the gap at the point of contact of the particles under the effect of the surface tension forces (Figure 6.6). Therefore, coagulation of such particles leads to formation of extra interfaces between the gas cavity, molten metal, and solid non-metallic inclusions. Thus, Equation (6.13) describing a change in the free energy for the given case can be written in the following form:

$$-\mathrm{d}G = [\sigma_{m-s} - (\sigma_{s-s} + \sigma_{m-g} + \sigma_{s-g})]\,\mathrm{d}S, \tag{6.14}$$

where σ_{m-s} and σ_{s-s} are the interphase tensions on the boundary of the metal and solid inclusion and between the solid inclusions, respectively, σ_{m-g} and σ_{s-g} are respectively the surface tensions of the metal melt and solid inclusion, and $\mathrm{d}S$ is the change of the interface in the merging of the inclusions.

It follows from Equation (6.14), that coagulation of inclusions, which are poorly wetted with metal, may take place spontaneously only if the following inequality holds:

$$\sigma_{m-s}\,\mathrm{d}S_{m-s} > \sigma_{s-s}\,\mathrm{d}S_{s-s} + \sigma_{s-g}\,\mathrm{d}S_{s-g} + \sigma_{m-g}\,\mathrm{d}S_{m-g}$$

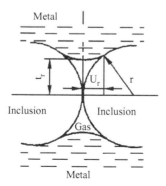

FIGURE 6.6 Flow diagram of the process of coagulation of solid non-metallic inclusions of a spherical shape.

For spherical particles, $dS_{s-g} \to 0$ and $dS_{s-g} \to dS_{n-s}$. Then the condition of the probability of spontaneous coagulation can be written as follows:

$$(\sigma_{m-s} - \sigma_{s-g})\,dS_{m-s} > \sigma_{m-g}\,dS_{m-g} \qquad (6.15)$$

or, allowing for Equation (2.12):

$$- \cos \theta \, dS_{m-s} > dS_{m-g} \qquad (6.16)$$

The value of σ_{m-s} at the interface with the iron melt for silica is equal to $1180\ \mathrm{mJ/m^2}$ and for alumina, $1400\ \mathrm{mJ/m^2}$.[396] Surface tension of solid Al_2O_3 is approximately $600\ \mathrm{mJ/m^2}$ and that of solid SiO_2 is about $400\ \mathrm{mJ/m^2}$.

6.3.3.1 The Role of the Gaseous Environment

The value of σ_{m-g} depends upon the compositions of the metal and the gaseous environment in the formed cavity. When the dispersion environment is low-carbon steel, it can be assumed, that $\sigma_{m-g} = 1100\ \mathrm{mJ/m^2}$. This value is very close to the value of surface tension of steel Sv-08 at the boundary with N_2, H_2, and CO,[397] that is, those gases which are most probably contained in the gas cavity. Substitution of the above values of surface energies into Equation (6.15) yields $dS_{m-s} > (1.4-1.5)\,dS_{m-g}$. Therefore, the process of coagulation leads to a decrease in the free energy of the system. Henceforth, it will occur spontaneously when the change in the interface between metal and inclusion is (as a minimum) higher by a factor of 1.4 than the corresponding change between metal and gas. Therefore, the worse is the wettability of solid non-metallic inclusions with the melt, the higher is the probability of their coagulation.

So, coalescence of liquid inclusions and coagulation of solid inclusions with liquid ones and of solid with solid inclusions may occur spontaneously. However, it is known from practice that dispersed systems, having a substantial free energy, can persist without changes for a long time. Therefore, based on the thermodynamic analysis, the possibilities of coarsening of non-metallic inclusions cannot always be realized.

It is absolutely apparent that the intensity of the process of coarsening of non-metallic inclusions in the weld pool depends upon how often they collide with each other. Depending upon the efficiency of collisions, coagulation may be fast, when all the collisions are efficient, or slow, when only part of the collisions leads to merging of the particles.

6.3.4 The Effect of Stirring

The effect of stirring of the liquid on the probability of collision of non-metallic inclusions can be seen from the following expression[274]:

$$\frac{W_{\mathrm{lam}}}{W_0} = \frac{\eta_m R_{ij}^3 (dv/dz)}{2kT}, \qquad (6.17)$$

where W_{lam} and W_0 are the probabilities of collision of spherical particles in a laminar motion of molten metal and in a quiescent metal melt, respectively, η_m is the viscosity of the metal melt, R_{ij} is the radius of a coagulated particle equal to r_i r_j, where r_i and r_j are the radii of the ith and jth particles, dv/dz is the velocity gradient, k is the Boltzmann's constant, and T is the temperature in K.

It follows from Equation (6.17), that if $dv/dz = 1 \ sec^{-1}$ and the diameter of at least some of the particles is more than 2 μm, then the value of W_{lam}/W_0 becomes more than 1, and stirring of the metal melt should not be ignored. If the metal melt contains non-metallic inclusions, whose size is close to 10 μm, and such inclusions are also found in the weld pool, then $W_{lam}/W_0 > 10^4$. Therefore, under welding conditions, the coagulation caused by stirring of the weld-pool metal will play a decisive role.

The probability of collision of particles in a liquid also grows, if at least part of them have a non-spherical shape. According to Muller,[398] the shape factor is equal approximately to $\ln (2a/b)$, where a and b are respectively the long and short axes of a particle, which is an ellipsoid of revolution. Elongated particles in the weld pool may be both solid inclusions and liquid deformed droplets.

It is a known fact[399] that a liquid droplet remains spherical as long as the following inequality is met:

$$\left[\frac{3v\eta_m(2\eta_m + 3\eta_{in})}{4\sigma_{m-in}(\eta_m + \eta_{in})} \right] \ll 1, \qquad (6.18)$$

where v is the velocity of movement of a droplet; σ_{m-in} is the interphase tension at the metal melt–inclusion boundary; η_{in} is the viscosity of the inclusion.

It follows from the inequality (Equation (6.18)), that for the deformation of a droplet containing a substantial amount of Fe and MnO, or sulfides characterized by a low interphase tension at the interface with the metal melt, the velocities at which the droplet is deformed should be 2–2.5 times lower than those needed for deformation of inclusions containing SiO_2, Al_2O_3, CaO, and MgO.

It should be noted that under actual conditions, not all the collisions are efficient. Under identical collision conditions, the efficiency of collisions of particles depends in many respects upon the state of their surfaces. It is very difficult to establish a relationship between the efficiency of merging of particles and the state of their surfaces. However, we can consider factors which prevent or favor the merging of non-metallic inclusions contained in the weld pool.

6.3.5 Disjoining Pressure

In the merging of fine particles ($\approx 10^{-9}$ m), the main role is played by forces of electrostatic repulsion of like charged particles and Van der Waals forces of attraction. In the merging of coarser particles (10^{-7}–10^{-4} m), the existing charges exert no marked effect on their merging. In this case, there are other causes which prevent their joining. One of such causes may be the presence of a disjoining pressure P_d

between the approaching particles, the value of which is equal to:[400]

$$P_d = \frac{-dG}{d\delta_g},$$

where G is the Gibbs energy, and δ_g is the width of the gap between the approaching particles.

When particles approach each other, the free energy of the system grows until the interlayer becomes $10^{-10}-10^{-9}$ m in thickness, after which it begins decreasing. Therefore, the disjoining pressure first grows and then passes through zero to become negative, which is indicative of prevalence of the adhesion forces. Formation of the disjoining pressure is caused primarily by interaction of diffusion-controlled electric double layers. No diffusion layers are present, if the molten metal contains oxide non-metallic inclusions on the side of the metal, at least the deoxidized one.[401] Therefore, the approaching particles in the weld pool will not be exposed to any disjoining pressure in those welding methods that provide a sufficiently reliable protection of the molten metal from air oxygen. The same is true metals containing a sufficient amount of d-eoxidizing elements.

However, if the metal contains much oxygen, which may be the case of welding and surfacing of low-carbon steels in oxidizing atmospheres (CO_2 or air), the diffusion part of the double layer may exist in the molten metal near the boundary with an oxide particle.[402] In this case, the structure of the electric double layer at the interface of the oxidized molten metal and the oxide is determined by an abrupt change in chemical potential of oxygen ions. The value of chemical potential of the oxygen ion can be characterized by ionicity of the bond, which is 50, 63, 75, and 78% for oxides SiO_2, Al_2O_3, MgO, and CaO, respectively.[403] The larger the difference between the ionicity of the Me$-$O bond (between the metal and oxide), the larger the abrupt change in chemical potential and the more developed the electric double layer is. The value of ionicity of the Me$-$O bond in a molten metal can be characterized by the ionicity of the electrostatic bond, which is equal to 78% in iron.[403]

Therefore, a developed double layer should be formed when the inclusions contain a substantial amount of SiO_2. This may lead to a marked growth of repulsive forces caused by the presence of the disjoining pressure. This seems to explain the fact that fine dispersed inclusions found in the weld are characterized by an increased content of SiO_2 (see Table 5.1).

Joining of non-metallic inclusions in the weld pool may also be prevented by increase in structural$-$mechanical properties of adsorption layers on the surface of non-metallic particles. Increase in structural viscosity and mechanical strength of the adsorption layers, resulting from adsorption of surface-active components, leads to the fact that liquid droplets behave as solid particles. They become less mobile, because of decrease in the velocity of their electrocapillary motion.

In addition, merging of such particles is hampered, as the rate of joining of liquid particles is inversely proportional to their viscosity.[404] It should be noted, that increase in the viscosity and mechanical strength of the surface layer of a liquid non-metallic inclusion may also take place if the molten metal contains the finest solid particles characterized by a selective wetting of an emulsion with an ambient atmosphere. When such particles get on to the surface of a liquid inclusion, they remain in the surface layer, causing an increase in the structural–mechanical properties of the surface of the particle. This event eventually hampers joining of the inclusions.

Joining of non-metallic inclusions may also be hampered by a change in interphase tension on the regenerated surface. When two particles approach each other, decrease in the thickness of the metal film between them occurs primarily due to adsorption of the liquid by thicker regions of the film. If the film on both sides is covered by adsorption layers of a surface-active material, whose concentration is far from the saturation one, the surface of the film increases with a rapid decrease in its thickness, although the concentration of the surface-active material in this case has no time to level. This causes a difference in the interface tensions $\sigma' - \sigma$, which in turn leads to the formation of metal flows which make up a thin region of the film. As this factor of stabilization of thickness of the film is associated with diffusion of the surface-active material, it plays a marked role only in rapid approaching of the particles.

Under welding conditions characterized by an intensive stirring of the weld-pool metal, the particles approach each other at a high velocity. That is why, in welding, this factor may exert a pronounced effect on the process of coarsening of non-metallic inclusions.

However, stirring of the weld-pool metal can lead not only to coarsening of non-metallic inclusions, but also can cause disruption of the bonds between the particles, that have already joined together. This is the most probable case of joining of solid non-metallic inclusions.

6.3.6 DETERMINATION OF ADHESION FORCES

The force of adhesion of two solid spherical particles (Figure 6.6) poorly wetted with the metal melt can be determined from the following formula[395]:

$$P = 2\pi\sigma_{m-g} \cdot r_{in} t_r$$

Here σ_{m-g} is the surface tension of metal, r_{in} is the radius of a non-metallic inclusion, and t_r is the dimensionless coordinate determined by jointly solving the following two equations:

$$t_r = 2\cos\theta - \left(\frac{r_{in}}{\sigma_{m-g}}\right)\left[\left(\frac{t_r^2}{2}\right) + \left[\text{sh}\left(\frac{2U_r}{t_r}\right)\right]\frac{t_r^3}{4} + \frac{U_r^3}{3} - U_r\right]$$

$$\times (P_{ms} - P_i) - \left[\text{sh}\left(\frac{2U_r}{t_r}\right)\right]\frac{t_r^2}{2U_r},$$

$$U_r = 1 - \sqrt{1 - t_r^2\,\text{ch}\left(\frac{t_r}{U_r}\right)},$$

where P_{ms} is the pressure that hinders movement of the metal surface and P_i is the pressure of gases that escaped to the cavity.

Therefore, the force of adhesion of two solid non-metallic inclusions depends greatly upon the wettability of non-metallic inclusion by the metal melt. The worse the wettability of non-metallic inclusions by the metal melt, the higher the adhesion force, provided sizes of the inclusions and value of surface tension of the melt remain unchanged. As shown by the calculations,[405] the velocity of the flow should be 13.5 m/sec and the velocity gradient should be $2.7 \cdot 10^6$ s^{-1} to disrupt the joined solid particles of a diameter of 5 μm. As such velocities and gradients can hardly be achieved in the weld pool, despite a considerable metal stirring, it is likely that welding can create conditions for the formation of sufficiently strong bonds between solid inclusions.

More favorable conditions for the joining of solid inclusions can be formed when the shell of an oxide, sulfide, or oxysulfide melt covers both the inclusions or at least one of them. A collar (Figure 6.7) is formed between such inclusions in contact, which leads to formation of extra capillary forces tending to keep and draw the particles together. The value of these forces can be found from the following equation[406]:

$$F_{cap} = \sigma_{m-c}\left[\left(\frac{1}{r'_1}\right) + \left(\frac{1}{r'_2}\right)\right]S_{m-c} + \sigma_{m-cp}\Pi, \qquad (6.19)$$

where σ_{m-c} is the interphase tension at the liquid collar–metal melt boundary; r'_1 and r'_2 are the main radii of curvature of the collar; S_{m-c} is the projection of the collar–inclusion interface on the surface normal to the direction of force F_{cap}; σ_{m-cp} is the projection of the interface tension vector on the axis; that connects the non-metallic inclusions; Π is the wetting perimeter length.

As shown by investigations of the capillary forces,[407–409] these forces are rather intense. So, they can have a marked effect on the joining of non-metallic inclusions.

Therefore, coarsening of non-metallic inclusions depends in many respects upon the surface properties and phenomena.

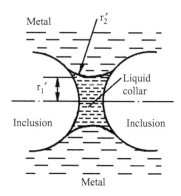

FIGURE 6.7 Schematic of formation of a liquid collar between solid non-metallic inclusions.

6.4 REMOVAL OF NON-METALLIC INCLUSIONS FROM THE WELD POOL

As the process of formation and coarsening of sulfide non-metallic inclusions occurs when the volume of the melt, its mobility, and the mobility of inclusions are small, these inclusions remain as a rule in the weld metal. Primarily the oxide non-metallic inclusions are removed from the weld pool.

6.4.1 THE BASICS OF REMOVAL

The process of removal of oxide non-metallic inclusions from the weld pool is a multi-stage one and consists of the following stages: (1) approach of an inclusion to the metal–gas or metal–slag interface, depending upon the shielding method employed; (2) transfer of the inclusion through the metal–gas or metal–slag interface; (3) escape of particles into the bulk of the slag, if the flux is involved in the process. It is apparent, that the velocity of the entire process will be determined by the velocity of the slowest stage. Therefore, let us consider peculiarities of each of the stages and role of the surface phenomena at each of the stages.

In intensive stirring of a small volume of the metal (which is the case of the majority of the fusion welding processes), the velocity at which the non-metallic inclusions arrive to the interface, does not limit the velocity at which they are removed from the metal. It should be noted, that sizes of the inclusions and their density will exert no marked effect on the velocity at which particles approach the interface, because as a result of circulation of the metal melt, the velocities of the movement of the particles will be many times as high as the velocities of their rising, as determined from the Stokes and Rybchinsky–Hadamard formulas. The velocity of movement of the particles and difference in the values of their adhesion to the metal will not be affected. This is because the inclusions move in the metal melt without slip.[410]

6.4.2 THEORETICAL CONSIDERATIONS

According to Khlynov, Sorokin, and Stratonovich,[411] constants of velocities of convection K_c and sedimentation K_s rises of inclusions in the metal in contact with slag are equal, respectively, to $K_c = U_f r_{in}/l_f H$ and $K_s = v/H$, where U_f is the velocity of flow of the metal melt at the interface with the slag; l_f is the width of the flow at this interface; H is the depth of the metal pool; v is the velocity of removal of a non-metallic inclusion determined from the Stokes formula.

For the arc welding methods, l_f is approximately equal to the weld width, and H is equal to the penetration depth.

The K_c/K_s ratio, which characterizes the effect of metal stirring on the delivery of inclusions to the metal–slag interface, is as follows: $K_c/K_s = U_f r_{in}/v l_f$. For oxide inclusions, $K_c/K_s = U_f/5.5 \cdot 10^3 l_f r_{in}$.[411] As is seen from Table 6.6, the effect of stirring of the metal melt on the velocity of the delivery of non-metallic inclusions to the interface is enhanced with increase in the velocity of metal stirring, decrease in size of the particles, and decrease of the width of the weld pool. Increase of depth of the

TABLE 6.6
Effect of Metal Stirring, Sizes of the Weld Pool, and Inclusions on the Velocity of Their Delivery to the Interface

l_f, 10^{-2} m	H, 10^{-2} m	r_{in}, 10^{-6} m	U_f, 10^{-2} n/s	K_c/K_s
1.0	0.5	1.0	1.0	1.82
2.0	0.5	1.0	1.0	0.91
1.0	1.0	1.0	1.0	1.82
2.0	1.0	1.0	1.0	0.91
1.0	1.0	0.1	1.0	18.2
1.0	1.0	0.01	1.0	182.0
1.0	1.0	1.0	2.0	3.64
1.0	1.0	1.0	5.0	9.1
1.0	1.0	1.0	10.0	18.2
1.0	1.0	1.0	100.0	182.0

weld pool diminishes the role of metal stirring in the process of delivery of the inclusions to the interface. Therefore, metal stirring will have the highest effect with the arc welding methods and the lowest effect with the electron beam welding method, where the weld pool has a large depth and a small width.

If the velocity of stirring of the metal melt amounts to a value[373] of $v = 9.38\sqrt{g\, r_{in}(\rho_{in} - \rho_m)/\rho_m}$, it is possible to remove from the melt the inclusions with a density exceeding that of the metal.

Velocity of delivery of a non-metallic inclusion to the interface may grow, if it joins a gas bubble, which is located in the weld pool or is formed on the surface of the inclusion. The possibility of formation of the gas bubble at the metal melt–inclusion interface will be considered in Chapter 7, while the former case is discussed below.

Spontaneous joining of the bubbles to the inclusion will occur only if this leads to a decrease in the free energy of the system. This is probable when the following inequality is met:

$$\sigma_{m-g}S^b_{m-g} + \sigma_{m-in}(S^b_{m-in} - S^e_{m-in}) > \sigma_{m-g}S^e_{m-g} + \sigma_{in-g}S_{in-g}, \qquad (6.20)$$

where σ is the surface energy at the specified interface; S^b_{m-g} and S^e_{m-g} are the areas of contact of the gas bubble with metal, respectively, before and after joining to the inclusion; S^b_{m-in} and S^e_{m-in} are the areas of contact of the inclusion with the metal melt, respectively, before and after joining to the bubble; S_{in-g} is the area of contact of the inclusion with the gas bubble.

Given that $S_{in-g} = S^b_{in-m} - S^e_{in-m} = S^b_{m-g} - S^e_{m-g}$ and $\sigma_{m-in} - \sigma_{in-g} = \sigma_{m-g}\cos\theta$, the inequality (Equation(6.20)) can be written in the following form:

$$\sigma_{m-g}(1 + \cos\theta)S_{in-g} > 0. \qquad (6.21)$$

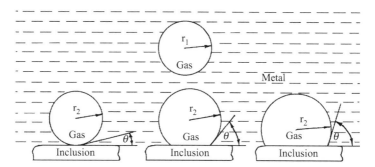

FIGURE 6.8 Schematic of the approach of a gas bubble to a coarse solid non-metallic inclusion and their joining at different wettability of inclusions with metal melt.

As follows from Equation (6.21), the higher the value of σ_{m-g}, the higher the probability of adhering of the bubbles to a non-metallic inclusion. In addition, in the joining of a bubble to a non-metallic inclusion, its size and shape depend upon the wettability of the inclusion with the metal melt. Besides, this determines also the value of S_{in-g} (Figure 6.8). It is apparent that the value of S_{in-g} will change to a larger degree with contact angle, θ than with cos θ.

6.4.3 THE COMBINED EFFECT OF A GAS BUBBLE AND THE INCLUSION

As the metal of the weld pool is intensively stirred, it is important to know what affects the force of adhesion between the bubble and the non-metallic inclusion, which determines reliability of contact between them. Factors that affect the approach of the bubble to the non-metallic inclusion, and conditions that influence their adhesion strength were studied by R.S. Nelson.[412] Assuming, that the gas in the bubble is an ideal gas, equations were derived for the estimation of the value of the energy of binding of the bubble and the non-metallic inclusion. This value is equal to a change in the free energy of the gas, caused by a change in size of the bubble upon contact with the non-metallic inclusion. If the bubble joins a flat surface of the solid inclusion, an equation of the following form holds good:

$$E_{free} = -8\pi r_1^2 \sigma_{m-g} \ln (r_2/r_1)/3 \ . \tag{6.22}$$

If the volume of the gas bubble after it has merged with a non-metallic inclusion remains unchanged, the ratio of radii of the bubble before (r_1) and after (r_2) merging with the inclusion will depend upon angle, θ (Figure 6.8). Figure 6.9 shows variations in the value of $\ln(r_2/r_1)$ with, θ. As is seen, the value of E_{free} grows very rapidly with increase in θ. Therefore, merging of the bubble with the inclusion is possible only in incomplete wetting of inclusions with metal or in the absence of wetting.

If the gas bubble joins a small solid or liquid non-metallic inclusion of a spherical shape (Figure 6.10), the radius of the bubble will change insignificantly. That is

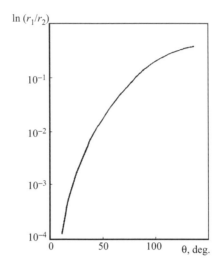

FIGURE 6.9 Dependence of the value of the ratio, r_1/r_2 upon the value of contact angle, θ.

why, the ratio, r_2/r_1 will be close to 1, and the force of binding of the bubble with such an inclusion will be low. If the bubble joins a large molten non-metallic inclusion (Figure 6.11), it will change its shape and size to a much larger degree. In this case, because of a difference in the values of surface tension of the metal and the inclusion material, the equilibrium shape of the gas bubble located on the surface of the inclusion will differ from the spherical one, and the radius of curvature will be much larger than the initial one. Therefore, in this case, the force of binding of the inclusion and gas bubble will also be larger.

The process of transfer of non-metallic inclusions from metal to the environment in contact is induced (like the process of coarsening of particles), by a decrease in the free energy of the system. Under the fusion welding conditions, it is possible that non-metallic particles transfer through two types of interfaces: metal–gas or

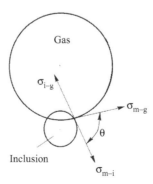

FIGURE 6.10 Schematic of the joining of a gas bubble with a fine non-metallic inclusion of a spherical shape.

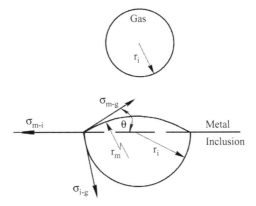

FIGURE 6.11 Schematic of the joining of a gas bubble with a course liquid non-metallic inclusion.

metal–slag. In both the cases, transfer of the particles through the interface occurs under the effect of gravity and surface forces,[413] and is determined by the sum of both the forces. In the transfer of a particle of a spherical shape through the interface the energy of the surface, A_{sur}, and gravity, A_{gr}, forces are as follows:

$$A_{sur} = \Delta\sigma 2\pi r_{in} dh, \quad A_{gr} = \frac{4}{3}(\pi r_{in}^3 \Delta\rho g \; dh),$$

where $\Delta\sigma = \sigma_{in-s} - \sigma_{m-in} - \sigma_{m-s}$, or $\Delta\sigma = \sigma_{in-g} - \sigma_{m-in} - \sigma_{m-g}$; $\Delta\rho$ is the difference in densities of the contacting phases; dh is the length of the path of movement of an inclusion from the metal to the contacting environment.

As is seen from Figure 6.12, the A_{sur} to A_{gr} ratio depends to a large extent upon the size of a moving non-metallic inclusion at its transfer through both metal–slag and metal–gas interfaces.

By comparing the values of A_{sur} and A_{gr} for the transfer of spherical particles through the metal–slag interface, one may notice that the effect of gravity forces on the transfer of non-metallic inclusions, the radius of which is not in excess of 10^{-4} m, is very low and thus, can be neglected.

In the transfer of inclusions through the metal–gas interface, if we assume that $\sigma_{in-g} = 600$ mJ/m^2, $\sigma_{m-in} = 1000$ mJ/m^2, $\sigma_{m-g} = 1100$ mJ/m^2, and $\Delta\rho = 7800$ kg/m^3, the A_{sur} to A_{gr} ratio is close to 1 for particles with a radius of $0.6 \cdot 10^{-2}$ m. Given that the size of non-metallic inclusions located in the weld pool most often varies from 10^{-7} to 10^{-8} m, the effect exerted by the gravity forces on their transfer through the interface can be neglected in this case as well.

6.4.4 DEFORMATION AND DISRUPTION

In the transfer of particles through the metal–slag interface, the energy of the surface forces decreases (Figure 6.13) with decrease in the value of interphase tension at the metal–slag and metal–non-metallic inclusion interfaces.

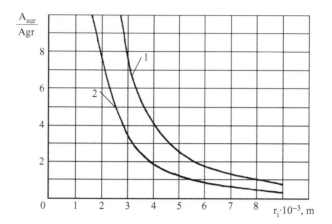

FIGURE 6.12 Variations in the A_{sur}/A_{gr} ratio depending upon the size of a non-metallic inclusion at its transfer throught the metal–slag (1) and metal–gas (2) interfaces: $\sigma_{i-g} = 600\,\text{mJ/m}^2$; $\sigma_{m-i} = 1000\,\text{mJ/m}^2$; $\sigma_{m-g} = 1100\,\text{mJ/m}^2$; $\sigma_{m-s} = 800\,\text{mJ/m}^2$; $\sigma_{i-s} = 0$; $\Delta\rho_{m-s} = 4300\,\text{kg/m}^3$; $\Delta\rho_{m-g} = 7800\,\text{kg/m}^3$.

According to Grigoryan, Shvindlerman, and Aleev,[413] transfer of any particle through the interface is possible only, if $\Delta\sigma < 0$. Otherwise, only the particles with a size exceeding some critical value will alone be able to transfer through the interface.

As $\sigma_{in-s} \approx 0$, for all values of σ_{in-m} and σ_{m-s}, the value of $\Delta\sigma < 0$ and, therefore, oxide inclusions can spontaneously transfer through the metal–slag interface. It can be seen from the above values of σ_{in-g}, σ_{in-m}, and σ_{m-g} that inclusions can transfer also spontaneously through the metal–gas interface.

Transfer of non-metallic inclusions through the metal–slag and metal–gas interfaces is preceded by deformation and disruption of a thin metal film.[414] As the escape of particles to the metal surface leads to the formation of a dome, it is likely that the stability of the film is determined by a value of excessive pressure $P = 2\sigma/r_{in}$ caused by the curvature of the interface. Therefore, the smaller the radius of an inclusion, the

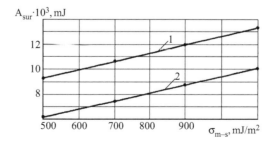

FIGURE 6.13 Change in the value of A_{sur} depending upon the values σ_{m-s}: (1) $\sigma_{m-i} = 1000\,\text{mJ/m}^2$; (2) $\sigma_{m-i} = 500\,\text{mJ/m}^2$.

higher the value of P and the lower the stability of the metal film. That is why, the smaller the size of liquid non-metallic inclusions, the easier they can be removed. For solid inclusions, $r_{in} \to 0$ and $P \to \infty$, which is attributable to the acute-angled microroughness inclusions present on the surface. Therefore, transfer of such inclusions through the interface will occur particularly rapidly, which is also proved by the experimental data. According to a published report,[414] the time of transfer through the metal–gas interface of solid non-metallic inclusions is less than 10^{-4} sec, and that of liquid inclusions is from fractions of a second to a minute.

The barrier effect exerted by the films in the removal of non-metallic inclusions from the molten metal is proved by the published results.[93]

However, non-metallic inclusions, that escaped to the metal surface, can be again carried along by the metal flows into the bulk of the weld pool, as in the case of gas-shielded welding in particular. In this case, one may not expect a marked removal of non-metallic inclusions from the weld-pool metal.

Inclusions, that escaped to the interface with the slag, will separate from the metal only when they are absorbed by the molten slag. It is important, that the rate of absorption of particles by the slag should be not lower than the rate of metal flows. Otherwise, the inclusions could be again carried along into the bulk of the weld pool.

6.4.5 Some Important Calculations

The efficiency of absorption of non-metallic inclusions can be evaluated from the rate of slag spreading over their surfaces.

As shown by the results of investigations of wettability of solid oxides by different compositions of slags, variations in the radius of the slag spot, r_s, during spreading in the viscous mode are described by the Frenkel' formula[380]:

$$r_s^2 \approx \tau_{sp} r_c \Delta\sigma / \eta_{in} \qquad (6.23)$$

and in the inertia mode, by the following formula:

$$r_s^2 \approx \tau_{sp} \sqrt{r_c \Delta\sigma / \rho_{in}}, \qquad (6.24)$$

where r_c is the radius of curvature at the apex of the spreading droplet; ρ_{in} is the density of the oxide melt; η_{in} is the viscosity of the oxide melt; τ_{sp} is the time of spreading.

The pulling force on the perimeter of the droplet, $\Delta\sigma$, for the inclusion–slag–gas system is $\Delta\sigma = \sigma_{in-g} - \sigma_{s-in} - \sigma_{s-g} \cos\theta$, while for the metal–inclusion–slag system it is $\Delta\sigma' = \sigma_{in-m} - \sigma_{s-in} - \sigma_{m-s} \cos\theta'$. Then the $\Delta\sigma' - \Delta\sigma$ difference will be approximately equal to $\sigma_{in-m} - \sigma_{in-g} - \sigma_{m-s} \cos\theta' - \sigma_{s-g} \cos\theta$.[415]

The value of σ_{in-m} varies from 400 to 1100 mJ/m^2, and σ_{in-g} varies from 400 to 900 mJ/m^2. Therefore, at $\theta \approx \theta'$, the $\Delta\sigma' - \Delta\sigma$ difference is not great. So, the results of the experiments on the wetting the oxide particles with the slag at the interface with gases can also be utilized for the investigation of the transfer of inclusions from metal to slag.

The calculations from Equation (6.23) and Equation (6.24) show that at $\Delta\sigma = 2000 \text{ mJ/m}^2$ and $\eta_{\text{in}} = 0.1 \text{ Pa}\cdot\text{s}$ the time of wetting the particles 10^{-4} m in size for a midship section is 10^{-5} sec, and that of the particles 10^{-6} m in size is 4.10^{-8} sec. Therefore, the process of transfer of non-metallic inclusions from the metal to the slag, allowing for their small sizes, should proceed very rapidly.

According to the data of Gurevich,[416] the initial velocity of the transfer of solid prismatic inclusions from the metal to the slag is as follows:

$$\left(\frac{\mathrm{d}x}{\mathrm{d}\tau}\right)_{x\to 0} \approx k_1 \frac{[(\sigma_{\text{m-g}} + \sigma_{\text{s-g}}) + (W_{\text{s-in}} - W_{\text{m-in}})]}{6m_0\pi\eta_{\text{m}}}. \tag{6.25}$$

Here, k_1 is the coefficient that allows for the size of inclusions and curvature of the surface of liquids, $\sigma_{\text{m-g}}$ and $\sigma_{\text{s-g}}$ are the surface tensions of the metal and the slag, respectively, $W_{\text{m-in}}$ and $W_{\text{s-in}}$ are the adhesions of an inclusion to the metal and the slag, respectively, and m_0 is the coefficient equal to $\sqrt[3]{3n_1/4\pi}$, where n_1 is the inclusion length to width ratio.

The final velocity of transfer of the solid prismatic inclusions is determined from the following expressions:

$$\left(\frac{\mathrm{d}x}{\mathrm{d}\tau}\right)_{x\to A} \approx \frac{k_1[(\sigma_{\text{m-g}} - \sigma_{\text{s-g}}) + (W_{\text{s-in}} - W_{\text{m-in}})]}{6m_0\pi\eta_{\text{s}}}, \tag{6.26}$$

where η_{s} is the slag viscosity.

Therefore, the rate of absorption of solid non-metallic inclusions by the slag depends upon the size of the inclusions and their shape, upon the value of surface tension of metal and slag, difference in the adhesion of an inclusion to the slag and metal, as well as upon the viscosity of the slag. It can be seen from Equation (6.25) and Equation (6.26) that the velocity of transfer of solid particles at the initial moment of time is much higher than the final velocity of their transfer, as the viscosity of slag can be tens, hundreds, and even thousands of times higher than the viscosity of the metal melt.

As the value of adhesion is equal to $\sigma_{\text{l-g}}(1 + \cos\theta)$, transfer of particles from the metal to the slag depends upon the wettability of a non-metallic inclusion by the metal and the slag. The better the wettability of the inclusion by the slag and the worse the wettability of the inclusion by the metal, the more rapid is the transfer of the particles from the weld pool to the slag. However, the transfer of the particles from the metal to the slag is affected to a larger degree by the difference in the viscosity of the metal and the slag. The effect of adhesion of inclusions to the slag and the metal, as well as the effect of viscosity of the slag on the rate of absorption of non-metallic inclusions by slag are shown in Figure 6.14.

The effect of different factors on the process of absorption of liquid non-metallic inclusions by a slag can be evaluated with a certain error from the data on merging of

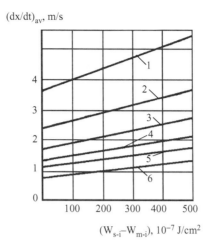

FIGURE 6.14 Effect of adhesion of solid non-metallic inclusions to the slag and the metal on the rate of their absorption by the slag of a different viscosity η_s(Pa·s): (1) 0.2; (2) 0.3; (3) 0.4; (4) 0.5; (5) 0.6; (6) 0.8. (From Gurevich, Yu.G., *Izv. Vuzov. Chyorn. Metall.*, 8, 5, 1968.)

the liquid slag droplets. In the merging of droplets of identical composition and size, the value of the neck r_n formed between the droplets will be as follows[417]:

$$r_n^2 = \frac{3\sigma_d r \tau_m}{2\pi\eta_d}$$

where σ_d is the surface tension of the droplet material; η_d is the viscosity of the droplet material; r is the initial radius of the droplet, τ_m is the time of merging of the droplets.

In the case of merging of droplets of different sizes and composition, the value of the neck is as follows[404]:

$$r_n^2 \approx \frac{\Delta\sigma r_{ne} \tau_m}{2\eta_n}, \tag{6.27}$$

where $\Delta\sigma = \sigma_1 + \sigma_2 - \sigma_{12}$ (σ_1 and σ_2 are the surface tensions of the droplet materials, and σ_{12} is the interphase tension at the boundary of the two droplets); $\eta_n = \eta_1\eta_2/(\eta_1 + \eta_2)$; $r_{ne} = r_1 r_2/(r_1 + r_2)$.

The time necessary for the particles to join each other increases with increase in their size ($\tau_{1/2} = \pi r^2/4v$, $v = r_n^2/\tau$). Therefore, absorption of fine inclusions is more probable than that of coarse ones. In addition, as is seen from Equation (6.27), increase in the viscosity of inclusions or slag also reduces the probability of assimilation of liquid inclusions by the slag.

Note that Equation (6.25) and Equation (6.26) are valid for the case of removal of solid non-metallic inclusions from a quiescent metal. The presence of flows in the metal melt should affect the transfer of non-metallic inclusions to the slag. Transfer of particles to the slag can be either facilitated or hampered due to the force of inertia caused by the presence of the flows. If at the moment when a particle approaches the metal–slag interface, the vector of its velocity is tangential to the interface, the centripetal forces will carry the particle back into the bulk of the metal. If the velocity vector crosses the interface, the velocity of transfer of particles will be higher than that of transfer of the same particles from the quiescent liquid. In this case, the presence of flows in the molten metal promotes removal of non-metallic inclusions from the metal.

The third stage, that is, transfer of inclusions deep into the slag seems to cause no difficulties in the removal of non-metallic inclusions from the weld pool, as in welding involving fluxes, especially with the arc welding methods, the slag is stirred as intensively as the metal. Intensified stirring of the slag pool in electroslag welding facilitates assimilation[418] of oxide inclusions by the slag.

Therefore the peculiarities of each of the stages of transfer of non-metallic particles from the metal to the slag provide a conclusion that under welding conditions, the rate of the entire process of removal of non-metallic inclusions is limited by the rate of the second stage. Meanwhile, the escape of non-metallic inclusions to the metal–slag interface and their absorption by the slag are determined in many respects by the surface properties of the metal, the slag and the inclusion, as well as by the surface phenomena (wettability and adsorption) at the interfaces of the contacting phases. The effect of the surface properties of the inclusions is especially pronounced[419] with intensive stirring of metal.

It should be again noted that a marked removal of non-metallic inclusions from the weld is possible only when the metal of the weld pool is covered by a slag layer or an oxide film of a sufficient thickness. This is possible in welding using covered electrodes, in flux- or flux paste-assisted welding, and welding in oxidizing gas atmospheres, where an oxide film is formed on the weld-pool surface, as well as in electroslag processes.

With inert-gas welding, it is impossible to achieve a marked removal of non-metallic inclusions. In this case, it is possible just to alter the shape and size of the non-metallic inclusions, as well as their location in the weld.

In electron beam welding, where the process occurs in a vacuum chamber, oxide non-metallic inclusions present in the weld pool may be dissolved due to increase in the oxidizing ability of carbon contained in the metal (see Section 1.4).

7 Porosity in Welds

Pores are the most common defects in welds made by different fusion welding methods. Therefore, the majority of researchers involved in welding or surfacing technologies studied the process of formation of pores in the welds to different extents. Experimental and theoretical information on this process, accumulated up to now, is very extensive.

However, many of the issues relating to the process of pore formation remain unclear so far or are controversial. For example, no unanimous opinion exists on the role of individual gases and interfaces, polarity of the welding current, composition of flux, metal, etc., in the process of pore formation.

In our opinion, this situation is attributable primarily to the following factors. First, the process itself is complex, that is, multi-factorial, and second, the process of pore formation is wrongly considered to consist of one stage. We think the process of pore formation should be subdivided into three stages: (1) formation of a stable nucleus, (2) growth of a gas bubble, and (3) escape of gas bubbles from the weld pool. Third, conclusions on peculiarities of the process are most often made on the basis of technological experiments, where shielding atmospheres and metal proper contain some amount of N_2, H_2, O_2, unaccounted inclusions, etc.

In recent studies, however, the process of pore formation is considered from the standpoint of physics of surface phenomena. This approach, based on the energy notions, provides a much better understanding of the process of pore formation in welds.

7.1 PORES AND PERFORMANCE OF A WELDED JOINT

The majority of researchers involved in studies of the effect of pores on performance of welded joints are of the opinion that pores present in the weld metal, their content being below a certain limit, which does not depend on the type of a material welded, have almost no effect on static strength of the weld metal. This limit is approximately 10% of the cross-sectional area for low-carbon steel,[420] 6–8% for pearlitic steels,[420–422] and 3.6% for aluminum alloys.[423]

According to the data of,[424] for aluminum alloy AMg6 the pores with a diameter of less than $0.5 \cdot 10^{-3}$, containing no oxide films, hardly have any effect on strength properties of a welded joint. If the diameter of such pores is within a range of $(1.8–2.0) \cdot 10^{-3}$ m, static strength of the weld without a reinforcement decreases by about 6%. Chains of partially merged pores ($d_p = 2 \cdot 10^{-3}$ m) of a total length of no more than 30% of that of the weld decrease its static strength by 11–15%.

The effect of pores on the load-carrying capacity of weld can be evaluated from the following equations[425]:

1. For individual pores

$$P_p = \frac{[\sigma]B_p\{(B_w/2) - d_p\}}{[2 + 1.18(1 - d_p/B_w)^4]};$$

2. For a chain and network of pores

$$P_p = \frac{[\sigma]B_p\left\{(B_w/2) - \sum_{i=1}^{p} d_{pi}\right\}}{[1 + 2(1 - d_p/a_p)^{1.35}]},$$

where d_p is the diameter of a pore (for chains and clusters it is the mean diameter); B_w is the maximum width of the weld; a_p is the smallest distance between the neighboring pores; B_p is the height of a through pore or the pore depth.

Therefore, the effect of pores on the mechanical properties of a welded joint depends on their size, quantity, and location in the weld.

It is likely that the presence of pores in the metal induces substantial stresses around them, because the pressure of gases in the pores amounts to tens and even hundreds of atmospheres. This pressure is especially high, if the metal melt contains hydrogen, which transfers to the gas bubbles during solidification.

Therefore, pores contained in the weld decrease not only the static strength of a welded joint; acting as stress-raisers,[426,427] they can decrease the fatigue life of the welded joint as well. A substantial effect of pores on fatigue resistance of welded joints for different metals is noted in a number of studies.[424,428–430]

Pores located in zones of high tensile residual stresses exert a particularly marked effect on fatigue life of welded joints. In this case, even individual pores may present a hazard.

Since tensile residual stresses are especially high in surface layers of the weld metal, the risk of fracture of a welded structure increases, if pores are located close to the surface. This is proved by the data of Sirgo,[431] according to which the value of the stress intensity factor for gas pores grows from 2.05 to 5.0, when a pore from the bulk approaches the surface to a distance equal to the pore diameter. As the effect of pores on the mechanical properties of a welded joint is related to the value of tensile residual stresses, it is apparent that the presence of pores causes a more intensive deterioration of the mechanical properties of long longitudinal welds, in which tensile residual stresses amount usually to high values. The presence of pores in short transverse welds, where the values of residual stresses are comparatively small, is less dangerous. Pores may cause fatigue fracture of a structure, if they are located in the connecting fillet and the butt welds. According to published data,[1] in the welding of low-carbon and low-alloyed steels, fatigue limit of the welds containing pores in the zones of high tensile stresses is 107 MPa with a pulsed cycle and 40 MPa with a symmetric cycle.

However, in addition to affecting the mechanical properties of the welds, pores may decrease their corrosion-resistant properties as well. Also, the presence of pores in the weld may be absolutely inadmissible for some items, for example structures operating under a high pressure or under deep vacuum conditions. In this connection, the issue of admissibility of pores in the weld is handled depending on the service conditions of a weldment. Nevertheless, in all the cases, the presence of pores in the weld is as a rule undesirable. Therefore, let us consider the process of formation of pores in more detail starting from the formation of stable nuclei of gas bubbles.

7.2 FORMATION OF GAS BUBBLE NUCLEI

The metal of the weld pool always contains some amount of gases, which are trapped during the production process or because of violation of shielding during welding, or they may be formed in the metal pool as a result of chemical reactions, which is the case, for example, of CO. There are conditions which may lead to oversaturation of the metal with the gas, that is, the weld pool or its part under such conditions is in a metastable state. Transition of the system from the metastable to stable state may take place either as a result of escape of the gases into the atmosphere or as a result of formation of gas bubbles in the bulk of the metal. Consider now the second process, which is of the highest interest, because it is in this case that defects, such as pores are formed in the weld.

Intensity of formation of stable nuclei of gas bubbles in the metal melt and their critical radius are determined from the same formulas as in the case of formation of non-metallic inclusions. The exception is that the value of surface tension of metal (σ_{m-g}) should be substituted into Equation (6.4) and Equation (6.5) instead of the value of interphase tension at the metal–slag interface (σ_{m-s}).

Peculiarity of formation of gas nuclei, compared with nucleation of non-metallic inclusions, is that at comparatively close values of molecular masses (CO–28, N_2–28, H_2–2, MnO–71, SiO_2–60, etc.), the densities of gases are thousands of times lower. Therefore, the intensity of nucleation of the gas bubbles is much lower than that of nucleation of non-metallic inclusions, and the critical size of a gas nucleus is larger than the radius of the stable nucleus of a non-metallic inclusion, the degree of oversaturation being identical.

As the energy of formation of the gas nucleus in a homogeneous environment, W_{hom}, is equal to a change in the free energy, (Equation (6.3)), the value of I is in exponential dependence on W_{hom}. Besides, the higher the value of W_{hom}, the higher is the value of I.

The value of pre-exponential factor A, according to the data by different researchers,[241,432,433] varies from 10^{10} to 10^{36}. These discrepancies, being so high, are attributable to the fact that it is very difficult to calculate this value correctly, because to do it one should take into account all the relationships and degrees of freedom for a nucleus consisting of a large number of molecules. Therefore, it is likely that the probability of formation of the stable nucleus of a new phase in the melt could be estimated more reliably from the value of W_{hom}.

The values of r_{cr} and W_{hom} determined from Equation (6.4) and Equation (6.3), depending on the metal composition, degree of oversaturation of the metal melt with gas, and the type of the gas are shown in Figure 7.1 and Figure 7.2. The values of r_{cr}

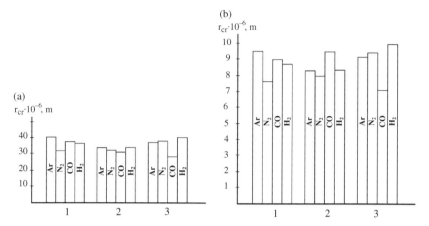

FIGURE 7.1   Values of critical radius of the gas bubble nucleus formed from different gases in a homogeneous metal melt ($T = 1803$ K). C/C_S value: (a) 1.1; (b) 1.5; The samples used are: [1] armco-iron; [2] Sv-10GS steel; [3] 08Kh20N10G6 steel.

and W_{hom} were also determined for the formation of a gas nucleus from argon, which for sure cannot be the cause of the formation of pores in the welds. However, these data allow a comparison of the values of critical radii and energy of formation of stable nuclei from different gases. The data on the surface tension of metals, needed for calculations, were taken from our studies,[191-193] and the gas density values used were valid for a temperature of 293 K. Figure 7.1 and Figure 7.2 show the value of r_{cr} and W_{hom} for two values of oversaturations: 1.1 and 1.5. The two values have been chosen because it is in this range that a change in the over-saturation has a particularly marked effect on the sizes of a stable nucleus and the energy of its formation. Further growth of the C/C_s values leads to a comparatively little

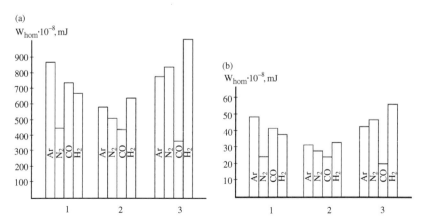

FIGURE 7.2   Values of the energy of formation of a stable gas nucleus in a homogeneous metal melt ($T = 1803$ K). C/C_S value: (a) 1.1; (b) 1.5; The samples used are: [1] armco-iron; [2] Sv-10GS steel; [3] 08Kh20N10G6 steel.

change in r_{cr} and W_{hom}. Therefore, for armco-iron, at a melt temperature of 1803 K in the case of formation of a stable nucleus from nitrogen, $r_{cr} = 7.6 \cdot 10^{-6}$ m at $C/C_s = 1.5$, and $r_{cr} = 1.34 \cdot 10^{-6}$ m at $C/C_s = 10.0$. Such relationships between the oversaturation values (higher than 1.5) and the values of r_{cr} and W_{hom} hold also for other metals investigated.

As is seen from Figure 7.1 and Figure 7.2, the values of r_{cr} and W_{hom} decrease with increase of the degree of oversaturation of the metal melt with gas. In addition, the values of the critical radius of a gas bubble nucleus decreases to a smaller degree than the energy of formation of such a nucleus. However, at any oversaturation, the energy needed for the formation of the stable gas nucleus from different gases is different. Therefore, for every metal, the lowest energy of the nucleus formation is characteristic of a certain gas, the presence of which in the metal leads to the formation of the stable gas nucleus.[397] Thus, the values of surface tension of a metal at the interface with different gases being known, using Equation (6.3) and Equation (6.5), we can find a gas which will become the cause of formation of gas nuclei in a given metal. This gas will be the one, at the interface with which the metal will have the lowest value of surface tension.

7.2.1 SURFACE TENSION AND RADIUS OF CURVATURE

Surface tension is known to vary with radius of curvature of a fracture surface, and this variation can be described by the Tallman equation[39]:

$$\frac{\sigma_g}{\sigma_0} \approx \frac{1}{(1 + 2\delta/r_0)},$$

where σ_g and σ_0 are the surface tensions of a curvilinear and flat surfaces, respectively; δ is the distance between the surface tension with radius, r and the equimolecular separating surface with radius r_0, for which the surface density of a material is equal to zero.

In this connection, the value of σ_{m-g} with allowance for the interface curvature should have been substituted into Equation (6.4) and Equation (6.3) to determine r_{cr} and W_{hom}. However, as shown by Khollomon and Turnbull,[42] determining the surface tension of liquids from the intensities of formation of nuclei and comparing them with the experimental values of σ_{l-g}, for a number of liquids (water, ethyl alcohol, etc.), indicated that the difference between them was no more than 2–3%. Given this fact, as well as the fact, that the critical radius of a gas nucleus is as a rule not less than 10^{-7} m, (i.e., higher than the value of the radius, at which it is necessary to allow for its curvature,[39]) r_{cr} and W_{hom} were determined using the values of the surface tension, we had estimated for large droplets.

As shown by calculations, in all the cases, the size of a critical gas nucleus is within a range of $10^{-7}-10^{-6}$ m. Such nuclei should contain (10^6-10^{11}) gas molecules. Apparently, the probability of formation of such nuclei is low, as it requires that a large quantity of gas molecules be located at a given moment of time in a small volume of the metal. Calculation of the intensity of formation of stable gas nuclei from one gas in a homogeneous environment indicates that, because of high values of surface tension of metal melts and very large sizes of critical nuclei, their formation in the melt is probable only at the 20–30-fold oversaturation of the metal with the gas. Such oversaturations can hardly be found in practice, although pores are really formed in welds. One of the causes of formation of pores in the welds may be local oversaturations or fluctuations.

It follows from the laws of statistical physics[59] that fluctuations with an increased density of a precipitating material can be formed in the bulk of the melt. In this case, the fluctuations are stable and capable of further growing, if their size exceeds a certain critical size. The probability of formation of stable fluctuations in the bulk of the metal increases[59] with increase of the total concentration of a component in the melt and its diffusion coefficient, as well as with decrease of size of the critical nucleus.

The probability of formation of nuclei of the gas bubbles due to fluctuations suggests that they can be formed also in the cases, where there is no gas oversaturation of the entire volume of the weld-pool metal.

As the probability of formation of fluctuations depends on the mobility of a component which forms the gas nucleus, it is likely that in this process, the important role should be played by hydrogen, which is characterized by the highest mobility, compared with other components participating in the formation of the pore nuclei. However, hydrogen does not exhibit a substantial surface activity in metal melts.

Nevertheless, it is apparent that all the factors which promote increase of mobility of components, decrease of size of a stable fluctuation, and growth of the concentration of gases in metal will promote formation of nuclei in the weld pool. One of such factors is the weld-pool temperature. Increasing it leads to increase of the diffusion coefficient, decrease of the r_{cr} value, and growth of the concentration of gas. Therefore, it should favor formation of stable fluctuations. Apparently, it is this factor that is one of the causes of increase of porosity of the weld with increase of the welding current over a certain critical value noted in the study of Pokhodnya,[298] and that explains the observed formation of pores in the active-spot zone of the weld pool.[434]

7.2.2 EXISTENCE OF DIFFERENT GASES IN THE PORES

All the above considerations are valid for the formation of a gas nucleus from one gas. Meanwhile, investigations into a composition of gas in pores of the welds are indicative of the fact that they always contain several gases. For example, in welding low-carbon and low-alloy steels, the pores contain primarily N_2, H_2, and CO; in aluminum welding the main gases contained in the pores are hydrogen and nitrogen; in the welding of stainless steels, these gases are nitrogen, hydrogen, and CO. In the last case, argon was also found among the gases contained in the pores. Most probably, it did not participate in the formation and development of a gas nucleus, but it got into the molten metal and pores due to suction by the arc column. As the pores always contain several gases, it is highly probable that several gases simultaneously participate in the formation of the gas bubble nuclei.

There are only two studies,[435,436] which considered the formation of bubble nuclei from two gases. However, these studies are limited to consideration of the qualitative aspect of the process,[436] or use assumptions,[435] which could not be realized under actual conditions. For example, it was assumed that the variations in the chemical potential with the transfer of components from the liquid to the gas bubble are identical. Moreover, no consideration was given to the effect of composition of a mixed nucleus on the probability of its emergence.

We handled this problem in a more general form[274] for two cases: (1) the nucleus is formed from two gases I and II dissolved in the melt and (2) one of the gases is dissolved in the metal and the other is formed as a result of a chemical reaction.

Assuming that gases that form the nucleus, are ideal and the composition of the melt does not vary in the formation of the nucleus, the following expressions were derived:

$$\frac{N_{cr(B)}}{N_{cr(0)}^{I}} = \frac{[\ln(x_1^I/x_2^I)]^3}{[B\ln(x_1^I/Bx_2^I) + (1-B)\ln\{x_1^{II}/(1-B)x_2^{II}\}]^3} = K, \quad (7.1)$$

$$\frac{W[N_{cr}(B)]}{W(N_{cr}^I)} = \frac{[\ln(x_1^I/x_2^I)]^2}{[B\ln(x_1^I/Bx_2^I) + (1-B)\ln\{x_1^{II}/(1-B)x_2^{II}\}]^2} = C^0 \quad (7.2)$$

Here, $N_{cr(B)}$ is the quantity of molecules in the nucleus of a critical size formed from two gases, $N_{cr(0)}^I$ is the quantity of molecules in the nucleus of a critical size formed from one gas, x_1 and x_2 are the mole fractions of a component in the melt and gas nucleus, respectively, B is the fraction of gas 1 in the nucleus, and $W[N_{cr}(B)]$ and $W(N_{cr}^I)$ are the energies of nucleation of a stable gas nucleus from two gases and one gas, respectively.

Assuming that $x_1^{II}/x_2^{II} = 1$, using Equation (7.1) and Equation (7.2), we can find the values of K and C^0 at different degrees of oversaturation of the metal with the gas 1 and at different values of B. As is seen from the data obtained (Table 7.1), at a small degree of oversaturation (up to 1.4) and at any possible value of B, the quantity of particles in a two-component nucleus and the value of the energy of its formation are smaller than for a nucleus formed from one gas. However, in this case, there may be such values of B, at which these relationships are minimum.

At a high degree of oversaturation with one of the gases, but at a small share of its participation in the formation of the gas bubble nucleus, an increasingly large quantity of its molecules is required to form a stable gas nucleus, which leads to the growth of the value of the nucleation energy. Therefore, increase of the degree of oversaturation with one of the gases (at a constant degree of oversaturation with the other gas) leads to (1) a decrease of difference in the quantities of molecules that form the nucleus (2) a decrease in the values of the energy of formation of two- and one-component nuclei.

In the formation of the gas nucleus from three gases I, II, and III the energy required for its formation is as follows:

$$W(N, E, F, U) = \alpha_k N^{2/3}\sigma_{m-g} - NRT[E\ln(x_1^I/Ex_2^I) + F\ln(x_1^{II}/Fx_2^{II}) + U\ln(x_1^{III}/Ux_2^{III})],$$

TABLE 7.1

Effect of the Degree of Oversaturation of Metal with Gas and Gas Content of a Nucleus on the K and C^0 Values

B	1.1	1.2	1.3	(C/C_s) 1.4	1.5	2.0	3.0	4.0
0.1	0.0812	0.284	0.562	0.882	1.37	3.11	6.4	11.02
	0.023	0.151	0.422	0.828	1.61	5.49	16.19	36.59
0.2	0.038	0.1156	0.226	0.55	0.487	1.178	2.33	3.84
	0.0062	0.039	0.107	0.327	0.34	1.278	3.55	7.53
0.3	0.022	0.0745	0.0145	0.244	0.306	0.716	1.37	2.16
	0.0033	0.02	0.055	0.106	0.169	0.605	1.6	3.18
0.4	0.018	0.0595	0.113	0.173	0.234	0.53	0.972	1.49
	0.0024	0.0145	0.038	0.072	0.113	0.386	0.956	1.82
0.5	0.019	0.0533	0.101	0.153	0.205	0.445	0.783	1.15
	0.026	0.0125	0.032	0.0598	0.093	0.297	0.693	1.24
0.6	0.017	0.0543	0.099	0.148	0.196	0.406	0.682	0.95
	0.0022	0.0126	0.031	0.057	0.087	0.258	0.564	0.93
0.7	0.02	0.0615	0.109	0.158	0.206	0.399	0.632	0.857
	0.0028	0.0152	0.036	0.063	0.094	0.252	0.502	0.794
0.8	0.027	0.0795	0.137	0.191	0.248	0.432	0.432	0.774
	0.0045	0.0224	0.051	0.083	0.123	0.284	0.284	0.681
0.9	0.0538	0.14	0.221	0.292	0.343	0.534	0.7	0.83
	0.0125	0.0523	0.104	0.157	0.201	0.391	0.586	0.75

Note: Upper figures are the values of C^0 and lower figures are the values of K.

where E, F, and U are the shares of participation of gases in the formation of the nucleus; α_k is the coefficient, which depends on the shape of the nucleus and for a sphere, it is equal to $(36\pi)^{1/3} V_m^{2/3}$, mean ($V_{m \text{ mean}}$ is the mean volume of a molecule).

Since with increase of the number of gases participating in the formation of a stable nucleus, the share of participation of each of them decreases, this leads to a corresponding growth of the ratio of the degrees of oversaturation with gases. Therefore, the larger the number of gases participating in the process of formation of pores, the smaller the degree of oversaturation with each of them. In this case, the degree of oversaturation with each of the gases may be less than 1, as the shares of participation of the gases are also less than 1. What is important here is that the ratio of the degree of oversaturation with a gas to its share of participation should be more than 1. In addition, the probability of formation of a multi-component fluctuation is higher than that of a one-component fluctuation. It is determined as a product of the probabilities of fluctuations of each type of components.[59]

It was assumed for the second probable case of formation of a gas nucleus, that the gas dissolved in the metal, and CO formed as a result of occurrence of the $[C] + [O] = \{CO\}$ reaction participated in the formation of this nucleus. Assuming, as before, that gases that form the nucleus are ideal and that the composition of a melt does not change in the formation of the stable nucleus, equations were derived[274] for the determination of the quantity of molecules in a critical nucleus and the energy of formation of such a nucleus. At a small content of [O] and [C] in the metal, the activities of components can be replaced by their concentrations, and the constant of equilibrium of the CO formation reaction can be assumed to be equal to 400.[437] Then, the equations will take the following form:

$$
N_{cr}^{CO+1} = \left[\frac{2\alpha_k \sigma_{m-g}}{3RT\left\{B\ln\left(\frac{400[O][C]}{BP_{CO}}\right) + (1-B)\ln\left(\frac{x_1}{(1-B)x_2}\right)\right\}} \right]^3
$$

and

$$
W(N_{cr},B) = \frac{4\alpha_k^3 \sigma_{m-g}^3}{27R^2T^2\left[B\ln\left(\frac{400[O][C]}{BP_{CO}}\right) + (1-B)\ln\left(\frac{x_1}{(1-B)x_2}\right)\right]^2}
$$

Therefore, irrespective of the fact, whether both the gases are dissolved in the metal or whether one of them is formed as a result of a chemical reaction, formation of the gas nucleus from two gases requires a lower energy consumption and smaller quantity of molecules than the formation of the nucleus from one gas.

The above theoretical statements are well confirmed by the known experimental data. For example, according to the data of Sapiro,[28] in submerged-arc welding, the combined addition of nitrogen and hydrogen to the arc zone leads to increase of porosity of the weld. Besides, even a small amount of nitrogen suffices to form the pores, which is in agreement with the calculation data of Table 7.1. Increase of porosity of the weld for a case of several gases present in the arc zone was noted also in welding of titanium samples.[438]

This is also proved by our experiments on welding of steels both in the $H_2 + N_2$ mixture (Figure 7.3) and in the natural gas atmosphere with additions of oxygen (Figure 7.4), which led to an increased formation of CO in the bulk of the metal and hence, in the compositions of the pores.[439]

It follows from Equation (6.3), Equation (6.4), and Equation (6.5), that the intensity of formation of gas nuclei, their formation energy, and critical size depend on the value of surface tension of the metal melt and, hence, on the presence of surfactants in the metal. Thus, for example, in the welding of steels, an addition of such components as O_2, S, Si, P, Se, etc. to the metal, which decreases surface tension of iron and alloys on its base, facilitates formation of nuclei of the gas bubbles. The stimulating effect of sulfur on the formation of pores in weld was also noted in a study.[28]

Oxygen takes a special place among surfactants. It possesses a high surface activity, and shielding gas atmospheres containing oxygen additions have been increasingly used recently for welding the most diverse materials, and primarily steels.[1,23,440] Therefore, in the following discussions, we will consider in more detail the effect of oxygen on the nucleation of pores in welding of steels.

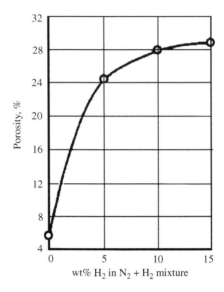

FIGURE 7.3   Dependence of porosity of the weld upon the hydrogen content of the $N_2 + H_2$ mixure used as a shielding gas.

7.2.3 EFFECT OF OXYGEN ON THE NUCLEATION OF PORES

It is a known fact[10,145] that, due to its high surface activity, oxygen present in the metal melt causes changes in the rate of transfer of nitrogen and hydrogen through the metal–gas interface. Therefore, the presence of oxygen in the metal may lead (see Section 1.4) to enrichment of the content of the gases (H_2 and N_2) and their over-saturation in the weld pool, thus promoting formation of the gas-bubble nuclei in the

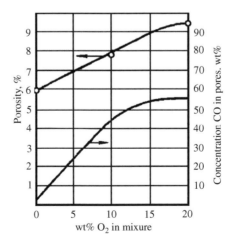

FIGURE 7.4   Dependence of porosity of the weld and concentration of CO in the pores upon the O_2 content of natural gas + oxygen mixture as a shielding gas.

weld pool. The presence of oxygen in the metal may also change the energy conditions for the formation of the bubble nuclei. As is established,[441] the presence of oxygen in the metal always leads to the formation of an electric double layer, and the formation of this layer does change the conditions for the formation and growth of a gas bubble.

Electrostatic part of the surface energy, σ_{el}, is always negative,[417] and can be determined from the following equation[442]:

$$\sigma_{el} = 4\pi\Gamma_{den}^2(ze)^2 l_d$$

where ze is the charge of a chemisorbed ion l_d is the length of a dipole, which can be assumed to be equal to a radius of the oxygen ion Γ_{den} is the specific density of chemisorbed ions $\Gamma_{den} = \Gamma_O \Omega$, Γ_O is the maximum adsorption of oxygen Ω is the portion of the surface occupied by the adsorbed oxygen.

Allowing for σ_{el}, the equations for the determination of the energy of formation of a stable gas nucleus in a homogeneous environment will be written as follows:

1. In formation of the nucleus from one gas:

$$W_{hom} = \frac{16\pi(\sigma_{m-g} - \sigma_{el})^3 \cdot V_m^2}{3[RT\ln(C/C_s)]^2}$$

2. In formation of the nucleus from two gases:

$$W_{hom} = \frac{4\alpha^3(\sigma_{m-g} - \sigma_{el})^3}{27[RT(B\ln(x_1^I/Bx_2^I) + (1-B)\ln(x_1^{II}/(1-B)x_2^{II})]^2}$$

Assuming[441] that $\Gamma_O = 1.29 \cdot 10^{19}$ atom/m,^{2} $l_d = 1.38 \cdot 10^{10}$ m, and $z = 2$, we find that $\sigma_{el} \approx 1630$ $\Omega.^2$ Therefore, if Ω has a sufficiently high value, σ_{el} will markedly decrease the value of surface tension of the metal and hence, facilitate the formation of stable gas nuclei.

Increase of porosity of the weld in the presence of oxygen and nitrogen in the weld-pool metal was noted in the study of Wei et al.[443] Addition of aluminum, which is a strong deoxidizer, to the metal leads to a complete elimination of porosity in the weld in the welding of low-carbon steels.

The intensity of formation of gas nuclei in the weld pool can be increased as a result of stirring of the metal melt. As is well known,[444] pressure in a stirred liquid decreases by a value of $\Gamma_c\rho_m/8\pi r_{cav}$. Here, Γ_c is the rate of circulation of liquid, ρ_m is the metal density, and r_{cav} is the radius of a cavity.

Therefore, nuclei of the gas bubbles may be formed in the bulk of the metal, where the rate of circulation of the melt is high, pressure is low, and the associated oversaturation of metal with gases is increased.

Similar phenomena take place also with feeding of ultrasonic oscillations to the weld pool. In this case, the pressure in different parts of the weld pool also changes, as well as the values of oversaturation of metal with gas do. If we designate the value of oversaturation in a certain volume at any moment of time as $C_o(\tau)$, the deviation of the degree of oversaturation from the initial value for a given volume at any moment of time will be $\Delta C_o(\tau) = C_o(\tau) - C_o$, and the maximum deviation of the degree of oversaturation will be $\Delta C_o^{max} = C_o^{max} - C_o$.

If we assume that at any point of the melt, the oversaturation changes under the effect of ultrasonic oscillations following the harmonic law, then

$$\Delta C_0(\tau) = \Delta C_0^{max} \cos(2\pi/z_\tau)\tau,$$

where z_τ is the duration of a full period of variation in the oversaturation degree.

As the $2\pi/z_\tau$ ratio is the angular frequency of oscillations of the oversaturation degree, ω, then $\Delta C_0(\tau) = \Delta C_0^{max} \cos\omega\tau$, and the intensity of formation of stable gas nuclei in the case of ultrasonic oscillations fed to the weld pool is as follows:

$$I_{us} = Ae^{-Y/\Psi}, \tag{7.3}$$

where $Y = 16\pi\sigma^3 V^2/3R^2T^3$ and $\Psi = [\ln C_0 + \ln(1 + \frac{\Delta C_0^{max}}{C_0} \cos\omega\tau)]^2$.

Analysis of Equation (7.3) allows the following conclusions: the intensity of formation of gas nuclei increases with ultrasonic oscillations fed to the weld pool, and the higher the value of C_0, the higher the above increase.

So, in formation of stable gas bubble nuclei in a homogeneous environment, the probability of their formation will decrease (1) in the absence of surfactant or at the decrease of the content of surfactants in the metal melt; (2) in the absence of stirring or at the decrease of the intensity of stirring of the weld-pool metal; (3) at the decrease of the amount of gases, which can participate in formation of such a nucleus in the metal.

7.3 FORMATION OF GAS BUBBLE NUCLEI IN A HETEROGENEOUS ENVIRONMENT

In the previous section, we considered the peculiarities of formation of gas nuclei in a homogeneous environment. However, under actual conditions, the weld pool always contains interfaces of different phases. These are the interfaces of molten metal with solid metal, non-metallic inclusions, or slag if the welding is performed by involving fluxes or using covered electrodes. It is known from the theory of transformations, that the presence of interfaces promotes formation of a new phase. To reveal the effect on the process of pore formation of the above interfaces, we will consider formation of stable gas nuclei at the interfaces, and first of all at the metal–slag and metal–molten non-metallic inclusion interfaces.

7.3.1 INTERMOLECULAR ADHESION FORCES

As nucleation of a bubble in a liquid is associated with overcoming of inter molecular adhesion forces, this bubble will develop more intensively toward that phase which offers less resistance to its growth. This phase is the slag, because of a lower force of adhesion between the atoms. The force of interatomic interaction is characterized by a value of surface tension, which is known to be the growth of the free energy of the system caused by partial disruption of bonds between the atoms located on the surface.

Therefore, if we ignore the hydrostatic pressure, the ratio of parts of the bubble volume in the slag and the metal will be in inversely proportional to the ratio of values of surface tension of the contacting phases, that is,

$$\frac{V_m}{V_s} = \frac{\sigma_{s-g}}{\sigma_{m-g}} \tag{7.4}$$

Here, V_m and V_s are the parts of volume of a gas nucleus in the metal melt and slag, respectively; σ_{s-g} and σ_{m-g} are the surface tensions of the metal and the slag, respectively.

Let as assume for simplicity that the shape of the nucleus of a gas bubble formed at the metal–slag interface is spherical (Figure 7.5). In the formation of such a bubble at the interface of the molten metal and the slag, at constant temperature and pressure, the value of variation in the free energy of the system will be as follows:

$$\Delta G = a\pi r^2 \sigma_{m-g} + b\pi r^2 \sigma_{s-g} - c\pi r^2 \sigma_{m-s} + \Delta G_{vol}, \tag{7.5}$$

where $a\pi r^2 = S_{m-g}$ and $b\pi r^2 = S_{s-g}$ are the areas of contact of the nucleus with metal and slag, respectively; $c\pi r^2 = S_{m-s}$ is the area of contact between the metal and the slag that disappeared; a, b, and c are the constant coefficients for each particular case. In addition, $a + b = 4$, because the total area of the sphere is equal to $4\pi r^2$ and $c = r_1^2/r^2 = \cos^2\alpha$.

Provided the gas bubble nucleus is formed from one gas, and that its chemical potentials in the slag and the metal are identical, the values of r_{cr} and ΔG_{max} can be found from the following expressions[445]:

$$r_{cr} = \frac{(a\sigma_{m-g} + b\sigma_{s-g} - c\sigma_{m-s})V_m}{2kT\ln x_\Sigma}, \tag{7.6}$$

$$\Delta G_{max} = 3W_{het} - 0.523\frac{(a\sigma_{m-g} + b\sigma_{s-g} - c\sigma_{m-s})^3 V_m^3}{k^2 T^2 (\ln x_\Sigma)^2}, \tag{7.7}$$

where W_{het} is the energy of formation of the new phase in a heterogeneous environment determined from the following formula:

$$W_{het} = (S_{m-g}\sigma_{m-g} + S_{s-g}\sigma_{s-g} - S_{m-s}\sigma_{m-s})/3, \tag{7.8}$$

x_Σ is the total mole fraction of gas in the metal and the slag.

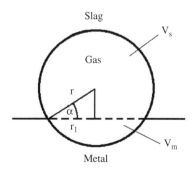

FIGURE 7.5 Diagram of formation of a gas bubble nucleus at the molten metal–slag interface.

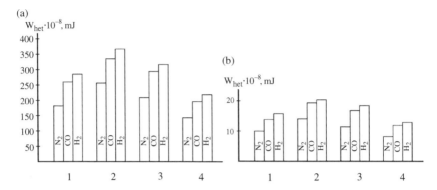

FIGURE 7.6 Values of the energy of formation of a stable bubble nucleus from different gases at the metal (Sv-08 steel)–slag interface ($T = 1803$ K). C/C_S: (a) 1.1; (b) 1.5. The composition of the slag (%) are; [1] 40 SiO_2 60 CaO; [2] 32 SiO_2 68 FeO; [3] 30 SiO_2 70 MnO; [4] 75 SiO_2 25 Na_2O.

It follows from Equation (7.7) and Equation (7.8) that formation of gas nuclei at the metal–slag interface depends on the surface properties of the metal and the slag, the value of interphase tension at the boundary between them, as well as on the gas content of metal and slag.

Values of the energy of formation of a stable gas nucleus at the interface between steel Sv-08 and slags of different composition are given in Figure 7.6. The required values of σ_{s-g} for the calculations were taken from reported studies,[90,292,446] and those of σ_{m-g} and σ_{m-s} were taken from Table 3.2 and Table 3.7. As the value of σ_{s-g} depends but slightly on the gas composition of the gas, it was necessary to simplify the calculations by assuming that the value of σ_{s-g} was constant in the formation of the gas nucleus from nitrogen, hydrogen, or CO. As seen from Figure 7.6, the energy of formation of the gas nucleus at the interface between the molten metal and slag is 2–3.4 times as low as the energy of formation of a similar nucleus in the bulk of the pure metal, the oversaturation values being identical. In addition, the energy of formation depends on the slag composition. CaO present in slag in an amount, that exceeds the σ_{m-s} value, makes formation of the gas nucleus easier, whereas FeO and MnO, which decrease the σ_{m-s} value, hamper the formation. An addition of Na_2O to the slag hardly changes the σ_{m-s} value, although it decreases σ_{s-g},[446] thus promoting also the formation of the gas nucleus at the metal–slag interface. The calculation data are in good agreement with the known experimental results. For example, according to the data of various studies,[1,31,447,448] the presence of Na_2O and CaO in the slag increases the porosity of the weld, whereas additions of MnO and FeO to the slag, as well as substitution of SiO_2 for CaO, lead to decrease of porosity.

7.3.2 The Influence of Electric Field

As noted previously, formation of nuclei of the gas bubbles at the metal–slag interface depends on the value of interphase tension at this interface and the gas content of the metal and the slag. Meanwhile, in the case of the submerged-arc and covered-electrode welding, as well as in electroslag welding, an external electric field is superposed on the metal–slag system, and the presence of this field can change the σ_{m-s} and x_{Σ} values and hence, affect the process of formation of pores in the weld.

As far back as 1951, it was established by D.M. Rabkin,[449] that in submerged-arc welding of low-carbon steel, the weld made at reverse polarity was less porous than that made at straight polarity. Similar facts were noted by other researchers as well, and they were related primarily to variations in the concentration of gas in the metal.

It is generally accepted now that nitrogen, hydrogen, and CO are the main gases, which cause porosity of the welds on steels. To our knowledge, no studies are available, dedicated to investigations of the effect of current polarity on the content of the above gases in the slag. What is known[29] is just that the current flowing through slag leads to increase of its nitrogen content. A large number of studies performed by different authors are dedicated to investigation into the relationship between the current polarity and the content nitrogen and hydrogen in the metal. However, since they consider primarily the processes, which occur in the metal–gas system, the established relationships must not be fully transmitted to the gas–slag–metal system. This is because in the gas–slag–metal system, a gas should first pass through the slag layer to penetrate into the metal.

Oxygen which participates in the formation of CO, as well as nitrogen and hydrogen are present in slag in the form of ions.[29] This allows interaction between gases dissolved in the slag and the metal to be regarded as an electrochemical interaction. Meanwhile, it is common knowledge that the rate and direction of electrochemical reactions are determined by the electrode (metal) potential, which depends on the amperage and direction of the current flowing. In this case, the flow of current is through the metal–slag interface. In this connection, superposition of an external electric field should either accelerate or decelerate occurrence of the electrochemical processes, depending on the electrode polarity. In otherwords, the applied electric field should change the gas content of metal, already proved by the experimental data.[450–454]

At present, the majority of researchers are of the opinion that hydrogen is contained in the slag in the form of ions OH^-, and that the transfer of hydrogen from the slag to the metal occurs[455] as a result of discharge of hydroxyl ions at the cathode:

$$(OH^-) + e = [H] + (O^{-2}).$$

Oxygen is contained in the slag in the form of ions O^{-2}, and its distribution between the metal and the slag can be described by the following diagram:

$$(O^{-2}) \rightleftharpoons [O] + 2e.$$

Opinions differ as to the form of nitrogen contained in the molten slag. Most often it is considered that nitrogen is present in the slag in the nitride N_3^- and cyanide CN^- forms, depending on the slag composition.[29]

In the gas–slag–metal system, transfer of gases to the metal melt occurs due to diffusion and convective flows. As shown by investigation of chemical composition of the slag crust,[456] it varies to a substantial degree through thickness, and this is indicative of a comparatively low stirring of the slag. Therefore, diffusion plays an important role in the processes of interaction of the slag with the metal.

The intensity of diffusion processes and the associated rate of occurrence of the above reactions increase with increase of the temperature of the molten slag, as this is accompanied by increase of mobility of the ions in the slag and current density through it. As the temperature of the slag at the interface with the molten electrode metal is always higher than that at the interface with the weld-pool metal, the total gas content of the weld metal should be determined primarily by the processes, which occur at the droplet stage. Therefore, one might expect that in straight-polarity welding, the hydrogen content of the weld metal would be slightly higher, and the nitrogen and oxygen content would be lower, compared to reverse-polarity welding. This is also proved by the experimental data[452–454] on the effect of polarity on the hydrogen, nitrogen, and oxygen contents of the deposited metal in sub-merged-arc welding and electroslag remelting.

The reaction of formation of CO is basically heterogeneous, and it might occur only if both carbon and oxygen are simultaneously present at the interface between the phases.

It was established,[457] that at a sufficiently high carbon content of the metal, the formation of CO takes place first at the metal–slag interface and then in the bulk of the metal, probably at the interfaces with oxide non-metallic inclusions. At a low content of carbon ([C] $\leq$ 0.1%) and other de-oxidizing elements in the metal, substantial amount of oxygen will go into the bulk of the metal, due to a low concentration of these elements in the surface layer of metal. This creates favorable conditions for the formation of CO at the interfaces with non-metallic inclusions. In this case, oxidation of carbon is determined mainly by the concentration of carbon at the interface and the rate of its delivery to the reaction zone.

It is well known,[458,459] that with DC flowing through a steel wire, the carbon content markedly decreases near the anode and increases near the cathode. It is likely that such changes in the concentration of carbon will take place also with the DC passing through a stationary metal melt. However, in the stirring of the metal melt, which is characteristic of the welding processes, the effect of the electric field on the transfer of carbon cations is less intensive. Therefore, the external electric field will hardly have a marked effect on the process of formation of CO under welding conditions. Nevertheless, as is seen from the above-said points, alternation of the current polarity does affect the gas content of metal. Besides, in straight-polarity welding, where porosity is higher, the amount of hydrogen in the weld metal grows to some degree, and the contents of nitrogen and the precipitating CO decrease, compared with reverse-polarity welding.

However, change in the current polarity affects not only the gas content of the metal, but also the value of surface tension at the metal–slag interface (see Section 3.6), which must lead to a change in the conditions for the formation of gas nuclei at the metal–slag interfaces.

7.3.2.1 Energy of Formation of the Gas Nucleus

To estimate the changes, that take place, let us determine the energy of formation of a stable gas nucleus at the metal–slag interface. In this approach, let us give allowance for a change in the value of the interface tension under the effect of the electric field, for which we will use Equation (7.8). Assuming, as before, that the gas phase has no effect on the value of surface tension of the slag, the nucleus has a spherical shape, and the V_m to V_s ratio is in inversely proportional to the ratio of surface tensions of the metal and the slag, let us additionally assume that the volume of a gas nucleus is equal to 1. This makes calculations simpler, however, the estimated values of the energy of formation of the gas nucleus will not be true. Nevertheless, the results obtained (Figure 7.7) make it possible to determine the nature of the effect exerted by the external electric field on the formation of the gas bubble nuclei at the metal–slag interface.

As is seen, when the metal has a positive charge, the energy of nucleation of bubbles at the interface between the metal and slags or liquid non-metallic inclusions consisting of $SiO_2–CaO$, $SiO_2–K_2O$, and $SiO_2–Na_2O$ is lower than that when a negative potential is applied to metal. In addition, this difference increases with increase of density of the current flowing through the metal–slag interface. If the slag consists of 60% SiO_2 and 40% MnO, the energy of nucleation of a gas bubble at the metal–slag interface will hardly depend on the polarity and current flowing through this interface.

Therefore, variation in the value of surface tension at the metal–slag interface with the electric current passing through it should affect the formation of pores in the weld.

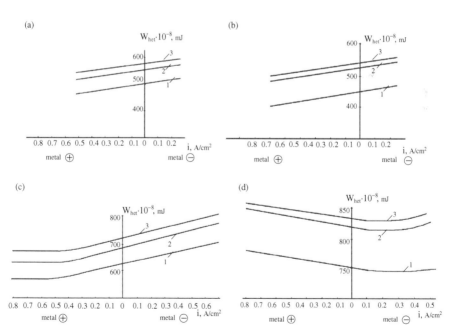

FIGURE 7.7 Change in the energy of nucleation of a stable gas bubble from different gases at the metal–slag interface depending upon the density and polarity of the current flowing through this interface ($T = 1803$ K). Gases: [1] N_2; [2] CO; [3] H_2. The composition of the slag (wt%): (a) 83 SiO_2 17 K_2O; (b) 83 SiO_2 17 Na_2O; (c) 50 SiO_2 50 CaO; (d) 60 SiO_2 40 MnO.

Relationship between the effect of the external electric field on surface tension in the metal–slag system and the pore formation process is proved also by the results of the following studies. It is well known,[1] that in A.C. welding, the amount of porosity of the weld takes an intermediate position between amounts of porosity in welding at straight and reverse polarities. Meanwhile, upon polarization of the metal–slag interface with A.C., σ_{m-s} will hardly change and also take the intermediate position between the σ_{m-s} values obtained on the metal at the positive and negative potentials.

7.3.3 THE INFLUENCE OF NON-METALLIC INCLUSIONS

In addition to the metal–slag and metal–molten non-metallic inclusion interfaces, the weld pool also contains interfaces between solid non-metallic inclusions and metal melt, as well as between the metal melt and crystalline grains.

Let us consider the formation of gas nuclei in the presence of solid non-metallic inclusions in the weld pool to reveal the mechanism of the effect of these inclusions on the process of pore formation.

It is apparent that formation of gas nuclei, their shape, and size will depend on the size of non-metallic inclusions and wettability of inclusions with the melt. Here three cases are possible: (1) the non-metallic inclusion is smaller than a critical gas nucleus and serves as its formation center (Figure 7.8); (2) the non-metallic inclusion is larger than the critical gas nucleus and is poorly wetted with metal melt, which leads to the formation on it of a gas nucleus in the form of a lens (Figure 7.8b); (3) the non-metallic inclusion which is larger than the critical gas nucleus, is well wetted with the melt, and thus, the nucleus has the form close to the spherical one (Figure 7.8c).

Let us consider the peculiarities of formation of the gas nuclei at the melt–solid non-metallic inclusion interface for each of the above cases.

Case 1. If in the formation of a gas nucleus, the temperature and external pressure remain constant, and the system is in equilibrium at the initial and final moments of time, a change in the free energy for the given system associated with formation of the nucleus can be described as follows:

$$\Delta G = (\sigma_{in-g} - \sigma_{in-l})4\pi r^2 + \sigma_{l-g}4\pi R^2 + \Delta G_p, \qquad (7.9)$$

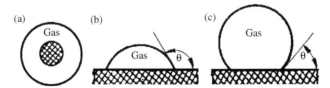

FIGURE 7.8 Formation of gas nuclei on solid non-metallic inclusions: (a) fine, poorly wetted with metal melt; (b) coarse, poorly wetted with metal melt: (c) coarse, well wetted with metal melt.

Here, σ is the specific surface energy; g and l are the subscripts designating an inclusion, gas, and molten metal, respectively; r is the radius of the inclusion; R is the radius of a gas nucleus; and ΔG_p is the change in the thermodynamic potential caused by the transfer of the gas from the molten metal to the nucleus.

It holds for the formation of the nucleus from one gas under the above conditions:

$$\Delta G_p = (\mu_2 - \mu_1)N, \tag{7.10}$$

where μ_1 and μ_2 are the chemical potentials of the gas in the melt and the nucleus, respectively, N is the number of moles of the gas in the nucleus determined from the following expression: $3\pi(R^3 - r^3)/4V_m$, where V_m is the mole volume of the evolved gas.

Chemical potentials of the gas in the melt and nucleus are, respectively, as follows:

$$\mu_1 - \mu_0 + R_g T \ln p_\infty$$
$$\mu_2 - \mu_0 + R_g T \ln p. \tag{7.11}$$

Here, μ_0 is the standard chemical potential; R_g is the gas constant; T is the absolute temperature; p is the gas pressure in the gas nucleus; p_∞ is the gas pressure over the flat surface of a liquid.

It is a known fact, that spontaneous transfer of particles from the melt to the gas nucleus is possible only when the chemical potential of a component in the molten metal is higher than in the nucleus, that is, μ_1 should be higher than μ_2. Allowing for the above-said facts, Equation (7.9) can be written as follows:

$$\Delta G = (\sigma_{in-g} - \sigma_{in-l})4\pi r^2 + \sigma_{l-g}4\pi R^2 - \left\{ \frac{[(\mu_1 - \mu_2)4\pi(R^3 - r^3)]}{3V_m} \right\},$$

Substitution of Equation (7.11) and transformations yield:

$$\Delta G = (\sigma_{in-g} - \sigma_{in-l})4\pi r^2 + \sigma_{l-g}4\pi R^2$$
$$- \left\{ \frac{[4\pi R_g T(R^3 - r^3)\ln(p/p_\infty)]}{3V_m} \right\}, \tag{7.12}$$

where $\ln p/p_\infty$ is the degree of gas oversaturation of the melt.

Replacing r by R/z ratio in Equation (7.12), where z characterizes the size of the inclusion and is a certain constant value for each particular case, Equation (7.12) will take the following form:

$$\Delta G = 4\pi R^2 \sigma_{l-g} + \frac{4\pi R^2(\sigma_{in-g} - \sigma_{in-l})}{z^2} - \left\{ \frac{[4\pi R^3 R_g T \ln(p/p_\infty)]}{3V_m} \right\}$$
$$+ \left\{ \frac{[4\pi R^3 R_g T \ln(p/p_\infty)]}{3V_m z^3} \right\}. \tag{7.13}$$

As is seen from Equation (7.13), ΔG is the function of the R value, and if $\Delta G_p < 0$, then ΔG will have a maximum at $R = R_{cr}$.[42] Therefore, to determine the value of R_{cr}, we will differentiate Equation (7.13), and equate the expression obtained to zero. Then, bearing in mind that $\sigma_{in-g} - \sigma_{in-l}/\sigma_{l-g} = \cos\theta$, where θ is the contact angle of a non-metallic inclusion, we will find that:

$$R_{cr}^1 = \frac{2\sigma_{l-g}(1 + \cos\theta/z^2)V_m}{R_gT(1 - 1/z^3)\ln(p/p_\infty)}. \qquad (7.14)$$

As can be seen, the value of R_{cr}^1 depends on the surface properties of the metal, σ_{l-g}; wettability of the non-metallic inclusion with the metal melt, $\cos\theta$; temperature of the system, T; oversaturation of the metal with the gas, $\ln(p/p_\infty)$; size of the non-metallic inclusion, z.

When the weld pool contains no non-metallic inclusions, that is, $r = 0$ or $z \rightarrow \infty$, Equation (7.14) will have the following form:

$$R_{cr}^1 = \frac{2\sigma_{l-g}V_m}{R_gT\ln(p/p_\infty)}, \qquad (7.15)$$

which is true for the case of formation of gas nuclei in the homogeneous environment.

Case 2. Assume for simplification, that the gas nucleus is formed on a flat surface. In this case, a change in the free energy of the system will be determined by the following equality:

$$\Delta G = (\sigma_{in-g} - \sigma_{in-l})\,\pi R^2 \cos^2(\theta - 90) + 2\pi R^2[1 - \sin(\theta - 90)]\sigma_{l-g}$$
$$- \{(\pi R^3 R_gT(2 - 3\sin(\theta - 90) + \sin^3(\theta - 90)]\ln(p/p_\infty)\}/3V_m. \qquad (7.16)$$

As $\Delta G = f(R)$ and $\Delta G_p < 0$, by differentiating Equation (7.16) and equating the resulting expression to zero, we can find that

$$R_{cr}^{11} = \frac{2\sigma_{l-g}V_m[2 + \cos^2(\theta - 90)\cos\theta - 2\sin(\theta - 90)]}{R_gT[2 - 3\sin(\theta - 90) + \sin^3(\theta - 90)]\ln(p/p_\infty)}. \qquad (7.17)$$

Case 3. In the formation of the gas nucleus on the surface of a coarse non-metallic inclusion, which is well wetted with the metal melt, a change in the free energy of the system will be equal to:

$$\Delta G = 2\pi R^2[1 + \sin(90 - \theta)]\sigma_{l-g} + (\sigma_{in-g} - \sigma_{in-l})\pi R^2 \cos(90 - \theta)$$
$$- \{\pi R^3 R_gT[2 + 3\sin(90 - \theta) - \sin^3(90 - \theta)]\ln(p/p_\infty)\}/3V_m. \qquad (7.18)$$

As before, by differentiating Equation (7.18) and equating it to zero, we find the value of a critical radius of the gas nucleus for the given case:

$$R_{cr}^{111} = \frac{2\sigma_{1-g}V_m[2 + 2\sin(90 - \theta) + \cos^2(90 - \theta)\cos\theta]}{R_g T[2 + 3\sin(90 - \theta) - \sin^3(90 - \theta)]\ln(p/p_\infty)}. \tag{7.19}$$

The last equation is valid also for the case of formation of a gas nucleus on the surface of a growing crystal, as it will be well wetted with the metal melt. Analysis of the resulting equations, namely, Equation (7.14), Equation (7.17), and Equation (7.19) is indicative of the following fact: with improvement in wettability (of a solid surface with the melt) the energy of formation of a stable gas nucleus at the interface becomes closer to the energy of formation of the nucleus in the homogeneous environment.

As the decrease of R_{cr} leads to facile of formation of the gas nucleus, let consider how the individual factors affect the value of R_{cr}. The results of solving Equation (7.14) at constant values of $\sigma_{m-g} = 1000 \text{ mJ/m},^2$ $T = 1803 \text{ K}$, $V = 22.4137$ m^3/mole and $R_g = 8.314 \text{ J/mole K}$ are shown in Figure 7.9. As can be seen, the worse the melt wets the fine non-metallic inclusion, the more readily is the gas nucleus formed on the surface of an inclusion of the same size. Increase of over-saturation of the metal melt with a gas, which forms a stable nucleus, leads to greater efficiency (Figure 7.10) of influence of non-metallic inclusions of any size on the pore-formation process.

7.3.3.1 Important Calculations

Calculations conducted in accordance with Equation (7.17) and Equation (7.19) at the same values of σ and T showed that for any θ, in the formation of nuclei on the surface of coarse non-metallic inclusions, their radius remains equal to that of

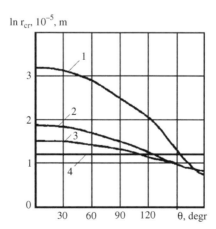

FIGURE 7.9 Dependence of variation in r_{cr} of the gas bubble nucleus upon the contact angle of wetting of a non-metallic inclusion with the metal melt. $C/C_s = 1.1$. Z: (1) 1.1; (2) 1.5; (3) 2.0; Homogeneous nucleation is represented in (4).

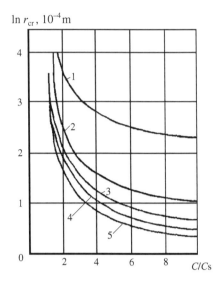

FIGURE 7.10 Dependence of r_{cr} of the gas bubble nucleus formed on a non-metallic inclusion upon the degree of oversaturation of the metal melt with a gas. Z: (1) 1.1; (2) 1.5; (3) 2.0; (4) 3.0. Homogeneous nucleation is represented in 5.

the nucleus formed in the homogeneous environment. However, when the volume of these nuclei and hence, the required quantity of gas molecules and energy of their formation will be smaller than those of the nuclei formed in the homogeneous environment (Figure 7.11). The presence of non-metallic inclusions in the weld pool promotes formation of pores also due to decrease of strength of the molten metal. It

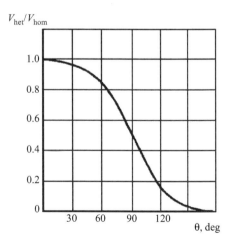

FIGURE 7.11 Dependence of the ratio of volumes of stable gas nuclei formed on a non-metallic inclusion, V_{het} to those in a homogeneous environment, V_{hom} upon the contact angle value.

follows from general equations of mechanics of liquids, that the distribution of pressures in a flow depends on the distribution of velocities at the steady-state motion of an incompressible liquid. In this case, the pressure may have even a negative value, provided the velocities are sufficiently high. As metal melts are incapable of taking up tensile stresses, formation of a negative pressure in some regions in the bulk of the weld pool should lead to violation of continuity of the flow and formation of gas cavities.

Extremely high pressures are required to violate continuity of a homogeneous liquid. However, liquids, including melts, always contain various impurities. The presence of soluble impurities, such as gases, in the melt hardly has any effect on the strength of the liquids, according to the theory of solutions. The presence of the finest bubbles of gases not dissolved in the metal, and especially non-metallic inclusions is the cause of decrease of strength of the melt.

According to the data by V.A. Zhuravlev, the strength of a homogeneous melt is determined as follows:

$$P_{\text{hom}} \approx P_{\text{s}} - 4 \left[\frac{\pi \sigma_{\text{m-g}}^3}{3kT\ln(A\tau)} \right]^{1/2}$$

and at the melt–non-metallic inclusion interface it is as follows:

$$P_{\text{het}} \approx P_{\text{s}} - 2 \left[\frac{\pi \sigma_{\text{m-g}}^3 (2 - \cos\theta)(1 + \cos\theta)^2}{3kT\ln(A'\tau)} \right]^{1/2},$$

where P_{s} is the pressure of the saturated vapors; A is the frequency factor; $A = 10^{36}\,\text{sec}^{-1} \cdot \text{cm}^{-3}$; $A' = 10^{24}\,\text{sec}^{-1} \cdot \text{cm}^{-3}$; τ is the mean time of expectation to formation of the first nucleus, k is the Botzmann's constant.

Therefore, the presence of non-metallic inclusions in the weld pool promotes formation of the gas nuclei also due to decrease of strength of the melt. As follows from the above formulas, the worse the melt wets the inclusion, the lower the strength of the metal melt, and the more readily the formation of a gas nucleus takes place.

The obtained theoretical data on the effect of non-metallic inclusions on the process of pore formation are confirmed experimentally.

Studies on surfacing of steel parts in shielding gas atmospheres with a different oxidation potential showed, that increase of the number of non-metallic inclusions led to increase of porosity of the deposited metal. The pores very often contact non-metallic inclusions, and at the interface with molten inclusions, they grow toward the inclusions (Figure 7.12).

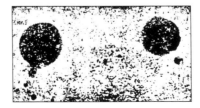

FIGURE 7.12 Appearance of pores contacting with a non-metallic inclusion.

7.3.4 OTHER SALIENT FEATURES

Our studies on the effect of solid non-metallic inclusions on the formation of pores in the weld showed that with the presence of corundum particles in the weld pool, the porosity was higher in welding of low-carbon steels, having a higher value of the contact angle; the porosity was lower in welding chrome–nickel steels, having a lower value of the contact angle. However, this difference is negligible. In the welding of low- and high-carbon materials, which have almost identical values of the contact angle, the porosity is higher (Figure 7.13) for high-carbon materials,

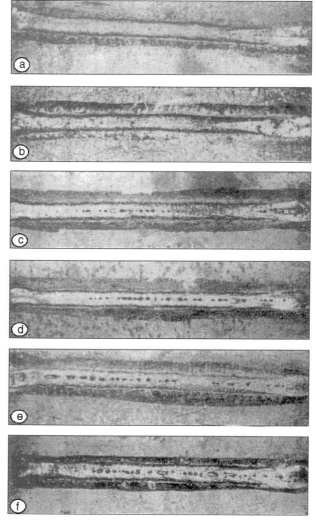

FIGURE 7.13   Appearance of beads deposited with wires: Sv-08 (a, b), 65G (c, d) и U8A (e, f) on steel St3 at oxide inclusions $0.5 \cdot 10^{-3}$ m (a, b,c), and $1.5 \cdot 10^{-3}$ m (b,d,f) in size.

and it grows with increase of sizes of the inclusions. This is attributable to a higher degree of oversaturation of the gas at the metal melt–inclusion interface, which is proved, in particular, by the data of Yavojsky[460] on an increased concentration of hydrogen near oxide and sulfide non-metallic inclusions. Increase of gas oversaturation at the interface with a non-metallic inclusion may be due to either a physical interaction of the non-metallic inclusion with the gases dissolved in the metal, or reactions occurring at the metal melt–inclusion interface.

7.3.4.1 Forces of Physical Interactions

Forces of physical interaction can be estimated from the value of dielectric constant of a non-metallic inclusion.[461] According to Mazurin,[462] at 1900 K, this value is equal to 15.4 for MgO, 3.8 for SiO_2, and 9.2 for Al_2O_3, that is, the interaction of gases dissolved in the metal will be most intensive at the interface with particles of MgO and least intensive at the interface with particles of SiO_2.

However, since physical interaction of gases dissolved in the metal and non-metallic inclusions is identical both for low-carbon and high-carbon steels, the noted growth of porosity in the welding of high-carbon materials cannot be explained by the physical interaction of inclusions with gases dissolved in the metal. It is likely that an increased porosity of the weld in welding high-carbon materials is determined first of all by a local oversaturation resulting from the formation of CO during the following reaction:

$$y[C] + (Me_xO_y) = x[Me] + y\{CO\}.$$

The possibility of occurrence of this reaction in the weld pool is confirmed by thermodynamic calculations and experimental results[439] of studies of the composition of gases in the pores formed at the interface with oxide non-metallic inclusions, as well as distribution of carbon at the interface with such pores. It should be taken into account that Al_2O_3 is a very stable oxide, whereas with the inclusions of SiO_2, MnO, and FeO present in the weld pool, the reactions of formation of CO will be more intensive.

To reveal the role of the crystal–metal melt interface in the pore-formation process, we will find relationship between the energy of nucleation of a gas bubble at this interface (W_{1-c}) and the energy of formation of a nucleus of the same volume in the melt (W_{hom}) using the Volmer's equation[463]:

$$\frac{W_{1-c}}{W_{hom}} = \frac{1}{2} - \frac{3}{4}\cos\theta + \frac{1}{4}\cos^3\theta, \qquad (7.20)$$

where θ is the contact angle.

It follows from the condition of equilibrium of the bubble formed on the crystal surface (Figure 7.14), that

$$\cos\theta = \frac{(\sigma_{s-1} - \sigma_{sol-g})}{\sigma_{1-g}},$$

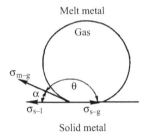

FIGURE 7.14 Formation of a gas bubble nucleus on the surface of the crystal.

where σ_{sol-g} and σ_{l-g} are the surface tensions of solid and liquid metals, respectively, and σ_{sol-l} is the surface tension at the liquid metal–crystal interface.
The value of σ_{sol-g} can be estimated from the following Equation[464]:

$$\sigma_{sol-g} = \sigma_{l-g}(\rho_s/\rho_l)^{2/3}(\gamma/\lambda),$$

where ρ_s and ρ_l are the densities of solid and molten metal, respectively, γ is the sublimation heat, and λ is the evaporation heat.

Let us assume for the calculations, that the values of ρ_s, ρ_l, γ, and λ are equal to the corresponding values for pure iron. The value of σ_{sol-l} can be determined by calculations using one of the formulas given in Section 1.4. However, the σ_{sol-l} values found from these formulas often differ from each other and from the experimentally found values by a factor of 2–5. Therefore, we will use the σ_{sol-l} value found experimentally, 204 mJ/m² for the case of iron liquid–crystal system to find $\cos\theta$.

As σ_{sol-l} is much lower than σ_{sol-g}, $\cos\theta$ has a negative value in all of the cases under consideration. Substituting the obtained values of $\cos\theta$ into Equation (7.20) yields that W_{l-c} is just a slightly lower than W_{hom} at identical oversaturations of the metal melt with gas. At the same time, the energy required for the formation of a stable gas nucleus at the metal melt–solid oxide interface (W_{l-so}) is no more than 5–12% of the energy required for the formation of a nucleus of the same volume in pure metal or at the molten metal–crystal interface (Table 7.2).

Therefore, of all the interfaces existing in the weld pool, the metal melt–solid oxide interface offers the most favorable conditions for the formation of stable nuclei of the gas bubbles. This is proved, in particular, by the results of our experiments and data obtained by other authors.

7.4 GROWTH OF GAS BUBBLES IN THE WELD POOL

The presence of coarse macro pores in the weld can be explained by three processes: (1) coalescence of micro bubbles contained in the weld pool; (2) evolution of gas cavities, which were present in the base metal, and in multilayer welding — in the previous layer, caused by pressure of gases contained in them; and (3) growth of nuclei of the gas bubbles as a result of diffusion of gases into them from metal.

TABLE 7.2
Ratios of Energies of Formation of Stable Gas Nuclei from Different Gases in a Homogeneous Environment and at the Interfaces of Molten Metal with Crystal and Solid Oxide

| Metal | W_{1-c}/W_{hom} (Numerator) and W_{1-so}/W_{hom} (Denominator) Ratios for Formation of Nucleus from Gases | | |
	Nitrogen	Carbon Oxides	Hydrogen
Sv-08	0.991	0.994	0.005
	0.05	0.07	0.08
Sv-10GS	0.99	0.99	0.919
	0.052	0.052	0.06
U8A	0.087	0.996	0.997
	0.053	0.07	0.1

7.4.1 THERMODYNAMIC CONSIDERATIONS

Let us consider peculiarities of each of these processes, and first of all the possibility of coalescence of the gas bubbles. It is a known fact, that the pressure of gas inside a bubble is larger than the pressure of the liquid, in which it is contained, by a value of $\Delta P = 2\sigma_{m-g}/r$. Therefore, if the melt contains two bubbles with radii r_1 and r_2, an increase of pressure for the first of them will be $\Delta P_1 = 2\sigma_{m-g}/r_1$ and that for the second will be $\Delta P_2 = 2\sigma_{m-g}/r_2$. In coalescence of such bubbles, the pressure inside a newly formed bubble due to the surface tension forces will be $\Delta P_\Sigma = 2\sigma_{m-g}/r_{12}$, where r_{12} is the radius of this bubble.

Change in the free energy of the dG system as a result of coalescence of the bubbles at constant P and T is

$$dG = \frac{4\pi\sigma_{m-g}(r_{12}^2 - r_1^2 - r_2^2)}{3}.$$

If the volume of the newly formed bubble, V_{12} is equal to the sum of volumes of the coalesced bubbles, that is, $V_{12} = V_1 + V_2$, then $r_{12} < r_1 + r_2$, and the expression in brackets will have a negative value. Thus, coalescence of two bubbles into one makes the system transform into a better equilibrium state. Therefore, it may occur spontaneously.

However, the process of coalescence is determined in many respects by the energy of interaction of the gas bubbles and strength of a molten film located between the approaching bubbles. Strength of the film depends on the kinetic, thermodynamic, and structural–mechanical factors acting in a way similar to the case of coalescence of non-metallic inclusions.

Disjoining pressure formed between the bubbles in their approach to each other, in contrast to the disjoining pressure in the approach of non-metallic inclusions to each other is equal to $P_d = \sigma_{m-g} (1/r_1 + 1/r_2)$. Moreover, since the presence of oxygen in the metal leads to formation of a electric double layer at the metal–bubble interface, it is apparent, that addition of oxygen to the metal melt will prevent coalescence of the gas bubbles. It is likely, that increase of thermodynamic stability of molten films is one of the factors, which explains the presence of a large number of fine pores in the weld metal during welding in oxidizing atmospheres.

In the case of collision of two bubbles of different sizes, the excessive pressure in a large bubble with radius r_1 will be lower than in a small one, because $r_1 > r_2$. Therefore, the film of the melt between the colliding bubbles will bend inside the large bubble as long as it can withstand this pressure. Thus, the probability of joining of the bubbles depends on the difference in the pressures:

$$\Delta P = \Delta P_2 - \Delta P_1 = 2\sigma_{m-g}\left(\frac{1}{r_2} - \frac{1}{r_1}\right) = 2\sigma_{m-g}\left[\frac{(r_1/r_2) - 1}{r_1}\right].$$

So, increase of the ratio of sizes of the bubbles that collided, leads to the growth of pressure acting on the film and, hence, it raises the probability of their coalescence. The probability of coalescence of bubbles of the same size is much lower.

7.4.2 Effect of Oscillations and Stirring

In addition to the above points, it should be noted that coalescence of the gas bubbles must grow with application of ultrasonic oscillations to the weld pool or electromagnetic stirring of the weld-pool metal, as in both cases, the film loses its stability.

Consider the following possible mechanisms of growth of the gas bubbles in the weld pool. There may be the case in welding, where the metal melted has already a gas cavity which was formed earlier. When such a cavity gets into the zone affected by the welding heat source (arc, electron beam, gas flame, etc.), the gas pressure in the cavity will grow due to an increase of temperature, thus leading to its growth. Meeting of the following condition is imperative for growth of the gas bubble:

$$P_{ex} \leq P_{in} - \frac{2\sigma_{m-g}}{r}, \tag{7.21}$$

where P_{ex} and P_{in} are the external and internal pressures acting on the wall of the bubble.

Under the fusion welding conditions, the external pressure preventing the growth of the gas bubble is made up of the atmospheric pressure P_{atm}, hydrostatic pressure and pressure caused by viscosity of the melt, $P_{vis} = (4\eta_m/r)\,dr/d\tau,$[380] while with the arc welding methods this includes also the arc pressure P_a.

Provided that the growth of the gas bubble occurs at a constant temperature, the value of P_{in} can be found from the $P_v V_0 = P_\tau V_\tau$ relationship, where P_v and V_0 are the gas pressure in the bubble and volume of the bubble at $\tau = 0$, and P_τ and V_τ are the gas pressure in the bubble and volume of the bubble at $\tau > 0$.

As $V_0 = 4\pi r_0^3/3$ and $V_\tau = 4\pi r_\tau^3$, then $P_\tau = P_v r_0^3/r_\tau^3$. In this case, the hydrostatic pressure being ignored, Equation (7.21) can be written in the following form:

$$\frac{dr}{d\tau} = \frac{[(P_v r_0^3/r_\tau^3) - P_{atm} - P_a - 2\sigma_{m-g}/r_\tau]r_\tau}{4\eta_m}. \tag{7.22}$$

The above equation is valid for the arc welding methods. In electron beam, flame, and electroslag welding, this equation will not contain the P_a value. In electron beam welding, this equation will have the following form:

$$\frac{dr}{d_\tau} = \frac{[(P_v r_0^3/r_\tau^3) - 2\sigma_{m-g}/r_\tau]r_\tau}{4\eta_m} \tag{7.23}$$

and the gas cavity will grow at the highest rate.

Note that with the arc welding methods the arc pressure will have a marked effect on the growth of the gas cavity, if it is located within the active spot zone.

Equation (7.22) and Equation (7.23) make it possible to determine the instantaneous rate of growth of the gas cavity which, as is seen from these equations, depends first of all on the viscosity of the metal melt.

7.4.3 DIFFUSION PHENOMENA

However, growth of the bubble due to diffusion of gases dissolved in the metal melt into this bubble is more characteristic of the welding processes. In this case, excess of the internal pressure over the external one, resulting from increase of mass of gas inside the bubble, can be found from the following form of the Mendeleev–Clapeyron equation:

$$\Delta P = \rho_g RT/\mu_g, \tag{7.24}$$

where ρ_g is the gas density; μ_g is the molar mass of the gas; R is the gas constant; T is the absolute temperature.

Solving jointly Equation (7.21) through Equation (7.24), assuming that μ_g, η_g, σ, T, R_{atm}, r_{cr}, and P_v are constantants, adding the constants $A_1 = 2\mu_g\sigma/RT$, $B_1 = 4\mu_g\eta_m/RT$, $C_1 = r_{cr}^3 P_v\mu_g/RT$, and $H_1 = \mu_g(P_{atm} + P_a)/RT$, and bearing in mind that the mass of the gas in the bubble is $m = 4\pi r^3\rho_g/3$, yield the following expression to describe the variation in the mass of the gas in the bubble with time:

$$dm/d\tau = \left(\frac{-4\pi r^2 H_1 dr}{d\tau}\right) - \left(8\,\pi r A_1 \frac{dr/d\tau}{3}\right)$$
$$- \frac{[8\,\pi r B_1(dr/d\tau)^2]}{3} - \frac{[4\,\pi r^2 B_1 d^2 r/d\tau^2]}{3}. \tag{7.25}$$

If variation in the mass of the gas in the bubble occurs only due to its diffusion from the metal, increase of mass of this gas in the bubble is equal to decrease of its mass in the metal[465]:

$$\frac{dm}{d\tau} = -DS\left(\frac{dC_g}{dx}\right),$$

where D is the gas diffusion coefficient, S is the surface, through which mass transfer takes place; dC_g/dx is the gradient of the concentration of gas dissolved in the melt in a direction normal to the transfer surface.

Variation in the concentration of the dissolved gas with time in the direction x in the absence of convective components of the flow can be determined from the $\partial C_g/\partial \tau = D(\partial^2 C/\partial x^2)$ equation, according to Fick's second law. $\partial C_g/\partial \tau = 0$ for a stationary process, then $\partial^2 Cg/\partial x^2 = 0$. The following equality holds for diffusion of a material through the film:

$$C_g = \left[\frac{(C_g' - C_g'')x}{\delta_1}\right] + C_g'',$$

where C_g' and C_g'' are the concentrations of the gas dissolved in the bulk of the melt at distance δ_1 from the phase boundary and at the interface, respectively, and δ_1 is the thickness of the diffusion layer.

Hence, the concentration gradient is $dC_g/dx = (C_g' - C_g'')/\delta_1$ and

$$dm/d\tau = -D4\ \pi r^2(C_g' - C_g'')/\delta_c. \tag{7.26}$$

Experimental data[466–468] indicate that variation in the radius of the gas bubble growing in the liquid is proportional to the square root of time, that is, $r = \phi(\sqrt{\tau})$. Thickness of the diffusion layer varies with time in the same way: $\delta_l = \phi(\sqrt{\tau})$.[469] Therefore, $r = k\delta_1$, where k is a number, which is a constant for the given case. Assuming,[298] that $k = 1$ and setting Equation (7.25) and Equation (7.26) equal to each other, we will have:

$$-\left[\frac{B_1 r(d^2 r/d\tau^2)}{3}\right] - \left[\frac{(2B_1(dr/d\tau)^2)}{3}\right] - \left[H_1 r + \frac{2A_1}{3}\right]\frac{dr}{d\tau} = -D(C_g' - C_g''),$$

or, designating $D\ (C_g' - C_g'')$ through K_1 and multiplying both parts of the equation by -1, we will have:

$$\left[\frac{B_1 r\ (d^2 r/d\tau^2)}{3}\right] + \left[\frac{2B_1(dr/d\tau)^2}{3}\right] + \left[H_1 r + \frac{2A_1}{3}\right]\frac{dr}{d\tau} = K_1. \tag{7.27}$$

Note that in this equation, H_1 value will vary depending on the welding method employed. For arc welding methods, $H_1 = \mu_g(P_{atm} + P_a)/RT$; in gas and laser beam welding, $H_1 = P_{atm}\mu_g/RT$; for electroslag welding, $H_1 = \mu_g(P_{atm} + P_{hs})/RT$, where $P_{hs} = \rho_s h_s$ is the hydrostatic pressure of the slag pool, which depends on the slag-pool depth, h_s and slag density, ρ_s.

Let us use an approximated method based on the Taylor's formula to solve Equation (7.27). By breaking the time interval under consideration into a sufficiently large number of equal parts, we will solve Equation (7.27) using numerical data of Gorshkov[470] and the value of the critical radius of the gas bubble nucleus, r_{cr} equal to 10^{-6} m. Results of the calculations thus made are indicative of the fact that the viscosity forces have almost no effect on the diffusion growth of the gas bubble. This coincides with the data of earlier studies.[471,472] The effect exerted by the surface tension forces is also very low (Figure 7.15). Even an unrealistic four-fold variation in the σ_{m-g} value had an insignificant effect on the rate of growth of the bubble. The major effect on the rate of growth of the bubble in the metal melt is exerted by the degree of oversaturation of metal with the gas, external pressure and mass-transfer rate (Figure 7.16).

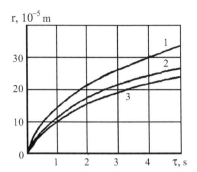

FIGURE 7.15 Variations in radius of a gas bubble with time. σ_{m-g} (mj/m^2): (1) 1510; (2) 6040; (3) 1510. $P_{atm} + P_a$ (MPa): (1) 0.103; (2) 0.103; (3) 0.206.

Meanwhile, it is a known fact, that an addition of surfactants to the melt has a marked effect on the mass-transfer rate. One of the causes of this effect may be adsorption of molecules on the interface, leading to reduction of the active contact surface or, in other words, to the growth of the surface resistance. Surface resistance, R_s is related to the concentration of surfactants on the bubble surface, C_s, through the following equation[373]:

$$\frac{5.4 \cdot 10^9 R_s}{[1 - \left(\frac{R_s}{7400}\right)]^{0.1}} = C_s,$$

Here

$$C_s = \frac{4 \cdot 10^{-5} A_0 \Gamma^\omega}{(1 - A_0 \Gamma^\omega)^{0.1}}.$$

Then,

$$\frac{1.35 \cdot 10^{14} R_s}{[1 - (R_s/7400)]^{0.1}} = \frac{A_0 \Gamma^\omega}{(1 - A_0 \Gamma^\omega)^{0.1}},$$

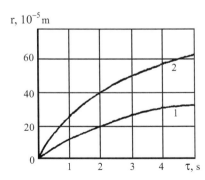

FIGURE 7.16 Variations in radius of a bubble with time at a gas content (g/cm^3) of metal: (1) $8 \cdot 10^{-6}$ (2) $32 \cdot 10^{-6}$.

where A_0 is the efficient cross-sectional area of adsorbed molecules; and Γ^ω is the concentration of molecules at the interface.

7.4.4 SURFACE VS. VOLUME CONCENTRATION OF COMPONENTS

Surface concentration of any element is related to its volume concentration, and this relationship is described by the following expression[473]: $\Gamma_i^\omega = F_i N_i / \Sigma F_i N_i$, where N_i is the volume concentration of a component, and F_i is the value, which characterizes surface activity of an impurity.

According to the data of Popel',[473] at a melt temperature of 1933 K, the value of F_i for different components is as follows:

Element	Surface Activity, *Fi*
Fe	1.0
C	2.0
Si	2.2
Mn	5.0
S	500
O	1000
H	1.0
N	150

Therefore, the rate of transfer of the gas from the melt to the bubble is affected to the highest degree by oxygen and sulfur. As noted earlier (Section 3.1), the simultaneous presence of several components in the melt may change the surface activity of each of them. For example, the presence of oxygen in the iron melt increases the surface activity of vanadium, phosphorus, and sulfur; the presence of carbon increases the activity of sulfur and manganese; and the presence of nitrogen increases the activity of carbon, silicon, and nickel, etc. Therefore, in multi-component systems, which are steels and many other alloys used in fabrication of welded structures, a blocking action of surfactants should show up to a greater degree than in the welding of relatively pure metals.

Ultrasonic oscillations applied to the weld pool, or electromagnetic stirring of the metal may affect the process of growth of the gas bubbles. In the distribution of ultrasonic oscillations in the weld-pool metal, the bubbles contained in it will make oscillatory movements, undergoing periodic contraction and expansion. These oscillations lead[474] to intensification of transfer of the gas from the melt to the bubble even when the solution is undersaturated with gas. The causes of this phenomenon are as follows. First, in the expansion of a bubble, its surface is larger than in contraction. That is why, the amount of gas entering into the bubble during expansion is larger than that leaving it during the contraction. Second, oscillations of the bubble lead to a periodic decrease of the thickness of the diffusion layer and increase of the concentration of the gas near the bubble. According to Equation (7.26), this should lead to an increase of mass of the gas entering into the bubble.

In addition, as shown by studies described in Kapustina,[474] micro-flows are formed at the surface of the bubble during oscillations, and these flows also lead to decrease of the thickness of the diffusion layer.

It is likely that intensification of degassing of the weld-pool metal seen in super-position of the external electric field is caused by a decrease of thickness of the diffusion layer and, probably, by the formation of the decreased pressure zones in the melt, where the gas bubbles have more favorable conditions for growth.

It should be noted that application of ultrasonic oscillations[474] and electromag-netic stirring of the metal melt increase the probability of collision of the gas bubbles, which favors their coalescence.

7.5 ESCAPE OF GAS BUBBLES FROM THE MOLTEN METAL

Gas bubbles formed and grown in the weld pool tend to escape to the melt surface, which is attributable to a difference in the metal and gas densities. The entire process of escape of a gas bubble from the weld pool may be subdivided into two stages: (1) approach of the bubble to the metal–gas or metal–slag interface, depending on the welding method employed; (2) transfer of the bubble through the interface.

Most often the movement of bubbles in a liquid is described by the well-known Stokes equation:

$$v = \frac{2(\rho_m - \rho_g)gr^2}{9\eta_m},\qquad(7.28)$$

where v is the velocity of movement of the bubble in the melt; g is the free fall accel-eration, ρ_m and ρ_g are the metal and gas densities, respectively; η_m is the viscosity of the metal melt; r is the bubble radius.

7.5.1 THE NEED FOR ADVANCED MATHEMATICAL TREATMENT

A relationship of the type shown in Equation (7.28) does not account for a number of factors characteristic of movement of bubbles in the weld pool. First of all, the Stokes equation yields sufficiently good results for floating up of the bubbles in a quiescent melt or for its laminar movement. According to Figurovsky,[475] increase of the intensity of stirring of the metal melt leads to increase of the error of esti-mation of the velocity, at which the bubble floats up in the case of using Equation (7.28). At Reynolds number Re $= 5.0$ this error amounts to 10%. Using this formula, the velocity of movement of the bubble can be calculated with a sufficient accuracy only at Re ≤ 2.0. For welding processes, especially when the weld-pool metal is subjected to intensive stirring, the use of the Stokes formula leads to incor-rect results. Also, it should be taken into account that Equation (7.28) is valid, if no circulation of gas takes place in an ascending bubble. However, actual conditions involve circulation of the gas in the bubble, which leads[274] to increase of the velocity of its movement. Note also that the Stokes formula can be applied for the description of the process of movement of spherical bubbles.

At the same time, an ascending bubble is affected by the friction forces and the liquid pressure gradient, which tend to deform the bubble, in addition to the surface tension forces, which tend to give it a spherical shape. Eventually, the shape of the bubble will be determined by the ratio of forces acting on it, the value of which depends probably on the size of the ascending bubble.

According to Nemchenko, Koz'min, and Popel',[476] the bubble will not deform in floating up, if its radius is $r = (324 \, \eta_{m}^2 \sigma_{m\text{-}g}/\rho_{m}^3 g^2)^{1/5}$. For steels, the shape of the ascending bubbles will remain spherical, if their radius is not in excess of $1\cdot10^{-4}$ m. Bubbles of a larger size will deform in floating up, and this deformation will be substantial, if $r \approx \sqrt{\sigma_{m\text{-}g}/\rho_{m}g}$.

As the weld may have pores, the sizes of which are by an order of magnitude larger, they may deform in stirring of the weld pool. Therefore, it seems necessary to use the Malenkov's formula[477] to describe movement of coarse deformed bubbles in the metal melt:

$$v = \sqrt{(\sigma_{m-g}g/\rho_{m}r_{e}) + gr_{e}},\qquad(7.29)$$

where r_e is the radius, which is equivalent to that of a spherical bubble.

Equation (7.29) is valid for the case of a turbulent floating up of bubbles having an elliptic or mushroom-like shape, which is also proved experimentally.[478] Comparison of values of the velocities of rising of the bubbles estimated from Equation (7.28) and Equation (7.29) shows, that deformation of a bubble leads to decrease of the velocity of its rising and hence, it causes an increase of the probability of the bubble being trapped in the weld metal.

The velocity of movement of a bubble in the metal melt will be affected by surfactants contained in metal, as well as by movement of the other bubbles. In the presence of the surfactants in the melt, the velocity of ascending of one bubble is as follows[479]:

$$v = \frac{v_{st}(3\eta_{m} + 3\eta_{g} + 3\gamma_{1})}{(2\eta_{m} + 3\eta_{g} + 3\gamma_{1})},\qquad(7.30)$$

where v_{st} is the velocity of ascending of the bubble estimated from the Stokes formula (Equation (7.28)),

$$\gamma_{1} = [2\Gamma\delta_{s}/3Dr]\frac{d\sigma_{m-g}}{dC};\qquad(7.31)$$

where Γ is the adsorption of a surface-active material, D is the molecular diffusion coefficient; C is the concentration of the surface-active material.

As $\eta_{g} \ll \eta_{m}$, viscosity of gas can be ignored. Then, Equation (7.30) becomes:

$$v = \frac{v_{st}[3(\eta_{m} + \gamma_{1})}{(2\eta_{m} + 3\gamma_{1})]}.\qquad(7.32)$$

Analysis of Equation (7.31) and Equation (7.32) shows that in the absence of surface-active materials in the melt or at their very low content ($C \rightarrow 0$) Equation (7.32) will have the following form:

$$v = \frac{(\rho_m - \rho_g)gr^2}{3\eta_m}. \tag{7.33}$$

Equation (7.33) is valid for the case of determination of the velocity, at which the bubble, that contains gas circulating in it, floats up.

If the content of surfactants in the melt is sufficiently high, the velocity of floating up of small bubbles should be described with a sufficient accuracy by the Stokes formula. This is associated with the fact, that in a liquid containing a substantial amount of surfactants, their adsorption may lead to the formation of films of a large thickness and high strength on the bubble–liquid surface. Formation of such films leads to the fact that bubbles start moving in a way similar to solid balls of a spherical shape, the movement of which in a liquid is described by Equation (7.28).

7.5.2 Providing Allowance for Many Bubbles

All the above expressions relate to the case of floating up of one bubble. Under actual conditions, several bubbles simultaneously move in the melt. In this case, each of the moving bubbles causes formation of flows which, while acting on the other bubbles, change their movement pattern. These changes depend in many respects on the shape and size of the weld pool.

In a system not restricted by walls, which is more characteristic of the electroslag welding process, where the weld pool is large in size, each of the bubbles while moving entrains a surrounding liquid with it. This leads to a decrease of resistance of the liquid and increase of the velocity of ascending of the bubbles. With a parallel movement of two spherical bubbles, the resistance to their movement, according to M. Smolukhovsky,[480] will decrease by $(9\pi r_1 r_2 \eta_m v \cos \gamma)/2\delta_b$ (where δ_b is the distance between the bubbles; γ is the angle between central lines and a direction of movement of the bubbles; r_1 and r_2 are the radii of the first and second bubbles).

In the movement of bubbles in a limited volume, the ascending bubbles involve a surrounding liquid into their movement, whereas the liquid located at a distance from them moves in an opposite direction. Therefore, in the case of movement of several bubbles in a limited volume, the velocity of their ascending will be lower than in the case of ascending of one bubble.

As noted earlier, the fusion welding systems are characterized by two types of interfaces through which the gas bubbles escape from the weld-pool transfer. These are the metal–gas and metal–slag interfaces. Let us consider first the process of transfer of the gas bubbles through the metal–gas interface.

As shown by experiments,[481] when a bubble approaches the free surface of a liquid, it causes the latter to bend. In this case, the velocity of ascending of the bubble decreases, while the liquid raised by the bubble flows down from the dome surface. Bending of the free surface of the liquid results in increase of the free energy of the system, which leads to a greater deviation of the system from equilibrium and decrease of its stability.

According to published data,[481] for steel, the film will remain stable, if its thickness is not less than $(0.01 - 0.02) \cdot 10^{-2}$ m, which is much larger than sizes of the molecules. Therefore, the model of polymolecular films suggested by De Fries seems suitable to describe the process of destruction of the metal film. According to this theory, destruction of the film begins from the formation of a hole in it which is energy consuming (Figure 7.17).

At the first moment of formation of a hole in the film, the surface of the system increases, and the value of this surface can be calculated from the following formula:

$$\Delta S = \pi^2 \delta_t (r_{min} + (\delta_t/2) - (\delta_t/\pi)) - 2\pi[r_{min} + (\delta_t/2)]^2,$$

where δ_t is the film thickness and r_{min} is the minimum hole radius.

The largest increase of the interface is achieved at $r = [(\pi/4) - (1/2)] \delta_t$. In this case:

$$\Delta S_{max} = [(\pi^3/8) - \pi]\delta_t^2 \approx 0.73\delta_t^2.$$

As increase of the free energy of the system in expansion of the hole, or the activation energy A, is equal to a product of ΔS by σ_{m-g}, then

$$A_{max} = \Delta S_{max}\sigma_{m-g} \approx 0.73\delta_t^2\sigma_{m-g}. \tag{7.34}$$

Therefore, the process of destruction of the film is affected to the largest degree by the thickness of this film and, hence by factors this thickness depends on.

When a bubble escapes to the surface, the melt located between the internal and external surfaces of the film will flow down to the weld pool, which will lead to a decrease of the film thickness.

Designate velocity of flowing of the melt at some point at a distance x from the film center as v_f. Gravity forces of the melt contained in between the two symmetric regions of the film, located at distance $2x$ from each other, are balanced by the viscosity forces[482]:

$$-\rho_m g x = \eta_m (\partial v_f/\partial x). \tag{7.35}$$

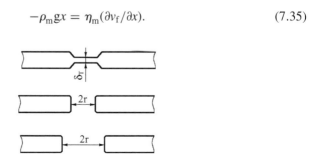

FIGURE 7.17   Formation of hole in the film according De Fries.

Integration of Equation (7.35) yields dependence of the velocity of flowing of metal on distance, x:

$$v_f = \rho_m g(\delta_t^2 - 4x^2)/8\eta_m. \tag{7.36}$$

As is seen, increase of viscosity and decrease of density of the melt lead to decrease of the velocity of flowing of the melt and hence, the rate of decrease of the film thickness.

7.5.3 DESTRUCTION OF THE FILM

Under welding conditions, the rate of destruction of the film, v_d has a special significance for the process of escape of the gas bubbles, which is attributable to a high rate of the process of weld formation. Assuming that the surface energy of the film fully transforms into the kinetic energy of the liquid, Dupre derived the following energy balance equation:

$$2\sigma_{m-g}S_f = mv_d^2/2 = S_f\delta_t\rho_m v_d/2,$$

where ρ_m is the density of the melt and S_f is the surface area of the film.

Hence, the rate of destruction of the film is as follows:

$$v_d = \sqrt{4\sigma_{m-g}/\rho_m\delta_t}. \tag{7.37}$$

However, the experimental results prove, that values of the rates of destruction of the film obtained by using Equation (7.37) are overestimated.

The process of destruction of the film can be described more accurately by the equation which was derived in the work of Mc Entee and Mysels[483] with an assumption, that the velocity of propagation of a circular wave, v_c, acting against the surface tension forces of the two-sided film, is a constant and equals:

$$v_c = \sqrt{2\sigma_{m-g}/\rho_m\delta_t}. \tag{7.38}$$

The difference in the values of the rates of destruction of the film determined from Equation (7.37) and Equation (7.38) can be explained as follows. It is known that destruction of the film, according to the De Fries diagram, begins from a round hole. A halo propagates ahead of the hole front in expansion,[483] this halo being a thickened part of the film. This leads to a loss in part of the kinetic energy, which may amount approximately to 50%.[484,485]

It follows from Equation (7.37) and Equation (7.38) that the activation energy in the expansion of the hole and the rate of destruction of the film depend on the values of surface tension of the melt and thickness of the film. The lower the surface tension of the melt, that is, in the case of surfactants present in the melt, the more marked is the decrease of values of the activation energy and the rate of the film destruction. The presence of surfactants in the metal melt makes the film more stable, thus hampering escape of bubbles from the weld pool into the atmosphere. This is proved by the results of experiments conducted on models,[486] which showed that in pure water, the film is destroyed immediately after the bubble has approached the surface. However, if a soap film is present on the water surface, the bubble persists on the surface for up to 10 sec.

7.5.3.1 Experimental Findings

As shown by investigations of the process of destruction of the gas bubbles on the surface of water using filming, destruction of the liquid film is accompanied by an outburst of fine liquid drops. Formation of these drops depends on the sizes of the bubbles escaping from the liquid. Thus, no outburst of liquid drops occurs in the destruction of bubbles up to $0.1 \cdot 10^{-3}$ m in diameter, whereas several hundreds of fine drops were formed in the destruction of bubbles more than $3 \cdot 10^{-3}$ m in diameter. Apparently, spattering of the film material may take place also on the weld-pool surface at the moment of destruction of the bubbles.

If we assume a gas contained in a bubble to be ideal, the following energy will release in the destruction of the bubble, that has escaped to the weld-pool surface:

$$A = (P_{atm} + P_\sigma)V_\tau \ln[1 + (P_\sigma/P_{atm})],$$

where V_τ is the volume of the escaping bubble, and $P_\sigma = 2\sigma_{m-g}/r$.

If we assume, that $P_{atm} = 0.1$ MPa and $\sigma_{m-g} = 1000$ mJ/m,^{2} as a result of destruction of the bubble with a radius of $r = 0.1 \cdot 10^{-3}$ m, the mass of metal equal to $0.41 \cdot 10^{-3}$ kg will acquire a velocity of 0.10 m/sec. Therefore, destruction of the gas bubbles escaping to the weld-pool metal – gas interface can be considered as one of the causes of formation of spatters on the surface of a part. This process depends on many factors. We studied[299] the effect of the welding current, arc voltage, presence of flux on the weld-pool surface, and surface properties of the weld-pool on sizes of the spatters and amount of the spattered metal.

The experiments were conducted as follows. Plates of low-carbon steel were melted for 20–22 sec. This was done using tungsten electrode with argon as a shielding atmosphere. Flux-cored wire 4 mm in diameter, filled with FeO or FeS, was added to the weld pool to change the value of surface tension of the metal. The wire was fed into the arc zone using an upgraded feeding mechanism of the ADSV-5 automatic welding device. The FeO and FeS content of the weld pool was varied by regulating the flux-cored wire feed speed. The effect of the welding current and arc voltage on spattering of the metal was studied by feeding only a sheath of the flux-cored wire to the weld pool. Specimens, 5 g in weight were cut from the weld metal after welding. The specimens were used to determine the value of the surface tension of metal, σ_{m-g}, by the sessile drop method at $T = 1803$ K in argon. After such melting of the plates, their surfaces were covered with spatters, the diameters of which were mostly not in excess of $0.8 \cdot 10^{-3}$ m, while the distance from the bead axis was $1 \cdot 10^{-1}$ m and more. All the spatters were collected and sieved into three groups: (1) with a diameter of more than 0.3 mm; (2) with a diameter of 0.2–0.3 mm; (3) with a diameter of less than 0.2 mm. Spatters with a diameter of less than 0.1 mm were ignored, as they consisted mostly of iron oxides. During the experiments the welding speed and gas-flow rate remained constant: $V_w = 0.0053$ m/sec; $Q_g = 17 \cdot 10^{-5}$ m^3/sec.

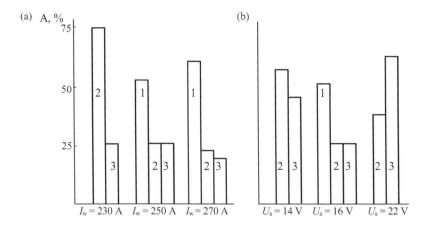

FIGURE 7.18 Size distribution of metal spatters depending upon (a) the welding current and (b) arc voltage. $U_a = 16$ V in (a); $I_w = 250$ A in (b); (1, 2, 3) group of spatters distributed in sizes (see in the text); A represents percentage of spatters of a given group.

The experiments showed that the amount and sizes of the spattered metal drops depend on the values of I_w and U_a (Figure 7.18). Increase of I_w and decrease of U_a lead to increase of size of the drops. Variations in the welding current affect the size of the spatters to a greater degree than variations in the arc voltage.

7.5.3.2 The Influence of FeS and FeO

Feeding of either FeO or FeS to the weld pool leads to decrease of metal spattering in addition to decrease of the σ_{m-g} value (Figure 7.19). This is attributable to the fact that decrease of surface tension is accompanied by increase of stability of the metal film,[482] reduction of its destruction rate,[483] and rise of the intensity of formation of gas nuclei which was discussed in Section 3.1. The latter leads to increase in the

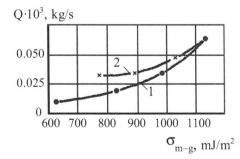

FIGURE 7.19 Influence of surface tension of the weld-pool metal ($T = 1803$ K) on the intensity of spattering of the weld-pool metal with the escaping gas bubbles. (1) Surface-active element: (1) oxygen; (2) sulfur $I_w = 250$ A; $U_a = 16$ V.

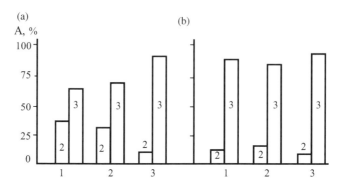

FIGURE 7.20   Effect of the welding current ($U_a = 16$ V) on sizes of spatters in the addition of oxygen (a) and sulfur (b) to the weld-pool metal. $I_w(A)$: (1) 230; (2) 250; (3) 270.

number of bubbles in the weld pool and, accordingly, to reduction of their sizes. As a result, the decrease of the σ_{m-g} value causes a decrease of the size of spatters (Figure 7.20). It should be noted that with sulfur added to the weld pool, the losses of metal and distance, to which spatters fly off, are larger than with an addition of oxygen. It is likely that decrease of spattering in the latter case results from the formation of the slag film on the weld-pool surface caused by the addition of FeO. This is proved by the data on the effect on spattering of the weld-pool metal by the thickness of the slag film (Figure 7.21).

Experiments on the investigation of the effect of the presence of flux on the losses of metal from the weld pool in arc welding were carried out as follows. The arc discharge was ignited between a low-carbon steel plate and a carbon electrode moving along the plate. The time, during which the arc acted on the plate, was $(16.4-23.9) \cdot 10$ sec. Experiments were conducted without and with a flux of the OSTs-45 grade. In the latter case, the thickness of the flux layer was $(3-6) \cdot 10^{-3}$ m, which provided the presence of a slag layer on the weld-pool surface.

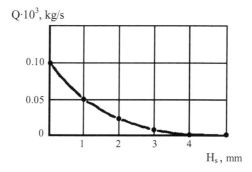

FIGURE 7.21   Effect of thickness of the slag layer present on the weld-pool surface on the weight of the metal spattered in a second $I_w = 270$ A; $U_a = 16$ V.

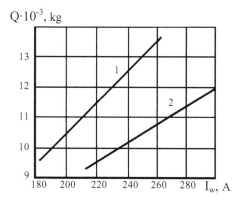

FIGURE 7.22 Dependence of weight of the spattered weld-pool metal upon the welding current in gas-shielded (1) and sumberged-arc (2) welding.

Losses of metal determined from the difference between the initial weight of the plate and that after welding and removal of spatters from it increased with growth of the welding current in welding with and without the flux (Figure 7.22).

In the presence of an oxide film or a slag layer on the weld-pool surface, a bubble, while transferring from the metal to the slag, involves a metal sheath with it, as shown by investigations conducted on models.[487] This sheath then transforms into a drop, which may descend into the weld pool, if during a time of its formation the slag and weld-pool metal have not yet solidified. The volume of the metal sheath carried away with the bubble is approximately twice as small as that of the bubble,[488] provided the distance between the bubble which floats up is not less than 4r.

It is likely that the process of transfer of bubbles from the metal melt to the slag depends on the relationship between the kinetic energy of an ascending bubble, energy of surface tension growing with increase of size of the dome (Figure 7.23), and the potential energy of the slag column and metal sheath of the dome. As the thickness of the metal sheath is very small, its effect can be ignored. Then the

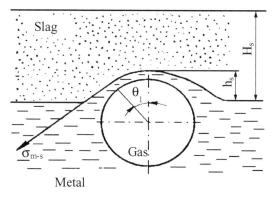

FIGURE 7.23 Transfer of a gas bubble through the metal–slag interface.

condition of transfer of the bubble from the metal to slag can be written as follows[299]:

$$(4r^3 \cdot \rho_g/6)[v_{st}(3\eta_m + 3\gamma_1)/(2\eta_m + 3\gamma_1)]^2$$

$$> [(r^2 \sin^2 \theta_1 + 2rh_{sl})\sigma_{m-sl}]$$

$$+ r^2 \sin^2 \theta_1 H_{sl} \cdot \rho_{sl} - [h_{sl}(3r^2 \sin^2 \theta_1 + h_{sl})\rho_{sl}/6],$$

where η_m is the viscosity of the metal melt, ρ_g and ρ_{sl} are the densities of the gas and slag, respectively, and σ_{m-s} is the tension at the metal–slag interface.

Increase of difference between the kinetic energy of the bubble and sum of the surface and potential energies raises the probability of transfer of the bubbles from the weld pool to the slag. Therefore, the velocity of transfer of the bubbles will increase with increase of their sizes and decrease of the values of σ_{m-s}, H_{sl}, η_m, and ρ_{sl}. Transfer of the bubbles from the metal to the slag seems to be favored by stirring of the weld-pool metal, as this leads to an increase of their kinetic energy.

It should be taken into account, that increase of the size of the dome lifted by a bubble will be prevented by the slag viscosity forces. Therefore, increase of the slag viscosity will probably hamper transfer of the bubble from metal to slag because of decrease of its ascending velocity.[489,490]

A gas bubble, which has escaped to the metal–slag interface and which is energetically incapable of transferring through it, may break up. In this case, the probability of its destruction, as in the case of destruction of the bubble at the metal–gas interface, will depend in many respects on the velocity of escape of the melt from the gap between metal and slag, which can be determined from the Equation (7.37).

Therefore, the escape of the gas bubbles from the weld pool and their destruction depend greatly on the stability of the film of the metal dome formed by the bubble at its escape to the interface. That is why, all the factors, which affect its stability, will also affect the process of escape of the gas bubbles from the metal melt. Thus, stirring of the weld-pool metal by electromagnetic fields, ultrasonic oscillations, and application of pulsed-arc welding promote the decrease of the porosity of the weld. In addition, stability of the metal film depends also on the spatial position of the welded joint. For example, in welding in the overhead or vertical position, the rate of flow of metal from the film is lower, compared to welding in the down-hand position, which is attributable to the effect of the gravity forces. In these cases, the kinetic energy of a bubble also decreases, as it depends on the presence and the rate of metal flows in the weld pool, especially in the overhead position. For these reasons, the degree of porosity of the welds made by welding in the overhead or vertical positions is higher than that in the downhand position.

So, surface properties of welding consumables and surface phenomena at the boundaries of the contacting phases characteristic of the welding systems affect the processes of formation of stable gas bubble nuclei in the weld pool, their growth, and escape from the metal melt. Consequently, they make it possible to control these processes. Therefore, it is necessary to know and allow for the surface properties of materials used for welding, and the various surface phenomena involed in welding so that one could eliminate or decrease the porosity of the welds or the deposited metal.

8 Solidification Cracking

As was noted in earlier chapters, cracks are the most dangerous defects of a welded joint. During fusion welding, the resulting welds may have either cold or hot cracks. A distinctive feature of hot cracks is their intercrystalline location, whereas cold cracks propagate mostly through the bulk of a crystalline grain, that is, they have a transcrystalline location.

Hot cracks are normally located in the weld metal and formed during solidification (solidification cracks) or at a temperature below the solidus temperature for a given metal (polygonization or subsolidus cracks). Formation of hot cracks has one factor in common, which is the presence of tensile stresses formed during the cooling of a welded joint. The main difference between them is that solidification cracks are formed in a solid–liquid metal, whereas the polygonization ones are formed in a solid metal. The solidification cracks are formed within a certain temperature range which, according to N.N. Prokhorov, is called the brittle temperature range (BTR).[511]

The value of BTR can be approximately determined from the constitutional diagram of a given alloy, assuming that increase of the solidification range is accompanied by increase of the effective solidification range and, hence, BTR. However, such an estimate is approximate, as it does not allow for the effect of impurities always present in commercial alloys on the properties of layers between the crystalline grains and on the solidus temperature.

Unlike the solidification cracks, the polygonization ones are characteristic only of a small class of metals (single-phase alloys, pure metals). Therefore, they are comparatively rarely found in welds.

Cold cracks are most often formed in the near-weld zone and, less often, in the weld metal. As a rule, they are formed in welding of parts of medium- and high-alloy steels of the pearlitic and martensitic grades. However, these cracks may be found also in the welded joints of low-alloy steels of the ferritic–pearlitic grades and high-alloy steels of the austenitic grade.

As the presence of cracks in a welded joint exerts a substantial effect on its performance, the process of formation of hot and cold cracks was studied by many researchers, who established factors affecting initiation of cracks and developed methods for estimation of the susceptibility of metals to formation of these cracks.

It should be noted that unlike the solidification cracks, the polygonization or subsolidus ones are formed not along the primary grain boundaries, but in regions characterized by a concentration of imperfections of the crystalline lattice (dislocation walls) formed under the effect of shrinkage or thermal stresses at very high temperatures close to the solidus. As the polygonization and cold cracks are formed in the solid metal, the effect of surface phenomena on the processes of their formation shows up to a much lesser degree than on the formation of solidification cracks. Therefore, in the following sections, we will consider the role of the surface phenomena only for the case of solidification cracks formed in a weld.

8.1 SOLIDIFICATION CRACKS: FACTORS AFFECTING THEIR FORMATION AND TEST METHODS

It is a well-known fact that the weld metal made by fusion welding is exposed to tensile stresses during solidification and subsequent cooling, which can be calculated from the following equation[491]:

$$\sigma_t = EaQ \frac{[(l/x) - (l/x + l)]}{2\pi c\rho h},\tag{8.1}$$

where x is the thermal diffusion (heat transfer) distance; a is the thermal expansion factor; E is the elasticity modulus; Q is the heat input, ρ is the material density; c is the specific heat of a material; h is the thickness of plates welded; l is the weld length.

It follows from Equation (8.1) that for one and the same base material, the value of tensile stresses increases with increase of the weld length and heat input, and with decrease of thickness of plates welded. Maximum values of stresses are found within a region located in the immediate vicinity of the weld pool.

As was noted earlier, solidification cracks are formed under the effect of welding stresses during a period of time when the weld pool is a two-phase system (crystalline grains–melt). Therefore, low-melting compounds, eutectics, or melts, being in a liquid state, may cause solidification cracks during the solidification of the weld metal.

Sulfur plays an important role in the process of formation of solidification cracks during the welding of carbon and low-alloy steels (see Section 3.5). The effect of sulfur depends on the type and content of the alloying elements in metal.

8.1.1 THE INFLUENCE OF CARBON CONTENT

As proved by a large amount of experimental data, the factor, which plays the most important role in the formation of solidification cracks in the weld, is the concentration of carbon in metal. This is attributable to the fact,[492] that increase of the carbon content of carbon and low-alloy steels leads to the following main changes: (1) increase of the degree of dendritic heterogeneity of distribution of sulfur in the weld metal; (2) coarsening and increase of number of sulfide inclusions; (3) decrease of the content of manganese sulfide in non-metallic inclusions. However, in carbon steel, having a low (0.01%) sulfur content, no solidification cracks are formed in the weld even at a high (0.6%) carbon content of metal.

In the welding of high-alloy steels and alloys, carbon may become a direct cause of formation of solidification cracks,[493] as this is the case of formation of low-melting carbides located along the grain boundaries.

The presence of some amount of strong carbide-forming components (Ti, Nb) in the metal may mitigate the harmful effect of carbon, and even increase resistance of the metal to solidification cracks due to variations in shape and composition of sulfide non-metallic inclusions.[494,495]

The process of formation of solidification cracks in the weld is affected to a lesser but still marked degree by the silicon content of the metal. Normally, susceptibility of the metal to solidification-cracking increases at a silicon content in the metal of more than 0.5%.[1] The effect of silicon on the formation of solidification cracks is stronger in the welding of austenitic steels. Susceptibility of the metal to solidification cracks grows in the presence of phosphorus contained in the metal.[1]

The risk of formation of solidification cracks in welding carbon and low-alloy steels decreases, if the metal contains oxygen, manganese, and chromium. The positive effect of oxygen is usually attributed to oxysulfide non-metallic inclusions formed in the metal instead of sulfide ones, which decreases the formation of a network of iron sulfides. By the way, it is to a decrease of the oxygen content in the metal by an addition of carbon and silicon to it; some authors relate the harmful effect of the latter nonmetals to the formation of solidification cracks.

The effect of Mn and Cr is normally related to the formation of their sulfides, the melting temperature of which is higher than the temperature of solidification of the weld-pool metal. Therefore, manganese and chromium sulfides are in a solid state in the metal melt and thus, cannot precipitate along the grain boundaries. In this case, the higher the concentration of manganese in the metal, the higher the content of manganese sulfide in the sulfide inclusions, which thus become more refractory and smaller in size.

In alloying of a metal with chromium, the solidification-crack resistance of the weld metal grows only to a certain level of Cr concentration.[496] If the chromium content exceeds this concentration, the effect of carbosulfide and carbide phases results. These phases are located in the form of chains along the grain boundaries, thus creating appropriate conditions for the formation of solidification cracks.

Experimental studies[497] made it possible to establish that chemical affinity of manganese for sulfur grows with decrease of temperature. Therefore, in the welding of multilayer welds, where the preceding layers are subjected to reheating, the amount of Mn in sulfide inclusions would increase.

No solidification cracks are usually formed in steel welds, if the manganese content of the metal is higher than the sulfur content by a factor of 4 to 6.

All the above-said facts prove the decisive role of sulfur in the formation of solidification cracks in the welding of carbon and low-alloy steels. The harmful effect of sulfur can be eliminated by transforming it to refractory sulfides through alloying the metal with rare-earth elements, as well as with Ti, Zr, Ca, Al, and some other elements.

8.1.2 THE ROLE OF THE METAL STRUCTURE

Metal structure can also have a substantial effect on the formation of solidification cracks. As is reported by Shorshorov et al.,[498] refining of crystalline grains at the same thickness of liquid interlayers leads to an increase of the ductility of two-phase systems following the hyperbolic law. In addition, refining of the primary structure of the weld metal leads to a decrease of the degree of chemical heterogeneity. In welding, the probability of formation of solidification cracks in the weld increases with increase of sizes of the base-metal grains,[499] which may become the centers of nucleation of crystalline grains in the weld pool.

Formation of cracks in the weld metal is also influenced by the shape of the weld pool, which determines the rate of solidification of the metal melt and the stressed state of the weld metal. As shown by John and Richards,[500] in the case where the weld pool has a shape close to that of a falling droplet, high tensile stresses will be formed in its tail edge, which facilitates formation of the cracks.

The critical carbon and silicon contents of metal, at which solidification cracks are formed, also depend on the shape of the weld pool. The optimal value of the form factor of the weld, which is the weld width to penetration depth ratio, is close to 6.0.[1]

For carbon and low-alloy steels, the relationship between chemical composition of the weld metal, its shape, and the susceptibility of the metal to formation of solidification cracks in it, characterized by the parameter H_C, can be described by the following expression[501]:

$$H_C = C + 0.75S - 0.03Mn - 0.07\psi_w,$$

where C, S, and Mn are the contents of these elements in the weld metal in wt.%, and ψ_w is the form factor of the weld.

This formula is valid for the content (wt.%) of alloying elements in a range of $C = 0.11-0.20$, $Mn = 0.4-1.5$ and $S = 0.015-0.06$, and for ψ_w ranging from 0.9 to 2.2. According to the data of Turi and El-Hebeary,[501] no solidification cracks would form in the weld, if $H_C < 0.05$.

Solidification cracks in the weld may be caused not only by low-melting sulfides, but also by low-melting metals located at the boundary with solid crystalline grains. Solidification cracks are formed most often[1] in the welding of copper to steel. In this case, the probability of formation of solidification cracks in the weld also depends on the chemical composition of the materials welded, magnitude of tensile stresses, type of the welded joint, etc.

As solidification cracks are very often formed in the weld metal, this makes it necessary to determine the probability of this phenomenon allowing for the factors which influence the process of their formation.

8.1.3 DETERMINATION OF SUSCEPTIBILITY TO SOLIDIFICATION CRACKING

Various methods for estimation of susceptibility of welded joints to solidification cracking, allowing for the effect of the base metal, filler metals, and the type of the welded structure, are available. All these methods can be conditionally subdivided into two groups: (1) methods providing for the use of the so-called "self-loading" samples; (2) methods, which involve samples loaded by external forces.

   In the first case, the use is made of the samples loaded during welding and sub-sequent cooling due to shrinkage and polymorphic transformations. Some samples utilized for these methods are shown in Figure 8.1. The criterion for estimation of susceptibility of the metal to the formation of solidification cracks in this case is as follows:"there are cracks" or "there are no cracks".

   Samples loaded by external forces are used with methods of the second group. This is the case, where bending or tensile deformation is induced in a welded joint using special loading devices. This requires very high forces, and the testing machines should have frames of a high rigidity. The estimation criteria with such tests are normally; the critical rate of deformation, at which cracks in the weld are formed; critical deformation; total crack length. These methods allow testing of single- and multilayer welds, welds with remelting of the previously deposited layers, and simple bead-on-plate welds. These methods are characterized by a high reproducibility of the experimental results.

   However, both groups of test methods are characterized by the necessity of using samples with a high metal consumption. They are labor-consuming and involve very sophisticated testing machines (the second group of the methods). Moreover, being purely technological in nature, they yield differing test results with variations in the test conditions, for example, welding parameters, sizes of samples, chemical composition of the metal, etc.

   Meanwhile, it is a known fact, that susceptibility of a metal to solidification cracking is determined in many respects by the nature of distribution and shape of the liquid phase at grain boundaries, which depend on the wettability of grain surfaces. Possible cases of distribution of the liquid phase along the grain boundaries

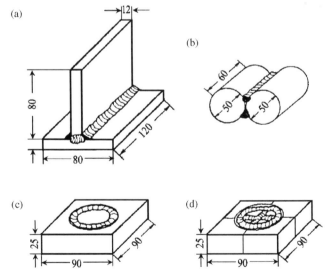

FIGURE 8.1 Specimens used to evaluate resistance of the weld metal to solidification cracking: (a) T-joint test specimen with two filled welds; (b) cylindrical specimen; (c) ring test specimen; (d) ring test specimen with previously made transverse weld.

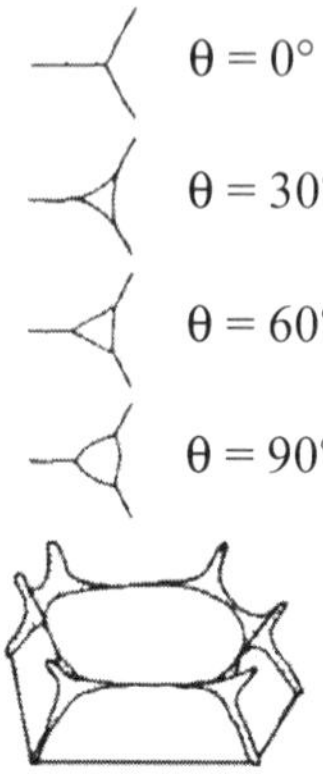

FIGURE 8.2 Diagram of possible distribution of liquid phase along the grain, boundaries depending upon the value of contact angle, θ: (1) 3D model at $\theta \approx 65°$ with part of the grains covered by a layer of liquid.

at different wettability of a solid metal with a low-melting melt are shown in Figure 8.2. The presence of liquid films formed in the case of good wettability of grain surfaces with the liquid phase favors formation of solidification cracks in the weld.[271,272] If the liquid phase is coagulated, and if at the triple angles it has a drop-like shape, close to the spherical one, then in such a case, no solidification cracking would occur as a rule. Therefore, formation of solidification cracks in the welds depends on the interaction between the solid metal and the liquid phase contacting it. This interaction may result in the Rebinder effect. As either a low-melting metal melt or the low-melting non-metallic inclusions may serve as the liquid phase during welding, we will separately consider the mechanisms of the effects of metallic and non-metallic melts on the fracture of the solid metal, resulting in the formation of solidification cracks.

8.2 MECHANISM OF THE EFFECT OF METAL MELTS ON THE FRACTURE OF SOLID METALS

The contact of liquid metals with solid wrought structural metals and alloys sometimes leads to a drastic decrease of strength and ductility of the latter. There are a lot of studies dedicated to the investigation of causes and principles of this phenomenon called the liquid-metal embrittlement. However, until now no generally recognized theory of this process has been available.

The presence of the metal melt at a boundary with the solid metal may promote formation of cracks in it as a result[252] of corrosion, diffusion, or adsorption effects. Each of these processes may occur spontaneously, leading to a decrease of free energy in the metal−melt system.[252]

8.2.1 Interaction of the Metal Atoms at the Crack Apex

To reveal peculiarities of the process of fracture of the solid metal in contact with a low-melting melt, as well as the role of the corrosion, diffusion, and adsorption effects, let us consider interaction of atoms of liquid and solid metals at the crack apex (Figure 8.3). Provided that the crack apex, as assumed by P.A. Rebinder[502] is wedge-shaped, stability of the crack will depend on the forces of binding of atoms at its apex. With a certain approximation, it can be considered[252] to depend on the binding forces f_{AA}^0 of that pair of atoms, A–A, the energy of interaction of

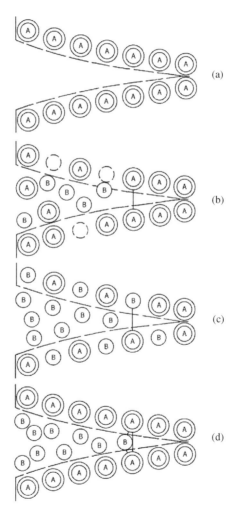

FIGURE 8.3 Diagram of interaction of metal atoms of the crack apex: (a) in the absence of melt; (b) corrosion effect exerted by the melt; (c) diffusion effect exerted by the melt; (d) adsorption effect exerted by the melt.

which is maximum. It follows from the wedge shape of the crack that intact interatomic bonds would gradually transform into the fully broken ones in a region from the apex of the crack to its ends upon their full divergence.

Disruption of bonds between two homogeneous atoms A–A at the crack apex occurs under the corrosion effect exerted by the melt on the solid metal as a result of its dissolution or formation of a chemical compound. This disruption is facilitated, if one of the atoms of the solid metal transfers itself to the melt. Condition of the corrosion effect of the low-melting metal medium on the solid metal can be written in the following form:

$$f_{AA}^{y} < f_{AA}^{0}, \tag{8.2}$$

where the superscript "y" indicates, that given forces are projections of vectors on axis y coinciding with the direction of action of an external force. It follows from Equation (8.2) that the lower the content of atoms A in a given melt and the higher the concentration of saturation of the liquid phase, the more pronounced is the corrosion effect caused by the low-melting metal melt on the solid metal.

With the diffusion effect, as in the case, where the melt diffuses into the solid metal, formation of a crack causes disruption of heterogeneous atoms B–A, rather than atoms A–A. Therefore, condition of the diffusion effect on the solid metal will have the following form:

$$f_{AB}^{y} < f_{AA}^{0}, \tag{8.3}$$

and the higher the value of inequality (Equation (8.3)), the more pronounced is this effect.

Penetration of atoms of an adsorption-active melt into a crack leads to weakening of the binding force f_{AA}^{y} down to f_{AA} and formation of new bonds f_{AB}. In the presence of the adsorption-active melt, propagation of the crack to one interatomic distance requires disruption of bonds f_{AA} and f_{AB}, rather than f°_{AA}. Therefore, a condition of the adsorption effect of the medium, which shows up in decrease of strength and ductility of a solid, can be written as follows:

$$f_{AB}^{y} + f_{AA}^{y} < f_{AA}^{0y}. \tag{8.4}$$

It should be noted that the effect of the adsorption decrease of strength and ductility of solids is found with any type of interactions between atoms: on metals, on ionic and molecular crystals, and on semiconductors. Strong surface-active media with respect to solid metals were proved to be the melts of: mercury for zinc and brass; gallium for aluminum and its alloys; zinc for iron and steels; cadmium for titanium.[252,503,504] At the same time, as noted in a number of cases, the higher the strength of a metal in an initial state, the stronger the effect of a melt on the mechanical properties of the solid metal.[252] Strength and ductility of a solid metal in contact with the metal melt are greatly affected by the temperature and strain rate used in experiments. With a substantial increase of temperature, if the solid metal is ductile enough even with a considerable decrease of surface energy, the effect of loss on the strength and ductility of the metal may disappear under the influence of the molten metal. Decrease of the strain rate for the solid metal may have an effect similar to that exerted by increase of temperature.[505] Therefore, embrittlement of a solid metal in contact with a low-melting metal melt is manifested within a certain temperature–rate range.[506] In addition, the temperature range of decreased ductility depends on the rate of deformation of the specimen, composition and structure of the metal under consideration, as well as geometry and size of the specimen. According to Popovich and Dmukhovskaya,[507] an experimental dependence of the upper value of the tough–brittle transition temperature (T_c) on the strain rate (ε) can be described by the following expression:

$$T_c = F(lgM - lg\varepsilon),$$

where F and M are constants.

For given test conditions, this dependence in a coordinate system of T versus $1/lg\,(126/\varepsilon)$ is a straight line (Figure 8.4). Similar dependence of the tough–brittle transition temperature upon the strain rate was derived in study of Shchukin.[508]

It should be noted that the temperature ranges of decrease of strength and that of ductility in liquid-metal embrittlement of the solid metal do not coincide.

8.2.1.1 The Complex Nature of the Mechanism

As was already noted, no single opinion of the mechanism of interaction of liquid metals with solid ones is available so far.[509] Complexity of the mechanism of interaction between solid and molten metals lies in the fact that in many cases, changes in the mechanical properties of the solid metal under the effect of the metal melt result from the simultaneous effect of all three factors (diffusion, corrosion, adsorption). It should be noted that the diffusion process is the slowest one, as it is related to diffusion of atoms of the melt into solid metal. The corrosion process is faster, as it is determined by diffusion of the atoms of the solid metal into the melt. However,

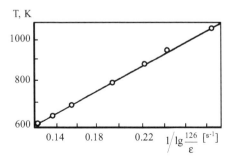

FIGURE 8.4 Dependence of the upper value of tough–brittle transition temperature upon the strain rate for St.3 steel in liquid-metal media.

in both the cases, a certain period of time is needed for the corrosion and diffusion effects to be manifested to ensure development of physical–chemical processes, which result in irreversible structural transformations of the metal.

The adsorption process is the fastest. Compared with the corrosion and diffusion effects of the melt on mechanical properties of solid metal, the adsorption effect has a number of specific features. Among them the most important for welding conditions are the following: (i) high rate of the effect of decrease of strength and ductility of a material; (ii) sufficiency of an extremely small amount of the adsorption-active melt. Another important factor inducing the Rebinder effect is the process of propagation of the surface-active melt, as it is associated with penetration of atoms or molecules of the melt to a new surface formed as a result of crack growth. Propagation of the adsorption-active medium is of the following basic types: (1) capillary flow inside a crack directed to its apex; (2) spread-over surfaces of the crack walls; (3) surface diffusion of single layers along the walls and into a narrow gap near the crack apex; (4) irregular diffusion at structural defects into the pre-fracture zone.

8.2.2 THE ROLE OF THE MELT

The role of the melt in the initiation and propagation of cracks is different and depends on the type of propagation of the surface-active medium. However, in any case, the movement of the adsorption-active medium is a necessary condition for the occurrence of the effect of an adsorption decrease of strength and ductility of a material. As is well known, this effect occurs in the presence of normal tensile and cleavage stresses, good wettability of solid metal with the metal melt, and low solubility of the melt in the solid metal.

As melts of copper and its alloys can readily wet solid steels, it might be expected, that their contact would lead to embrittlement of the solid metal.

8.2.2.1 Investigation of Mechanical Properties

To study the effect of melts of copper and its alloys on the mechanical properties of solid metals, we investigated the strength and ductility of St.3 and 12Kh18N10T steels in contact with the copper melts. Experiments (Figure 8.5) were conducted on specimens (1) of a tubular shape (outside diameter: 9 mm, gauge length: 64 mm). The internal working surface of the specimens was manually treated with a reamer, and the external one was polished. Prior to the test, the specimens were washed in ethanol and filled up with copper or copper alloy. Then openings in the specimens were closed with plugs (2) made from the same metal as the specimens, and the specimens were welded in a vacuum chamber filled with argon.

Uniaxial tension of the specimens was provided using the upgraded tensile testing machine UM-5 in a vacuum chamber at a residual pressure of 6 MPa or in high-grade argon. In the latter case, the three-fold purging of the chamber with argon eliminated the effect of active air elements on the experimental results. The experiments were conducted at a temperature of 1373 K.

The preliminary experiments made it possible to establish that the optimal thickness of the wall of a specimen should be 1.5 mm and weight of a low-melting metal

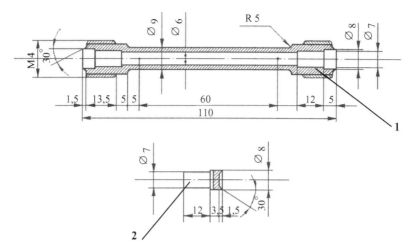

FIGURE 8.5 Types of specimens for investigation of the effect of low-melting point melts on mechanical properties of solid metal: (1) tubular specimen; (2) plug.

placed inside the specimen (copper or its alloys) should be 4 g. Temperature and deformation of the specimens were controlled during the tests using the plotting device N-306 and potentiometer KSP-4.

The effect of the rate of deformation of the St.3 steel specimens on the strength and ductility of the steel in contact with the melts of copper M1 and bronze BrKMts3–1 and without a contact with the copper melts is shown in Figure 8.6 and Figure 8.7. As can be seen, the contact of St.3 steel with the copper melts leads to a decrease both of strength and ductility. Ductility of solid steel decreases

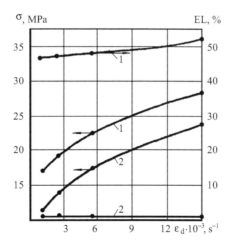

FIGURE 8.6 Change of St.3 steel strength and ductility depending on the straining rate ($T = 1373$ K): without (1) and with (2) M1 copper.

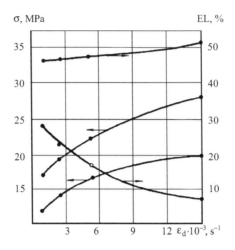

FIGURE 8.7 Change of strength and ductility of 12Kh18N10T steel depending on the straining rate ($T = 1373$ K): with (1) and without (2) BrKMts3-1.

to a greater degree compared with its strength. Such relationships were noted also in the case of the contact of St.3 and 12Kh18N10T steels with melts of copper M3r, brass LK62-0.5, and the copper alloy MNZhKT5-1-0.2-0.2. In all these cases, the decrease of strength of the metals was not high (up to 22% and 25% for St.3 and 12Kh18N10T steels respectively). Copper melts have a much higher effect on the ductility of the solid metal (Figure 8.6, Figure 8.7, and Figure 8.8). As seen from the figures, ductility of St.3 steel decreases to the greatest degree in contact with

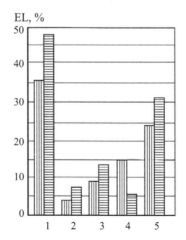

FIGURE 8.8 Ductility of steels (▦ St3; ▤ 12Kh18N10T) without and with copper melts ($T = 1373$ K): (1) without copper melt; (2) M3r copper; (3) BrKMts3-1 bronze; (4) MNZhKT5-1-0.2-0.2 alloy; (5) LK62-0.5 brass.

melts of copper M1, M3r, and bronze BrKMts 3-1, and to the smallest degree in contact with the melt of brass LK62-0.5. Ductility of 12Kh18N10T steel decreases most of all in contact with the melts of MNZhKT5-1-0.2-0.2 (8.5 times) and M3r (6 times).

8.2.2.2 The Corrosion Effect

One of the causes of decrease of strength and ductility of solid steels in contact with copper melts can be a corrosion effect of the melts on solid metal. Therefore, additional studies were carried out to estimate the effect of corrosion processes on the decrease of mechanical properties of solid steels in contact with copper melts. Experiments to study the corrosion effect of copper melts on solid steel were conducted as follows. Cylindrical steel specimens were placed together with a copper melt under consideration into sealed caps (Figure 8.9) made from stainless steel. The experiments were carried out in the Tamman furnace at a temperature of 1373 K. The time of holding specimens at this temperature was varied from 30 to 360 s. Comparative corrosion activity of copper melts with respect to steels was estimated from the size of the corrosion—affected zone measured in a fixed region 0.45-mm long.

A marked corrosion effect was noticed in the contact of St.3 steel with the melts of copper M3r and the alloy MNZhKT5-1-0.2-0.2, as well as in the contact of steel 12Kh18N10T with bronze BrKNts3-1. Copper alloys under consideration can be arranged in the following order by their corrosion activity with respect to steels: St.3: M3r (5.0), MNZhKT5-1-0.2-0.2 (3.0), BrKMts3-1 (1.5), and LK62-0.5 (1.0). The coefficient of comparative corrosion activity of a copper melt in contact with steel at a contact time of 360 s is given in brackets. In the case of contact of copper melts with 12Kh18N10T steel, the order of the corrosion activity is as

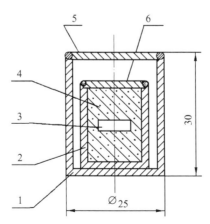

FIGURE 8.9 Appearance of ampoule used to study the corrosion effect of low-melting point melt on the solid metal: (1, 2) casing of external and internal cups; (3) solid metal specimen; (4) low-melting point medium; (5, 6) covers of the external and internal cups.

follows: BrKMts3-1 (10.0), LK62-0.5 (3.5), MNZhKT5-1-0.2-0.2 (3.0), and M3r (1.5). However, in all the cases, the corrosion effect of copper melts on solid steel is manifested at a considerable time of contact (60–180 s). Therefore, these processes will hardly have a marked effect on the solidification cracking of the welds.

Therefore, the melts of copper and its alloys can lead to a substantial decrease of strength and especially ductility of the solid metals in contact with the given melts. This has been proved by the results of investigations conducted by otherworkers. Moreover, this effect is associated with wettability and spreading of these melts over the solid-metal surfaces. This should be taken into account in the development of technologies for surfacing and welding of copper–steel parts.

As shown by our investigations (Section 3.5), the iron sulfide melt can wet well many metals. Thus, it is highly probable that its contact with the solid metal may also lead to embrittlement of the latter. Meanwhile, no data are available on the effect of low-melting non-metallic melts, and FeS in particular, on the mechanical properties of solid metals, while such studies can help understand the mechanism of formation of solidification cracks of a sulfide origin.

8.3 EFFECT OF LOW-MELTING NON-METALLIC MELTS ON THE MECHANICAL PROPERTIES OF SOLID METALS

This section gives results of investigations conducted to study the effect of melts of iron sulfide and alloy $SiO_2–Na_2O$ on the mechanical properties of solid metals in contact with the above non-metallic melts. The first series of the experiments was intended to study the effect of the FeS melt on the strength and ductility of armco-iron and 15KhMNF, 9KhF, 10Kh18N10T, 10Kh17N13M3T steels, and the transformer steel E3A.

8.3.1 Experimental Findings on Mechanical Properties

Experiments on the investigation of the effect of the melts of copper and its alloys on the mechanical properties of solid metals were carried out on specimens of a tubular shape having the same sizes as the specimens used to study the effect of copper melts on mechanical properties of solid steels. Preparation of specimens for the experiments and the experimental procedure remained unchanged, except that another testing machine of the MP-4G model was used in addition to the tensile testing machine UM-5. The experiments were conducted at a temperature of 1373 K, which was controlled using three chromel–alumel thermocouples, the hot junctions of which were secured to the center and edges of a specimen.

As before, in order to optimize the experimental conditions, the influence of the following factors on the pattern of variations in the mechanical properties of the solid metal was first studied. The effect of the amount of the FeS powder placed inside a tubular specimen, specimen wall thickness, and variation in the contact area between the FeS melt and metal studied.

As shown by the previous studies (Figure 8.10), starting from 3–4 g, any further increase of the amount of the powder placed inside the tubular specimen hardly had

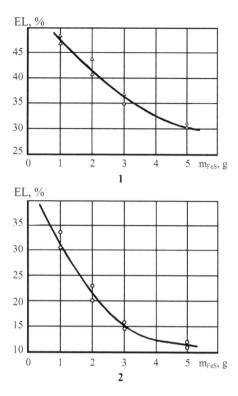

FIGURE 8.10 Effect of weight of powder FeS fed to a tubular specimen on the ductility of steels: (1) 9KhF; (2) 15KhMNF; $\varepsilon_d = 5 \cdot 10^{-3} \text{ s}^{-1}$.

any effect on variations in strength and ductility of metal. Variations in the specimen wall thickness from 1.0 to 2.5 mm showed (Figure 8.11), that increase of the specimen wall thickness was accompanied by a decline in the mechanical properties of the solid metal in contact with the non-metallic melt.

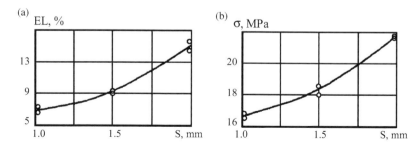

FIGURE 8.11 Effect of variation in thickness of the tubular specimen wall on the variation in the ductility and strength of armco-iron in contact with the solid metal with the FeS melt; $m_{\text{FeS}} = 5 \text{ g}$; $\varepsilon_d = 5 \cdot 10^{-3} \text{ s}^{-1}$.

8.3.1.1   The Effect of Comfort Area

Two series of experiments were conducted to study the effect of the contact area between FeS and solid metal. In the first series, FeS was fed only inside a tubular specimen, that is, only the inside surface of the specimen was wetted with the FeS melt. In the second series of the experiments, 1 g of the FeS powder was deposited on the outer surface of the specimen, in addition to the powder placed inside the specimen. In this case, the use was made of special split caps, which were put on the test specimen.

As is seen from Figure 8.12 and Figure 8.13, the simultaneous contact of the FeS melt with the outside and inside surfaces of a specimen leads to enhancement of the effect of decrease in the strength and ductility of the solid metal. This is also proved by the results of the tests of flat specimens immersed into the FeS melt, which were conducted using the ALA-TOO (IMASh-20-75) testing

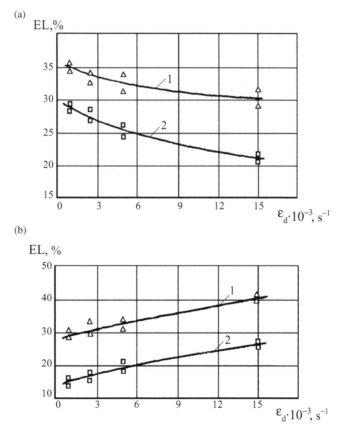

FIGURE 8.12 Variation in the ductility of steels 15KhMNF (a) and 10Kh17N13M3T (b) depending upon the strain rate of a specimen. Contact of FeS with the specimen at (1) the internal surface; (2) the internal and external surfaces.

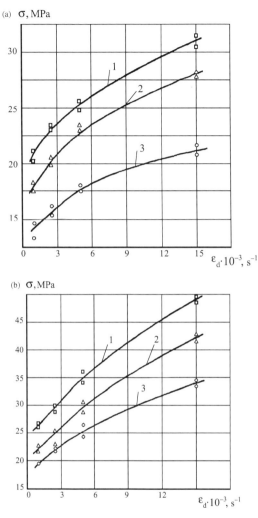

FIGURE 8.13 Variation in the strength of steels 15KhMNF (a) and 10Kh17N13M3T (b) depending upon the strain rate of a specimen: (1) without contact with FeS melt; (2) contact of the internal surface of the specimen with FeS; (3) contact of the internal and external surfaces of the specimen with FeS.

machine. Thus, for armco-iron specimens with a cross-sectional area of 1.84 to 1.86 mm^2, at $\varepsilon_d = 2.38 \cdot 10^{-2}$ s^{-1} and $T = 1373$ K, when not in contact with the FeS melt, the fracture stress is $\sigma_y = 39.5$ MPa, whereas in the presence of the melt, $\sigma_y = 9.9$ MPa. Elongation of the metal in this case is 38% and 8%, respectively for the two conditions.

Preliminary experiments allowed a conclusion, that the optimal thickness of the specimen wall should be 1.5 mm and weight of the iron sulfide powder poured into a

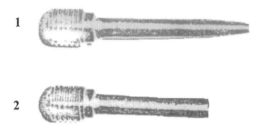

FIGURE 8.14 Appearance of specimens of armco-iron after the tests; $T = 1373$ K; $\varepsilon_d = 5 \cdot 10^{-3} s^{-1}$. (1) without contact with the FeS; (2) in contact with the FeS melt.

specimen should be 5 g. These parameters were set constant for all the further experiments.

As shown by the experiments conducted (Figure 8.14), in contact with the FeS melt, armco-iron loses its ductility at all the strain rates under consideration. Strength of the metal also markedly decreases, although to a lesser degree, with decrease of ε_d. The embrittlement effect of the FeS melt on armco-iron is confirmed by the physical appearance of the specimens (Figure 8.14) and by the metallography results (Figure 8.15). Contact melting and formation of FeS–Fe, eutectic, as well as its wetting of the specimen surface, take place at the test temperature. Because of heterogeneity of the structure and applied stresses, interaction of the melt with iron is of a selective nature, which promotes formation of cracks and accelerated fracture. Structure of the crack filler (Figure 8.15) is characteristic of the eutectic system, and is probably of the Fe–FeS eutectic.

8.3.1.2   The Effect of Ductility

Contact with FeS also leads to a marked decrease of ductility of the transformer steel E3A (Figure 8.16). This decrease accelerates to a certain extent with increase of the strain rate, i.e. with reduction in time of contact of the melt with the solid metal. Strength, like in the case of armco-iron, decreases to a greater degree at low values of the specimen strain rate (ε_d).

(a)   (b)   (c)

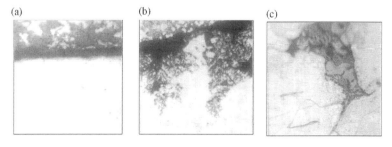

FIGURE 8.15 Microstructure of specimens of armco-iron after mechanical tests: (a) without FeS ($\times 70$); (b) with FeS ($\times 70$); (c) with FeS ($\times 600$).

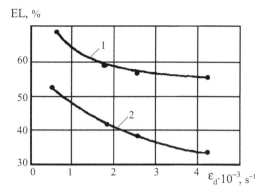

FIGURE 8.16 Variation in the ductility of steel E3A depending upon the strain rate of a specimen: (1) without contact with the FeS melt; (2) in contact with the FeS melt.

Decrease of ductility and strength of the metal is observed also in the cases of contact of the FeS melt with steels: 15KhMNF, 9KhF, 10Kh18N10T, and 10Kh17N13M3T (Figure 8.17). As is seen from the figure, 9KhF steel exhibits a more substantial decrease of ductility, compared to 10Kh18N10T and

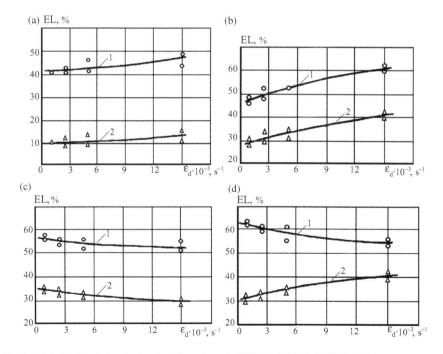

FIGURE 8.17 Variation in the ductility of steels 9KhF (a), 10Kh17N13M3T (b), 15KhMNF (c), and10Kh18N10T (d), depending upon the strain rate of a specimen: (1) without contact with the FeS melt; (2) in contact with the FeS melt.

15KhMNF steels. Increase of the strain rate is accompanied by a less-substantial decrease of ductility for steels 10Kh17N13M3T, and a more intensive decrease of ductility for steels 9KhF and 15KhMNF. All of these steels, although to a differing degree, exhibit a decrease of strength under the effect of the iron sulfide melt.

There is almost no change in strength of the metal, when 10Kh18N10T steel is in contact with the FeS melt. The effect of FeS on the ductility of this steel decreases with increase of the strain rate.

It can be seed from the results obtained that the better the wettability of this metal with the sulfide melt, the stronger is the effect of decrease of strength and especially ductility under the impact of FeS.

Additional experiments were conducted to verify the effect of a melt material on the strength and ductility of a solid metal. The melt used in these experiments was a binary alloy consisting of 80 wt.% SiO_2 and 20 wt.% Na_2O. This choice was based on the fact, that this alloy and FeS have close values of surface tension ($\sigma_{FeS} = 318$ mJ/m^2 and $\sigma_{SiO_2-Na_2O} = 324$ mJ/m^2 at $T = 1373$ K) and, hence have close cohesion strength, although the silicate melt cannot fully wet armco-iron ($\theta = 46°$). It was found, that in the contact of the silicate melt with armco-iron andl 15KhMNF stee, the values of strength and ductility of the solid metals remained unchanged over the entire range of variation of the strain rate of the specimens.

Therefore, decrease of strength and especially ductility of solid steels in contact with a non-metallic melt depends on how well it wets-solid metals, rather than on the cohesion strength of the melt. In contact with the FeS melt, which wets the solid metals readily, the latter change their strength and ductility.

Under welding conditions, since the time of contact of a molten medium with a solid metal is not in excess of a few seconds, most probably the process of formation of solidification cracks in the weld metal can be affected only by two processes characterized by the highest rate. These are the corrosion and adsorption processes. The corrosion effect of the FeS melt on the metal was studied to reveal the role of each of them in the fracture of solid metals.

8.3.2 Experimental Study of the Corrosion

The experimental procedure was as follows: cylindrical specimens 7 mm in diameter and 2.5 mm thick were cut from the metal studied; the specimens were polished and placed into stainless steel cups together with 3 g of FeS powder; then the cup was welded up in a vacuum chamber in argon atmosphere and put into a larger cup also made from stainless steel, which was also welded up. Such a preparation of specimens enabled avoiding any probability of contact of the specimens with air during the experiments. The air-tight ampoule thus prepared was placed into the Tamman furnace heated to 1373 K. The preliminary experiments showed that normally, the FeS powder was fully melted after holding at this temperature for 25–30 sec. Thus, the time of 30 sec was assumed to be the reference point. After the preset holding, the ampoule was taken out of the furnace, cooled in air to room temperature, and cut. Sections for metallographic examinations were made from the cylindrical specimen.

Corrosion effect of the FeS melt was studied for armco-iron and 9KhF, 15KhMNF, 10Kh18N10T, 10Kh17N13M3T, and E3A steels. It was found, that FeS exerts corrosion effect on all of the metals studied at a temperature of 1373 K, the minimum interaction time being about 15 sec.

This time of 15 s, is enough for the corrosion effect of FeS melt on armco-iron to be manifested (Figure 8.18). Longer time of contact of the FeS melt with armco-iron led to enhancement of the corrosion effect on the solid metal. Holding for 15 sec led to a clearly defined selective penetration of FeS (Figure 8.18e) to form an eutectic,

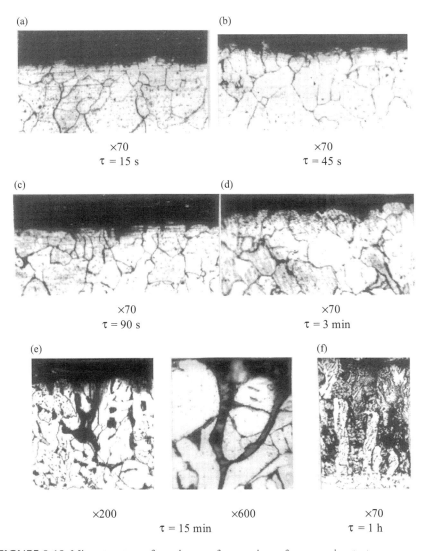

(a) ×70 τ = 15 s

(b) ×70 τ = 45 s

(c) ×70 τ = 90 s

(d) ×70 τ = 3 min

(e) ×200 ×600 τ = 15 min

(f) ×70 τ = 1 h

FIGURE 8.18 Microstructure of specimens of armco-iron after corrosion tests.

resulting from contact melting. Groups of grains with surface melting can be seen in Figure 8.18. Longer testing time led to an intensive corrosion, which can be regarded as more uniform, since local penetrations of FeS melt in the form of cracks disappeared (Figure 8.18f), and the entire surface layer of the metal was subjected to corrosion.

8.3.2.1 Carbon-Rich Steels

For steels containing a considerable amount of carbon (9KhF, 15KhMNF), the process of penetration of FeS into the metal slows down because of the opposing diffusion of carbon from the metal into molten sulfide (Figure 8.19 and Figure 8.20).

However, even in this case, soon after 15 sec, we can see the results of interaction of the FeS melt with the 9KhF steel. This leads to the formation of a zone on the solid metal surface, the structure of which differs from the initial one for it now contains the ferrite phase as a separate structural component. Thickness of this zone grows with increase of the holding time, and at $\tau = 5$ min, it amounts to 0.15 mm. At the same time, the content of ferrite in it also grows. It should be noted, that ferrite regions have as a rule a certain orientation, that is, they are oriented preferably normal to the specimen surface. Holding for 5 min leads to the formation of small wedge-shaped crack nuclei in some regions on the specimen surface. These nuclei are filled with the Fe–FeS eutectic formed as a result of interaction of the FeS melt with the steel.

Interaction of iron sulfide with 15KhMNF steel is of a slightly different nature. In addition to the formation of the zone with redundant ferrite, no later than after

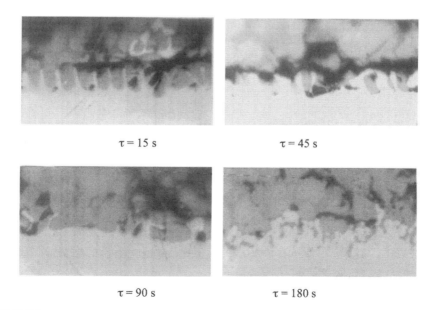

$\tau = 15$ s	$\tau = 45$ s
$\tau = 90$ s	$\tau = 180$ s

FIGURE 8.19 Microstructure of specimens of 9KhF steel after corrosion tests ($\times 500$).

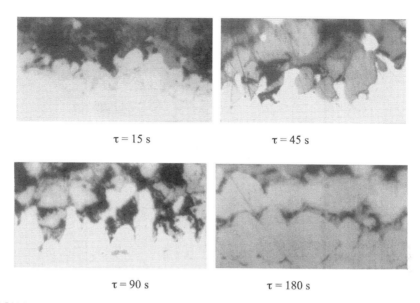

$\tau = 15$ s $\qquad\qquad$ $\tau = 45$ s

$\tau = 90$ s $\qquad\qquad$ $\tau = 180$ s

FIGURE 8.20 Microstructure of specimens of steel 15KhMNF after corrosion tests ($\times 500$).

15 sec of holding, FeS starts penetrating (Figure 8.20) into the base metal to form the eutectic along the boundaries of the prior austenitic grains. Longer time of contact of FeS melt with solid steel makes this penetration more pronounced. The zone with a changed structure also increases in size, although this increase is less intensive compared to 9KhF steel.

Therefore, contact of FeS melt with steels containing a considerable amount of carbon leads to decarburization of the surface layer and formation of a redundant ferrite phase. This process becomes more pronounced with increase of the carbon content of the metal.

8.3.2.2 Austenitic Steels

In austenitic steels, 10Kh18N10T and 10Kh17N13M3T with a comparatively low content of carbon and rather high content of chromium, which is a strong carbide-forming element, the process of decarburization of the surface layer of metal is not pronounced. The nature of interaction of the FeS melt with these steels is similar in many respects. After holding for no more than 15 sec, the eutectic formed as a result of the interaction of FeS with the steel starts penetrating into the metal primarily along the grain boundaries (Figure 8.21 and Figure 8.22).

8.3.3 Important Observations in Metallography

Longer time of contact of the solid metal with the iron sulfide melt leads to a marked increase of the number and especially the size of cracks filled with the eutectic. It can be well seen on the etched sections, that in addition to a direct penetration of the eutectic into the metal in the form of thin cracks, there is also an interaction

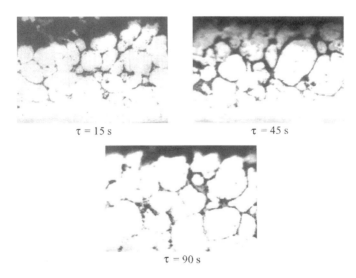

FIGURE 8.21 Microstructure of specimens of steel 10Kh18N10T after corrosion tests (×500).

of the sulfide with metal in the bulk of grains, leading to formation of redundant phases in the form of dispersed inclusions.

It is likely, that dissolution of the solid metal in FeS melt, associated with destruction of the crystalline lattice of solid metal, takes place in all of the systems studied. This type of corrosion is most characteristic of steel E3A. Examination of the transformer steel specimens after corrosion tests in FeS melt showed,

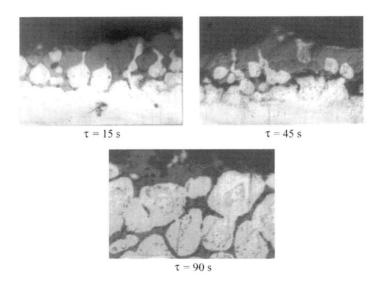

FIGURE 8.22 Microstructure of specimens of steel10Kh17N13M3T after corrosion tests (×500).

that interaction of the melt with the metal became pronounced only after a long-time holding (Figure 8.23). In this case, an increase of the time of contact of iron sulfide melt with steel enhances the corrosion effect. However, the rate of the process markedly decreases. It should be noted, that corrosion taking place on this steel is sufficiently uniform over the entire metal surface, unlike, for example, armco-iron. In the interaction of the transformer steel with FeS melt, the corrosion attack on steel is likely to occur due to formation of a binary (Fe–FeS) or ternary (Fe–FeS–FeSi) eutectic. In this steel containing a considerable amount of silicon (2.88%), which is characterized by a higher chemical affinity for sulfur than iron, the corrosion process is intensified by the following reaction of formation of SiS_2 with a melting point of 1363 K: $Si + 2FeS \rightarrow SiS_2 + 2Fe$.

Metallography of the specimens after tensile tests at a temperature of 1373 K showed, that in this case, the degree of the corrosion attack on the solid metal was very high. In contrast to armco-iron, steel E3A has almost no sharp cavities filled with eutectic and sulfides, which can be considered as cracks or their initiators (Figure 8.23). Thus, decrease of ductility of steel E3A in contact with the iron sulfide melt is caused to a large degree by its corrosion effect on the solid metal.

Therefore, the studies conducted showed that at a temperature of 1373 K, all the solid metals examined were subjected to corrosion, when in contact with the iron sulfide melt. The minimum time required for interaction of the melt with solid metal is approximately 15 sec.

As was noted earlier, in the interaction of FeS melt with solid metal, the fracture of the latter under welding conditions can be caused by the following main

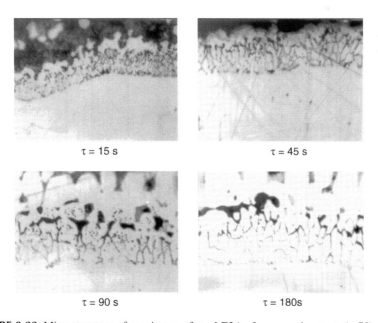

$\tau = 15$ s $\tau = 45$ s

$\tau = 90$ s $\tau = 180$s

FIGURE 8.23 Microstructure of specimens of steel E3A after corrosion tests ($\times 70$).

factors: the corrosion and adsorption effect of the melt. Naturally, the mode of fracture of metal depends on which of the processes is dominant.

8.3.4 THE MECHANISM OF THE CORROSION EFFECT

The corrosion effect can be caused by the following processes[252]: (1) dissolution of solid metal in melt; (2) thermal mass transfer; (3) isothermal mass transfer; (4) intercrystalline fracture; (5) formation of solid solutions and compounds; (6) interaction with impurities in the melt. Among the above processes the less significant is the thermal mass transfer because the FeS melt has no substantial temperature gradients. Isothermal mass transfer is also equally important, as it requires a simultaneous contact of the melt with two solid metals capable of forming intermetallics or solid solutions with each other. The rest of the four processes may occur in the interaction of the melt with metals during solidification of the weld metal.

It is clear that out of all types of the corrosion effect of the FeS melt on the solid metal, of the highest danger is not the general corrosion but the local and micro-local corrosion phenomenea. The latter may begin in non-equilibrium regions of the metal structure in terms of thermodynamics, such as grain boundaries, places where slip bands escape to the surface, etc. As there is always a concentration of stresses at the apex of the formed crack, atoms in this region will have a much higher potential energy than atoms located in the neighboring regions of the crack and on the rest of the metal surface. That is why the probability of transfer of atoms located at the crack apex to the melt is much higher than that for the rest of the atoms. Therefore, with a comparatively low rate of general corrosion, the possibility exists of an intensive occurrence of corrosion at the grain boundaries, that is, the intercrystalline corrosion will take place.

The fact that intercrystalline corrosion is much faster than the general one can be seen from the expression $(l_i/l_g)_{max} \approx \exp(\Delta Q_i/RT)$, where l_i and l_g are the depths of intercrystalline and general corrosions, respectively, and ΔQ is the free energy of the grains.[252] For γ-iron, according to the data of Mc Lin,[510] the mean value of $\Delta Q = 33.633 \cdot 10^3$ J/(mole deg). Then at $T = 1373$ K, the l_i/l_g ratio will be equal to about 19, that is, the depth of the intercrystalline corrosion is much greater than that of the general corrosion.

Fracture of solid metals at grain boundaries greatly depends on the ratio of the free surface energy at a boundary between two grains and the dihedral contact angle. It follows from the condition of equilibrium of the system that $\sigma_{s-g} = 2\sigma_{s-1} \cos \theta$. This equation is valid for a case of $180° > \theta > 0$, if $\sigma_{s-1} > \sigma_{s-s}/2$, and for $\theta = 120°$, if $\sigma_{s-1} = \sigma_{s-s}$. However, this equality does not hold for any of the θ values, if $\sigma_{s-1} < \sigma_{s-s}$. In this case, system equilibrium may be reached only if the melt penetrates along the grain boundaries. The higher the value of σ_{s-s} and the larger the decrease of this value caused by the presence of the melt, the more intensively the melt penetrates along the grain boundaries and the more intensive is the separation of the grains. It should be noted, that at low values, the contact angle is approximately equal to the dihedral contact angle.[67]

According to the data given in the study by W. Missol, for γ-iron at 1653 K, the energy of the grain boundary σ_{s-s} is 770 mJ/m^2, for the alloy Fe 3% Si at 1473 K, this energy is equal to 346 mJ/m^2, and for the alloy Fe 18% Cr 9% Ni at 1433 K it is 735 mJ/m^2.[87] Then, for the metals studied, because of a good wettability of armco-iron with FeS melt and high values of σ_{s-s}, the iron sulfide melt should penetrate most intensively along the grain boundaries in armco-iron. This is proved also by metallography. Comparatively worse wettability and lower values of σ_{s-s} for the transformer steel explain a decrease of intercrystalline corrosion for this steel. For steel 10Kh18N10T, the presence of intercrystalline corrosion is related to a high value of σ_{s-s} and, probably, formation of NiS, which seems to readily wet the grains of this metal. Intercrystalline corrosion of steel 10Kh17N13M3T is attributable to good wettability of this steel with FeS melt and, probably, to a very high value of σ_{s-s}, similar to steel 10Kh18N10T.

8.3.5 OTHER PARAMETERS

Intercrystalline corrosion of steels 9KhF and 15KhMNF, taking place after decarburization of the surface layer of the metal, is probably related also to a better wettability of the decarburized metal with FeS melt and growth of the value of σ_{s-s} of metal caused by its decreased carbon content.

Therefore, predominance of this or that type of the corrosion effect depends in many respects on the wettability of the solid metal with a melt and the value of the interface surface energy at grain boundaries. It should be noted that the data generated on the corrosion effect of the iron sulfide melt on metals are valid for a case where metal is in an unstressed state. Under actual conditions, where the metal is in the stressed state, corrosion processes would occur more intensively.

As noted above, the contact of FeS melt with solid metals leads to decrease of strength and ductility of the latter.

Strength of a metal in contact with an adsorption-active medium can be determined from the Griffits equation, in which the surface energy is expressed in terms of two components of the interface energy σ_{s-l} and energy σ_p consumed primarily for plastic deformation of the metal at the crack apex. Then the condition of fracture of the metal can be written as follows:

$$\sigma_p = \mu\sqrt{E(\sigma_{s-l} + \sigma_p)/l}, \tag{8.5}$$

where μ is the proportionality coefficient, which for a plane strain state is $\mu = \sqrt{2}/\pi$, and l is the size of a defect. It follows from Equation (8.5) that the more intensively the adsorption-active medium decreases the surface energy of metal, the more intensive is the decrease of the value of σ_p. If in the contact of solid metals with the adsorption-active medium, the decrease in the surface energy of the metals is identical, the value of σ_p will be lower for a more brittle metal.

However, the experiments have proved that in contact of solid metals with the iron sulfide melt, their ductile properties change much more intensively than strength. Therefore, let us consider in more detail the effect of the iron sulfide melt on the ductile properties of solid metals.

For armco-iron and steels 15KhMNF, 9KhF, and E3A (Figure 8.14, Figure 8.16, and Figure 8.17), an increase of the strain rate leads to a more marked decrease of ductility of the metal in contact with the iron sulfide melt. For steel 10Kh17N13M3T (Figure 8.17), a change in the ε_d value hardly changes the difference in the ductility of the metal in the presence or absence of FeS melt. Only for steel 10Kh18N10T (Figure 8.17), an increase of the strain rate leads to a decrease of the difference in ductility in tests with and without FeS.

As an increase of the strain rate causes a decrease of the time of contact of FeS melt with solid metal, the relationships noted are indicative of a dominant role of the adsorption effect of FeS on armco-iron and steels 9KhF, 15KhMNF, and E3A. The corrosion effect is dominant for steel 10Kh18N10T, whereas both effects are likely to be valid for steel 10Kh17N13M3T. This conclusion is supported also by the fact that ductility of the metals studied decreases in the presence of FeS, and the better is this metal wetted with the FeS melt, the more intensive is the above decrease. Meanwhile, good wettability is one of the conditions for the adsorption effect to be manifested in.

Therefore, depending on the composition of the solid metal in contact with FeS melt, the embrittlement of the former may be due either to the adsorption or corrosion effect, or the combined action of both of the processes occurring simultaneously. Let us consider how these processes affect solidification cracking of the welds.

8.4 EFFECT OF LOW-MELTING MELTS ON SOLIDIFICATION CRACKING

It is a known fact, that the adsorption effect of a melt shows up even during a short-time contact of this medium with the metal. However, for a crack to grow, it is important, that the medium has enough time to penetrate into the pre-fracture zone located at the crack apex. As the penetration of the adsorption-active medium over the walls of the crack is driven by capillary forces, sizes of the crack should depend on the rate of spread of the medium over the metal and on the quantity of the medium.

8.4.1 CRACK FORMATION

To verify this assumption, investigations were conducted to study the effect of FeS on the formation of cracks in the welding of armco-iron and 15KhMNF, 9KhF, 10Kh18N10T, 40Kh, 30KhGSA, St.3, and E3A steels. All the experiments on welding were carried out using the automatic welding unit ADSV-5. Welding was performed with a tungsten electrode in argon atmosphere under the following conditions: $I_w = 195$ A, $U_a = 13.5 - 14.5$ V, and $v_w = 0.0043 - 0.005$ m/sec. Plates measuring $120 \times 60 \times 2.5$ mm^3 were used as the samples for welding. Prior to welding, the plates were cleaned and degreased. Iron sulfide was added to the weld

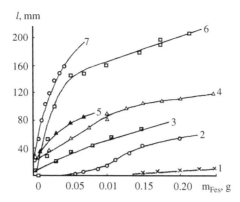

l, mm

FIGURE 8.24 Effect of the amount of FeS added to the weld pool on the sizes of solidification cracks in the weld. Steel samples: (1) 10Kh18N10T; (2) 30KhGSA; (3) 15KhMNF; (4) 40Kh; (5) 9KhF; (6) St3. (7) is armco-iron.

through the weld root and using a flux-cored wire filled with FeS. In the first case, welding was performed on a copper backing, which had a longitudinal slot 1-mm deep and 6 mm in radius. Prior to welding, the slot was filled with FeS powder uniformly distributed over the entire length. A sample of the chosen metal was placed on the backing so that the weld axis coincided with the slot axis.

As shown by the experiments (Figure 8.24 and Figure 8.25), with a continuous addition of FeS to the weld, the largest sizes of the cracks were in the case of armco-iron, and the smallest ones in steel 10Kh18N10T. In the latter case, the first cracks appeared at a FeS amount of about 0.15 g per 100 mm of the weld. The length of the cracks in the welds grew with increase of FeS amount.

Experiments with continuous and intermittent feed of FeS to the weld metal were conducted for steel E3A. In the latter case, FeS powder was placed into the

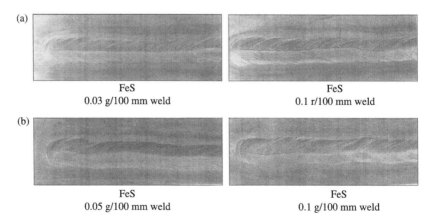

FIGURE 8.25 Appearance of welds with FeS added to the weld pool: (a) armco-iron; (b) steel 15KhMNF.

slot in regions 30-mm long with spacings of 20 mm. The choice of this size of a
spacing was based on the fact, that it should be longer than the length of the weld
pool under given welding conditions.

In welding with an intermittent feed of FeS powder, the cracks did not exceed
the limits of the region containing iron sulfide, and the length of the cracks was
not in excess of 20 mm (Figure 8.26 and Figure 8.27). With a continuous feed of
FeS, the cracks were much longer. It was experimentally established that increase
of the welding speed, while intensifying solidification of the metal, decreased the
sizes of the cracks (Figure 8.28). Although this led to a growth of thermal stresses
in the tail end of the weld pool,[500] such stresses should promote formation of solidification cracks. However, as complete wetting of this steel with FeS melt can be
achieved only after 12–14 sec, reduction of time of existence of the weld in solid–
liquid state due to increase of the welding speed reduces the adsorption effect of
decrease of the strength and ductility of steel E3A.

Addition of a low-melting binary alloy SiO_2–Na_2O to the weld pool in the
welding of steel 15KhMNF and armco-iron did not lead to cracking. The latter
results evidence that formation of solidification cracks in the presence of a low-
melting medium in the weld metal is associated not with a low strength of this
medium, which is often reported in the literature,[498,511] but with a physical–chemical interaction of the low-melting medium with the solid metal.

Analysis of the results obtained indicates that solidification cracks were formed
with an addition of FeS to the weld pool primarily in the welding of metals (armco-
iron and 40Kh, 15KhMNF, 9KhF, St. 3, E3A steels) which are well wetted with the
FeS melt, and embrittlement of which occurred as a result of the adsorption-induced
decrease of strength and ductility. In addition, the better the FeS melt wetted a given
metal, the larger were the cracks and the lower was the FeS content of the weld, at
which they were formed. In 10Kh18N10T steel (the embrittlement of which occurs
as a result of the corrosion effect of FeS melt), feeding FeS to the weld pool did not

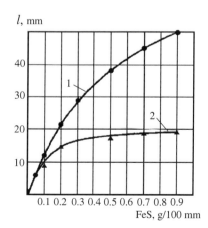

FIGURE 8.26 Sizes of cracks in welds on steel E3A depending upon the amount of FeS fed to
the weld pool: (1) continuous feeding; (2) intermittent feeding.

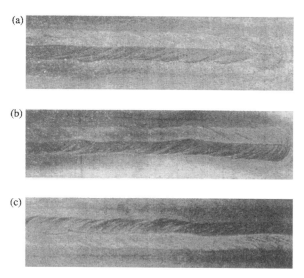

FIGURE 8.27 Appearance of welds on steel E3A depending upon the amount of FeS fed to the weld pool (g/100 m): (a) 0.05; (b) 0.12; (c) 0.25.

lead to cracking, since the time of existence of the weld in solid–liquid state seemed to be much shorter than that required for the corrosion embrittlement of the metal to take place. Formation of solidification cracks as a result of the corrosion embrittlement of the metal may occur only in the case, where the time of existence of the weld in solid–liquid state is sufficiently long. For example, this is the case of electroslag welding or surfacing.

8.4.2 THE ROLE OF THE SURFACE PHENOMENA

Allowance for the surface phenomena during the formation of solidification cracks makes it possible to explain many facts known from practice, such as, for example, the effect of different alloying elements on susceptibility of the metal to solidification cracking. Thus, as noted in Section 8.1, if the metal contains carbon, this

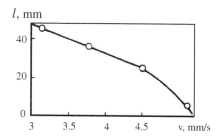

FIGURE 8.28 Effect of welding speed on the size of a crack formed in the weld on steel E3A. $I_w = 195$ A; $U_a = 13.5$ V; FeS $= 0.2$ g/100 mm.

increases the probability of cracking, if this metal also contains sulfur. Meanwhile, it is a known fact,[512] that the presence of carbon in Fe–C–S melt increases surface activity of sulfur. Knowing the dependence of the value of surface tension of the melt upon the concentration of a surface-active component, C, one can readily determine graphically the value of adsorption of this component. For this, the tangent lines until they intersect the axis of abscissas are drawn on the plot (Figure 8.29) at points, for which it is necessary to find the value of adsorption Γ_a. The slope formed by these lines will be equal to $d\sigma/dC$.[50] This value being known, it is possible to determine the value of Γ_a using the Gibbs equation:

$$\Gamma_a = -(C/RT)(d\sigma/dC), \tag{8.6}$$

where R is the universal gas constant.

It should be noted, that Equation (8.6) is valid only for the two-component systems. However, as in the system under consideration the interaction of carbon with sulfur is insignificant, the carbon content can be assumed to be constant, and Equation (8.6) can be used for calculations, assuming that the iron–carbon alloy serves as the solvent.

Calculations of the values of adsorption of sulfur at different concentrations of carbon in the metal show that increase of the carbon content of the metal leads to an increase of adsorption of sulfur. This should lead to an increase of the surface concentration of sulfur at a boundary with the crystal, in accordance with the following expressions:[10]

$$[S]_{surf} = [S]_{vol} + (M\Gamma_a\sqrt{\Gamma_a N_o}(100\%/\rho_m)). \tag{8.7}$$

Here, $[S]_{surf}$ and $[S]_{vol}$ are the surface and volume concentrations of sulfur in the melt; M is the molecular mass of an adsorbing component; ρ_m is the metal density; N_0 is the Avogadro number.

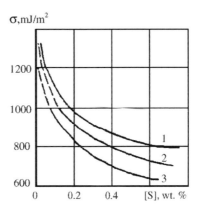

FIGURE 8.29 Variation in the surface tension of the Fe–C–S alloy depending upon its sulfur content (1723 K). Carbon content (wt.%): (1) 1.25; (2) 2.2; (3) 4.0.

The calculations made using Equation (8.7) show, that sulfur content of the surface layer is much in excess of that in the bulk of the metal (Figure 8.30), and the growth of the carbon content of metal is accompanied by increase of this difference. Increase of segregation of sulfur with the growth of carbon content of the metal is proved also by experiments.[55,513]

Therefore, the presence of carbon in the melt, while increasing the surface activity of sulfur, favors growth of its concentration at the boundaries with a crystal, which makes precipitation of sulfides at the grain surfaces much easier. Greater amount of FeS (as is seen from the data shown in Figure 8.10, Figure 8.11, and Figure 8.12), promotes decrease of ductility of the metal. In addition, increase of the carbon content of the metal makes it more brittle. In turn, while decreasing the value of energy for plastic deformation of the metal at the crack apex, σ_p, this should decrease also the strength of the metal, according to Equation (8.5), in the presence of the adsorption-active medium. This could be one of the causes for the increased susceptibility of the metal to formation of solidification cracks with increase of the carbon content of metal.

8.4.3 COMBINED EFFECT OF OXYGEN AND SULFUR

Decrease of the probability of formation of solidification cracks of the sulfide origin with an addition of oxygen to the metal can be also explained from these standpoints. To some extent, this can be associated with formation of oxide non-metallic inclusions and subsequent precipitation of sulfides on them, which was discussed in Chapter 6. In addition, as oxygen and sulfur increase the surface activity of

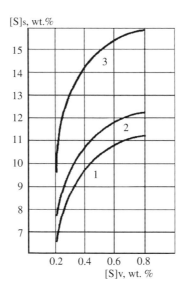

FIGURE 8.30 Variation in surface concentration (wt.%) of sulfur upon its content in the volume of the Fe–C–S melt: Carbon content (wt.%): (1) 0.15; (2) 0.25; (3) 0.54.

each other,[110] the simultaneous growth of their concentrations in the metal layer adjoining the surface of the crystal creates favorable conditions for the formation of two-component (oxysulfide) inclusions, which require less energy to appear.[385,386] In the contact of such inclusions with the solid metal, the probability of embrittlement of this metal will be much lower, as these inclusions wet the solid metal worse and have a higher melting point than pure iron sulfide, thus causing decrease of BTR.

Therefore, even such a brief consideration of the effect of individual components (C, O_2) on the segregation of sulfur in the metal allows some explanation of the effect of these components on the formation of solidification cracks of the sulfide origin, allowing for the surface activity of sulfur and the type of non-metallic inclusions formed.

Formation of solidification cracks in the welding of steel to copper is also associated with the adsorption effect of decrease of strength and ductility of the solid metal in contact with the copper melt. This is proved by the results of experiments on welding steel samples using filler wires of copper M1 and copper alloys (BrKMts3-1 and MNZhKT5-1-0.2-0.2). The following results were obtained in argon-arc surfacing under appropriate conditions ($I_w = 120$ A, $U_a = 14$ V, $v_w = 0.0011$ m/sec), which ensured the required penetration of plates 3-mm thick and good formation of the deposited bead. In surfacing of plates of steel St.3, the longest cracks (25 mm) resulted from using a filler wire of copper M1, that is the material, the melt of which decreases the ductility of the given steel to the greatest degree. In surfacing using wire BrKNts3-1, the cracks were 15-mm long, while in the case of the MNZhKT5-1-0.2-0.2 wire, the cracks were 20 mm long. In the surfacing of 12Kh18N10T steel, the crack length was 55 mm with a filler wire of copper M1, it was 70 mm with a filler wire of BrKMts3-1, and 65 mm with a wire MNZhKT5-1-0.2-0.2.

The results obtained are in good agreement with the data available (Section 3.5) on the wettability of steels with copper melts and decrease of ductility of steels in contact with the copper melts. This proves once more the importance of the adsorption embrittlement of a solid metal caused by a low-melting metal melt during the process of solidification cracking of the weld.

Therefore, susceptibility of a given metal to the formation of solidification cracks can be estimated on the basis of wettability of the metal with a low-melting metal or non-metal melt. With this method of estimation, it is possible to separately find the susceptibility of both base and electrode (filler) metals to solidification cracking. In addition, this method is more objective, as this criterion characterizes a physical–chemical process, which occurs at the interface between a solid metal and a low-melting melt in contact with it.

9 Development of Welding and Surfacing Technologies with Allowance for Surface Phenomena

As evidenced by the above data, surface properties of welding consumables used for the fabrication of a welded structure and surface phenomena occurring at interfaces between the contacting phases play an important role in the processes of fusion welding. Transfer of electrode and filler metals, formation of a weld and deposited bead, shape and size of the penetration zone, and formation of different types of defects (undercuts, burns-through, lack-of-penetration, craters, lack-of-fusion, non-metallic inclusions, pores, solidification cracks) depend on them and should be taken into account in the development of welding and surfacing technologies. This chapter describes the development of some welding and surfacing technologies with an account for the surface phenomena by an example of specific parts and some welding and surfacing methods, as well as conditions, under which these processes occur.

9.1 WELDING OF COMPOSITE FIBER-REINFORCED MATERIALS

Composite fiber-reinforced materials find an increasingly wide application recently, especially in aircraft and transport engineering. These materials allow the weight of many structures to be markedly decreased, their strength, rigidity, and impact load resistance remaining at the same level or even increasing. In this section, we consider peculiarities of fusion welding of composite metal-matrix fiber-reinforced materials. Metal matrix of these materials contains reinforcing fibers, the volume fraction of which may amount to 50–60%.

Various methods of pressure joining, such as spot and seam welding, diffusion bonding, etc., are used currently to join composite materials. This is associated with the fact that the above joining methods do not lead to a marked heating of pieces joined, which means that a composite material will not lose its strength. Fusion welding is employed to a lesser degree for this purpose so far. However, given that this method shows high promise for fabrication of structures of composite fiber-reinforced materials, studies in this area are required.

9.1.1 PROBLEMS IN THE FUSION WELDING OF COMPOSITE MATERIALS

Fusion welding of composite materials may involve two problems: production of a joint along the fibers or across the fibers. In the first case, it is necessary to produce a welded joint, the strength of which should be at a level of that of the matrix material. This could be achieved by properly selecting the welding conditions. The second problem is much more difficult. In this case, it is necessary to join the fibers, the diameter of which is not in excess of $0.15-0.3$ mm with a matrix material. Currently, the fibers are most often made from high-strength steels, boron, carbon, or tungsten. Welding of the high-strength steel fibers is accompanied by their weakening. Welding of tungsten and carbon fibers is a very complicated engineering problem, while welding of boron has been little studied as yet. Moreover, simultaneous joining of tens or hundreds of fibers is also extremely difficult.

However, the process of welding a composite material does not stop with joining the fibers to each other. It is also necessary to weld to it a matrix material. Here, the decisive processes during welding are wetting and impregnation of fiber spacing with the metal melt. However, these processes have never been studied in terms of welding of composite materials and matrix–fiber systems employed.

Wettability should have a pronounced effect on the production of a sound fusion welded joint between composite materials reinforced by continuous fibers. Spontaneous impregnation of the fiber spacing with the metal melt is possible only when this process is accompanied by a decrease of free energy, ΔF in the fibrous material–melt system, that is, if the following inequality is met:

$$\Delta F < 0$$

For isothermal conditions, and in the case of negligible changes in volume of the phases, decrease of the free energy of the system is determined by the following expression:

$$\Delta F = \sigma_{s-g}\Delta S_{s-g} + \sigma_{l-g}\Delta S_{l-g} + \sigma_{s-l}\Delta S_{s-l}, \qquad (9.1)$$

where ΔS_{s-g}, ΔS_{l-g}, and ΔS_{s-l} are the variations in the values of surface areas at the interfaces with the melt flowing between the fibers; σ is the value of the interface energy.

As the area of contact at the melt–gas interface hardly changes with the movement of the melt, it can be assumed that $\Delta S_{l-g} = 0$. In addition, after the melt fills up the fiber spacing, the solid–gas interface is replaced by the solid–liquid interface. Thus, $\Delta S_{s-g} = \Delta S_{s-l}$. Given this fact, the condition of impregnation of the fiber spacing with the metal melt can be written down as follows:

$$\sigma_{s-g} - \sigma_{s-l} > 0. \qquad (9.2)$$

Allowing for the Young's equation, the condition for spontaneous impregnation will have the following form:

$$\sigma_{s-g}\cos\theta > 0. \qquad (9.3)$$

Therefore, the efficiency of using fusion welding for joining parts of composite fiber-reinforced materials depends in many respects on the wettability of fibers with the metal melt. Moreover, spontaneous penetration of the molten metal into the fiber spacing is possible only at $\theta < 90°$.

The processes of wetting and spreading of the melts occurring during welding of conventional metals are of a very complicated nature. They are even more complicated during welding of composite materials, where spreading of a metal melt occurs over the surface of fibers. In this case, the spreading process may be described with a certain inaccuracy as spreading of a liquid over a rough surface, the principles of which have been considered in chapter 3 (see Section 3.3). The effect of roughness on the wetting and spreading processes is estimated using roughness factor, k which is defined as the ratio of an actual surface area to the area of projection of the actual surface on a horizontal plane, that is, $k = F_{ac}/F_p$. It holds for a layer of fibers with diameter, d_f:

$$F_{ac} = \pi d_f/2; \quad F_p = d_f \qquad (9.4)$$

then the roughness factor is $k = \pi/2 = 1.571$.

According to the Ventsel–Deryagin equation, $\cos \theta_r = k \cos \theta_0$, where θ_r and θ_0 are the contact angles of the rough and smooth surfaces, respectively. It follows from this equation that at a poor wetting ($\theta > 90°$), an increase of the surface roughness leads to an increase of the macro contact angle. If a liquid wets a solid body ($\theta < 90°$), increase of roughness leads to a decrease of the contact angle. If $k < 1/\cos\theta_0$, the condition of complete wetting is met (at $\theta_0 < 90°$). It follows therefrom that complete wetting of a layer of fibers across their orientation can be achieved when the equilibrium contact angle is $\theta = \text{arc cos}\,(1/k) = \text{arc cos}(2/\pi) = 50.5°$.

Therefore, the presence of roughness on the surface of a part can provide complete wetting, whereas wetting of the smooth surface for the same materials will be limited. This is proved, for example, by the experimental data on the spreading of mercury over the surface of polycrystalline zinc. Spreading of mercury over the rough surface was complete, whereas on the smooth surface the contact angle was $10°$.[229]

It should be noted that wetting and spreading of a liquid over the smooth and rough surfaces of a solid body are characterized by fundamental differences. When a liquid spreads over the rough surface of a solid, on reaching a protrusion, this liquid will continue spreading only if it has a sufficient energy to overcome this protrusion. This overcoming can be facilitated by oscillations of the liquid, which can be either natural or artificially induced in the liquid, for example, ultrasonic oscillations. The presence of roughness also leads to a change in the contact angle. As is seen (Figure 9.1), the contact angle on the rough surface is substantially different from the equilibrium contact angle on the smooth surface. All this should be taken into account, while studying the processes of wetting and spreading of a liquid over the rough surface.

9.1.2 The Heterogeneous Nature of the Surface

Composite materials are classed with solids having a heterogeneous surface. The presence of large heterogeneous regions leads to the formation of additional energy

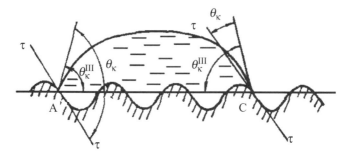

FIGURE 9.1 Drop of melt on a rough surface of solid body.

barriers between them and, thus, the state of a metastable equilibrium. A requirement for producing a continuous welded joint makes it necessary to completely fill up the gaps between the fibers with matrix material. In addition, it is important that the time of contact of the melt with fibers be as short as possible, as this determines the degree of weakening of a composite material taking place during fusion welding.

It should be noted that publications dedicated to investigations of peculiarities of welding composite materials are very few. Besides, all the authors note that fusion welding of composite materials induces irreversible processes accompanied by dissolution of fibers in the matrix to form intermetallic compounds. Thus, argon-arc welding of a composite material of the aluminum (matrix)–boron (fibers) system leads to complete fracture of fibers within the welding zone. A resulting weld is characterized by high brittleness. Negative results were obtained also in an attempt to use fusion welding to join the parts of a composite material, where titanium was used as the matrix and sapphire was used as single-crystal fibers. Welding of such a material resulted in a complete dissolution of sapphire in titanium, which led to the formation of a coarse brittle structure in the weld metal. Such welds, consisting of continuous cracks, fractured during cooling. Similar results were obtained also in fusion welding of composite materials of the aluminum (matrix)–steel (fibers) and aluminum (matrix)–tungsten (fibers) systems. Attempts to weld nickel reinforced with dispersed particles of thorium were unsuccessful either. As demonstrated by our experiments in the argon-arc welding of the composite material KAS-1A, where aluminum is used as the matrix and fibers are made from steel VNS-9, increase of the welding current and decrease of the welding speed lead to weakening of the composite material (Figure 9.2). This is associated to a considerable degree with fracture of the reinforcing fibers, occurring as a result of their interaction with the metal melt of the matrix (Figure 9.3). Therefore, decrease of time of contact of the matrix melt with the fibers should promote improvement in the quality of a welded joint.

9.1.2.1 The Capillary Forces

The time of contact of the metal melt with fibers can be reduced by increasing the rate of impregnation of fiber spacing with the metal melt. As the gap between the fibers in a composite material is very small, it can be considered as a capillary.

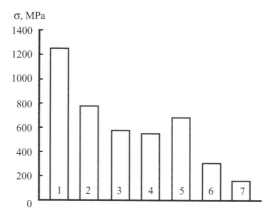

FIGURE 9.2 Effect of parameters of tungsten argon-arc welding on the strength of composite material KAS-1A. (1) material in initial state. I_w(A): (2) 25; (3,4) 45; (5–7) 45. V_w (m/s): (2,7) 0.003; (3,5) 0.005; (4,6) 0.005.

The process of movement of a liquid in capillaries has been sufficiently well studied, especially for cylindrical capillaries. A concave meniscus is formed in such capillaries, which are uniform in length, the capillary walls being well wetted with a liquid. Formation of the meniscus leads to the formation of a capillary pressure, which is equal to $2\sigma \cos \theta/R_c$. Along with the hydrostatic pressure, this pressure determines the movement of a liquid in a capillary. Here, σ is the surface tension of the liquid; θ is the contact angle; R_c is the capillary radius.

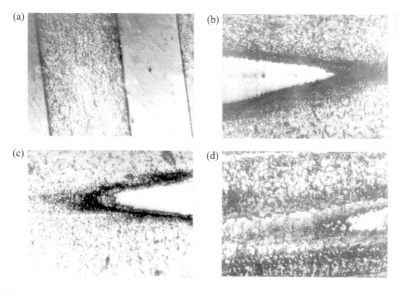

FIGURE 9.3 Fracture of fibers under the effect of the direct arc on composite material KAS-1A ($\times$ 500): (a) material in initial state. I_w (A): (b) 25; (c) 35; (d) 45.

We studied the movement of the metal melt through a slot capillary, as its shape is closest to the shape of a gap between the fibers. In this case, no account was made of the chemical interaction between the components of the composite material.

According to Maslov,[514] the time of penetration of a liquid via channels in its laminar movement to distance, L at low values of the hydrostatic pressure is as follows:

1. For a cylindrical capillary

$$\tau = \frac{2L^2 \eta_1}{\sigma R_c \cos\theta};$$ (9.5)

2. For a slot capillary

$$\tau = \frac{3L^2 \eta_1}{\sigma \delta \cos\theta}.$$ (9.6)

Here, η_1 is the viscosity of the liquid; δ is the width of the slot capillary.

It follows from Equation (9.5) and Equation (9.6) that irrespective of the shape of a capillary, the rate of impregnation is determined by viscosity and surface tension of the melt, as well as its capability of wetting a solid. All these parameters depend on the temperature, chemical composition of the substrate, and the melt and composition of the gas phase, while wettability additionally depends on the cleanliness of the substrate surface.

Distinctive feature of actual welding processes consists in the fact that the melt will move in a gap provided that a temperature gradient is present in this gap along its length. This may change to some extent the velocity of movement of the melt, as it was mentioned earlier. In addition, the values of σ, η_1, and θ would also change along the length of the channel, as they all depend on the temperature.

The value of surface tension would change to the least degree. According to our data, the values of σ for the aluminum melt measured in an argon atmosphere decreased from 867 to 792 mJ/m^2 with an increase of temperature from 973 to 1573 K. The values of η_1 and θ change to a larger degree. Therefore, it is these values that could have a substantial effect on the process of impregnation of a fiber gap with the metal melt.

Consider first of all how the presence of temperature gradient and the associated change in viscosity of the melt affect the velocity, at which it flows through the slot capillary.

The integral mean of viscosity of the melt along the length of the slot capillary equals[515]

$$\eta = (1/L) \int_0^L \eta(x) dx.$$ (9.7)

For a composite material, because of its small thickness, it can be assumed that variation in temperature in a direction of motion of the melt is of a linear nature.

Then

$$T = T_0 + \beta L, \quad \text{or} \quad T = T_0(1 + \alpha L), \quad (9.8)$$

where $\alpha = \beta/T_0$.

Out of the known dependencies relating variations in viscosity to temperature, the expression which is most often used is that derived in the study of Frenkel',[516] according to which

$$\eta = A \exp(E/RT), \quad (9.9)$$

where A is the pre-exponential factor; E is the activation energy of a viscous flow of liquid; R is the universal gas constant; and T is the temperature.

As for pure aluminum, $R = 20962 \text{ J/mol}$,[517] and at a temperature of $T = 1217$ K, viscosity $(\eta_0) = 0.7922 \cdot 10^{-2}$ П,[518] variation in viscosity of molten aluminum depending on the temperature can be written as follows:

$$\eta = 1.07 \cdot 10^{-3} \exp(2521/T). \quad (9.10)$$

By performing transformations similar to those made in the study of Ivanov study,[515] we find that, when the melt moves in a narrow (slot) gap at a temperature and viscosity varying along its length, the time of movement of the melt to distance, L is equal to:

$$\tau = 3L^2 \eta_0 \frac{\left(1 + \sum_{m=1}^{\infty} y_m L^m\right)}{\sigma \delta \cos \theta} \quad (9.11)$$

Coefficients y_m can be found from the following expressions:

$$y_1 = -\alpha a/3; \; y_2 = \frac{\alpha^2(a + a^2/2)}{6}; \; y_3 = \frac{\alpha^3(a + a^2 + a^3/6)}{10}, \quad (9.12)$$

etc., where

$$a = A/T_0 \; (A = E/R).$$

Comparison of conditions of penetration of the aluminum melt into the narrow gap (cm): $\beta_1 = 0$, $\beta_2 = -65$, $\beta_3 = -150$, $\beta_4 = -300$, and $\beta_5 = -500$ (where β_1 is the temperature gradient) yields that:

$$\tau_1 = 3\eta_0 L^2 / \sigma \delta \cos \theta;$$

$$\tau_2 = \tau_1(1 + 0.025L + 0.0015L^2 + 0.0000634L^3);$$

$$\tau_3 = \tau_1(1 + 0.581L + 0.00055L^2 + 0.00079L^3); \quad (9.13)$$

$$\tau_4 = \tau_1(1 + 0.116L + 0.0022L^2 + 0.0064L^3);$$

$$\tau_5 = \tau_1(1 + 0.194L + 0.063L^2 + 0.031L^3).$$

9.1.3 Experimental Observations

As seen above, the time necessary for the melt to penetrate to the same depth grows by a factor of K with increase of temperature gradient. The values of the factor K, depending on the temperature gradient β and depth of penetration of the melt, are given in Table 9.1 Therefore, variations in the viscosity of the melt occurring due to the presence of the temperature gradient along the length of the channel will have a substantial effect on penetration of the melt only, if the penetration depth and temperature gradient have sufficiently high values.

Variations in the value of contact angle, θ must have a greater effect on the penetration of the melt. For example, with the contact angle decreasing from $80°$ to $20°$ the time of penetration of the melt into the narrow gap decreases almost by a factor of 5.5, other parameters included into Equation (9.6) being constant.

We conducted experiments to study the effect exerted by the arc discharge on the process of penetration of the aluminum melt into a narrow gap for titanium and steel capillaries.

The experiments were conducted on samples with a varied narrow gap simulating the fiber spacing in a composite material. A setup described in Section 2.5 was used in the experiments. The experiments were carried out to study the effect of the substrate material, its initial temperature (T_{sub}), presence of ultrasonic oscillations and their parameters on the velocity of penetration of the aluminum melt into the narrow gap. The velocity of penetration (V_p) of the aluminum melt was determined as the L_{max}/τ_{arc} ratio, where L_{max} is the maximum depth of penetration of the melt into the gap, and τ_{arc} is the arcing time.

The experiments were carried out as follows. The substrate assembled with a fixed gap was placed on a mount. An aluminum sample 180–200 mg in weight was placed on the substrate. The arc burning in the argon atmosphere was ignited between the sample and a tungsten electrode. Temperature of the substrate was measured using a tungsten–rhenium thermocouple (VR 5/20). Ultrasonic oscillations, the frequency of which was kept constant during the experiments at 24 kHz, were excited using the URSK-7-N-18 generator. The ultrasonic oscillations were applied to the substrate.

The first experiments showed that in arc melting of aluminum placed on the steel substrate the melt did not penetrate into the gap even when the welding current was

TABLE 9.1
Values of Factor K Depending on the Values of Temperature Gradient and Melt Penetration Depth

Melt Penetration	Values of Factor K at β			
Depth, cm	$-65°/cm$	$-150°/cm$	$-300°/cm$	$-500°/cm$
0.2	1.003	1.012	1.024	1.04
0.5	1.013	1.030	1.100	1.12
1.0	1.030	1.060	1.150	1.29

increased to 100 A. This is attributable to the fact that an oxide film remained on the surface of a drop in contact with the substrate surface, and that this film prevented penetration of the melt into the gap. In the experiments that followed, the substrate surface was covered with 10–15 mg of flux AF-4a. This led to the removal of the oxide film from the rear side of the drop and provided penetration of the aluminum melt into the gap.

As can be seen (Figure 9.4 and Figure 9.5), the velocity of penetration of the aluminum melt through the narrow gap depends on the substrate material. This velocity is higher in the case of a titanium substrate, and lower in the case of substrates of St.3 and 12Kh18N10T steels. Variation in the size of the gap from 0.3 to 1.0 mm leads to an insignificant variation in the velocity of penetration of the melt. Variations in V_p are affected to a much greater degree by the welding current. Thus, increase of I_w by a factor of 3.5 leads to increase of V_p by a factor of 4–5.

Preheating of the substrate leads to an increase of V_p (Figure 9.6). However, the effect of the welding current on variations in the V_p value remains almost the same as in the case, where the melt flows through a slot capillary without preheating.

Comparing the data on the wettability of the substrate materials used in these experiments by the aluminum melt with the results obtained from the evaluation of the velocities of penetration of the melt into the narrow gap shows that the velocity of penetration increases with decrease of the contact angle.

The velocity of impregnation substantially grows under the effect of ultrasonic oscillations. For example, in the case of using 12Kh18N10T steel as the substrate, at

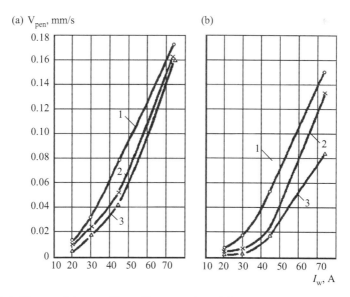

FIGURE 9.4 Change of the rate V_{pen} of impregnation of the through-slot capillary in steels: St.3 (a) and 12Kh18N10T (b) by aluminum melt depending on the welding current value ($T_m = 293$ K). δ (mm): [1] 0.3; [2] 0.7; [3] 1.0.

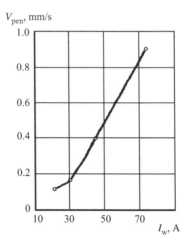

FIGURE 9.5 Change of the rate V_{pen} of impregnation of the through-slot capillary in titatinm by aluminum melt depending on the current value. $T_m = 293$ K; $\delta = 0.7$ mm.

a gap of 0.3 mm and an oscillation amplitude of 50 μm, the value of V_p increased by a factor of 6 (Figure 9.7). Increase of the amplitude of ultrasonic oscillations leads to increase of the velocity of penetration of the melt into the narrow gap. This is especially pronounced in the case of small gaps.

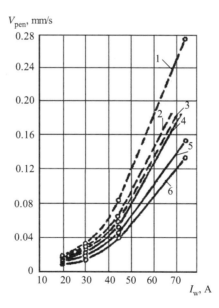

FIGURE 9.6 Change of the rate of impregnation of the through-slot capillary by aluminum melt depending on the welding current value ($T_m = 473$ K): (——) without substrate preheating; (– – –) with substrate preheating. δ (mm): (1), (4) 0.3; (2), (5) 0.7; (3), (6) 1.0.

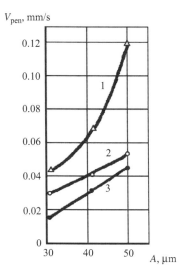

FIGURE 9.7 Change of the rate of impregnation of the through-slot capillary in 12Kh18N10T steel depending on ultrasonic oscillations amplitude, A ($T_m = 293$ K; $I_w = 30$ A). δ (mm): (1) 0.3; (2) 0.7; (3) 1.0.

It is of interest to note that Equation (9.5) and Equation (9.6) fail to comprehensively describe the obtained experimental data. This is attributable to the fact that these equations are valid, if no chemical interaction occurs between the melt and the capillary material. When the aluminum melt flows through steel capillaries, this induces an intensive chemical interaction between them. This may result[519] in a substantial increase of the viscosity of the melt. In turn, this leads to a decrease of the velocity of its movement. In addition, the arc discharge also has its influence. When the arc acts on the melt, it penetrates to a greater degree in the arc center, that is; where the arc temperature and pressure are higher, and to a lesser degree on the periphery, where these parameters are lower.

The experimental data obtained were used for development of the technology for welding composite material KAS-1A. In the case of welding, for a composite material with a volume content of fibers of 30%, the strength of a welded joint amounted to 70–82% of the initial strength of the material.

9.2 ELECTRON BEAM SURFACING

Electron beam surfacing using traditional standard units with vacuum chambers has a limited commercial application. First of all, this is associated with high costs of such works and a low productivity of the process, as prior to surfacing it is necessary to create vacuum in the chamber, and after surfacing the chamber should be depressurized.

Meanwhile, high-power (≥ 50 kW) electron accelerators have been developed in the last years. They allow the beam of electrons to be ejected into the atmosphere,

which enables surfacing of the parts to be performed in open air. In turn, this eliminates restrictions on sizes of workpieces and leads to rise in the process productivity.
The feasibility of using a beam of relativistic electrons for welding and surfacing was
proved in studies by A.N. Skrinsky, L.P. Fominsly, and V.A. Gorbunov,[520] who noted a
high productivity of the process. However, publications available on this subject com-
prise no systematic analysis of the effect exerted by individual parameters of the
process on quality and formation of a deposited bead. No experimental data are available either on saturation of metal with gases, which is important because in this case the
surfacing process is conducted with no shielding of the molten metal.

9.2.1 Surfacing—Experimental Findings

We carried out experiments on the surfacing of copper and steel plates using the ELV-4
electron accelerator. The block diagram of this accelerator is shown in Figure 9.8. The
scanning electron beam with an energy of $E = 1.5$ meV and power of $P = 30-50$ kW
processed a sample moving normal to the path of movement of the beam.

Similar surfacing consumables as discussed in Chapter 5 (see Section 5.4) were
used for surfacing: copper plates 4.5- and 12-mm thick (δ_{pl}), such as a copper
powder; copper plates $\delta_{pl} = 2.5$ mm; powders PR-N80Kh12S2R, PR-N70
Kh17S4R4, and PR-KhN80SR2.

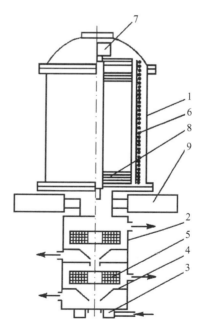

FIGURE 9.8 ELB-4 type electron beam accelerator: (1) accelerator; (2) device for
concentrated release of the electron beam into the atmosphere; (3) blower nozzle;
(4) diaphragms; (5) electromagnetic lenses; (6) primary winding; (7) injector; (8) secondary
winding; (9) magnetic discharge pumps.

9.2.1.1 Effect of Composition

First of all, the investigations were conducted to study the effect of composition of a surfacing consumable applied, thickness of the powder layer and surfacing speed on the depth of penetration of a copper plate. As is seen (Figure 9.9), with the copper powder used, the penetration depth (H_{pen}) is slightly larger than in surfacing using the chromium–nickel powders. For all of the chromium–nickel powders, the penetration depth was almost identical. Increase of thickness of the powder layer for all materials investigated (Figure 9.10) led to decrease of the penetration depth. The penetration depth also decreased with increase of the surfacing speed. The effect of the surfacing speed on the depth of penetration of base metal was more pronounced (Figure 9.10) at small thickness of the powder layer deposited on the surface of a sample processed.

9.2.1.2 Effect of Gases

As the surfacing process was performed without extra shielding of the molten metal, it is highly possible that in this case, a considerable amount of gases would be transferred to the deposited metal. This may lead to deterioration of the mechanical properties of the deposited metal and to porosity. The gas (O_2 and N_2) content of the deposited metal was estimated using a gas analyzer EAO-202 by the vacuum melting method. It was established that electron beam surfacing performed in open air provided (Table 9.2) sufficiently good shielding of the molten metal from interaction with the atmosphere. This is proved by a low nitrogen content of the deposited metal (0.002–0.005 wt%). It is likely that this occurs because of formation of metal vapors due to the electron beam acting on this metal

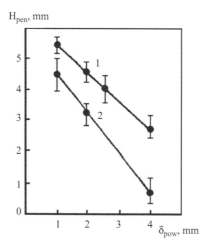

FIGURE 9.9 Dependence of sample penetration depth on the thickness of deposited layer at $I_w = 30$ mA, $v_s = 0.008–0.01$ mm/s, beam scaning frequency $v = 8–9$ Hz, and $\delta_{pl} = 12$ mm: (1) M1; (2) PG-KhN80SR2, PR-N70Kh17S4R4, PR-N80Kh13S2R.

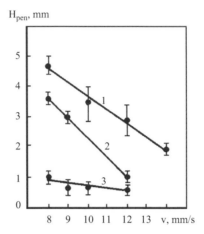

FIGURE 9.10 Dependence of penetration depth on surfacing speed for PR-N80Kh13S2R, PR-N70Kh17S4R4, and PG-KhN80SR2 powders: $I_w = 30$ mA; thickness of copper plate, $\delta_{pl} = 12$ mm; beam scanning frequency, $\nu = 8-9$ Hz; δ_{pow} values are 1, 2, 4 respectively for 1, 2, 3.

(see Section 2.2), and then the vapors press out air from the metal melt. In addition, the chrome–nickel powders used for surfacing are self-fluxing, which also promote shielding of metal from interaction with the atmosphere.

9.2.1.3 Effect of Deposited Metal

In addition, the effect of the deposited and base metals on the formation of pores and cracks in the deposited metal was studied visually and by the x-ray inspection method. It was found that the deposited metal usually contained one or two cracks (small in size), per 100 mm of the deposited layer.

TABLE 9.2
Dependence of the Gas Content and Porosity of the Deposited Metal on the Base and Deposited Metal Compositions

Base Metal	Powder Grade	Gas Content of Deposited Metal (wt%) O_2, N_2	Volume of Pores (mm³) Per 100 mm of Deposited Metal
Steel St. 3	PR-N80Kh13S2R	0.01–0.02, 0.002–0.004	0
	PR-N70Kh17S4R4	0.075–0.09, 0.013–0.005	8
Steel 30KhGSA	PR-N80Kh13S2R	0.02–0.03, 0.005–0.006	4
	PR-N70Kh17S4R4	0.02–0.03, 0.003–0.005	3
Copper M1	PR-N70Kh17S4R4	0.015–0.02, 0.003–0.005	6

As pores present in the deposited metal were of different size, porosity of the deposited metal was estimated from the total volume of pores determined from the following formula:

$$V = \sum_{d_{min}}^{d_{max}} = (\pi d^3/6)n,$$

where V is the volume of pores per 100 mm of the deposited metal, mm^3; d is the diameter of a pore, mm; and n is the number of pores of a given diameter per 100 mm of the deposited bead.

The low gas content of the deposited metal provides a pore-free deposited bead. As seen from Table 9.2, porosity is higher in those cases, where the deposited metal contains more oxygen and nitrogen. This coincides with the data given in Chapter 7.

Therefore, surfacing of the materials studied using the electron beam directed to the atmosphere provides the deposited metal with a low content of O_2 and N_2, low porosity, and good crack resistance.

The picture observed in surfacing of St. 3 steel with the iron-based powder M6F3 is different. In this case, the deposited metal contains a large number of pores ($V > 50$ mm^3), as well as undercuts, rolls, and hot cracks. This is attributable to the fact, that this powder is not self-fluxing, which leads to a more pronounced transfer of gases from air to the metal.

9.2.1.4 Effect of Beam Power

Results of experiments conducted to study the effect of the beam power on the deposited metal formation (Figure 9.11 and Figure 9.12) are indicative of the fact that a good formation may take place only in a certain power range, which provides sufficient preheating of the base metal. Optimal values of the power, which provide a good formation of the deposited layer, depend on the amount of the powder melted and thickness and composition of plates treated.

As is seen from Figure 9.11, surfacing of copper plates at low values of power leads to the formation of drops of the deposited metal. Increase of power eliminates formation of drops, still leaving bridges in the deposited layer which can be removed by further increasing the power.

Thus, surfacing using chromium–nickel powders provided a good formation of the deposited metal on copper plates 12-mm thick at the following values of the electron beam power, Q for a corresponding thickness of the powder layer applied to the surface of a copper plate:

Q, J/M^3	Thickness (S), mm
$116 \cdot 10^{12} - 140 \cdot 10^{12}$	1.0–1.5
$48 \cdot 10^{12} - 77 \cdot 10^{12}$	2.0
$29 \cdot 10^{12} - 36 \cdot 10^{12}$	4

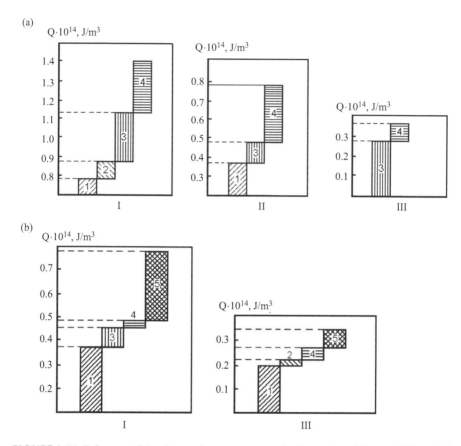

FIGURE 9.11 Influence of the electron beam energy on the formation of the metal deposited on copper of thickness 2 mm (a) and 5 mm (b): [I] PR-N80Kh13S2R, PR-N70Kh17S4R4, PG-KhN80S2R; [II] PG-KhN80S2R, PR-N80Kh13S2R; [III] PR-N80Kh13S2R, PR-N70Kh17S4R4. [1] area of drop formation; [2] area occurence of bridges; [3] area narrowing of deposited bead; [4] area of good formation of deposited bead; [5] area of deep penetration; δ_{pow} (mm): (I) 1.0–1.5; (II) 2; (III) 4.

In surfacing of copper plates 12-mm thick with a copper powder (layer 1–2 mm thick) or copper plate 2.5-mm thick, a good formation of the deposited layer can be provided within a wider power range, that is, $Q = 40 \cdot 10^{12} - 128 \cdot 10^{12}$ J/m^3.

For a thinner copper plate ($\delta = 4.5$ mm) the range of optimal power values is narrower. For example, with a thickness of the powder layer equal to 1–1.5 mm and $Q = 46 \cdot 10^{12} - 49 \cdot 10^{12}$ J/m^3, and at a thickness of the powder layer equal to 3.5–4 mm, $Q = 22 \cdot 10^{12} - 25 \cdot 10^{12}$ J/m^3. If the power exceeds the optimal range, it leads to burns-through of the base metal.

Therefore, formation of the deposited metal in surfacing of copper parts using powdered materials and plate electrodes depends on wettability of the base metal

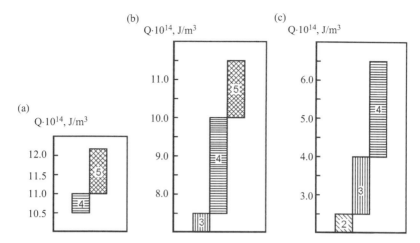

FIGURE 9.12 Influence of the electron beam energy on the formation of the metal deposited on St.3 steel ($\delta_{pl} = 5.0$ mm). δ_{pow} (mm): (a) 1.0 (b) 1.5–2.0, (c) 4.0–4.5. [2] area occurence of bridges; [3] area narrowing of deposited bead; [4] area of good formation of deposited bead; (5) area of deep penetration.

with the metal melt. The required wettability can be provided by proper selection of surfacing conditions.

Studies of surfacing of steel parts (St.3) using powdered materials showed that in this case the deposited metal formation also depends on the value of the electron beam power (Figure 9.12). However, the required values of power in this case are 2–3 orders of magnitude lower than in surfacing of copper parts. This is associated primarily with lower thermal conductivity of steel, compared to copper. Therefore, less heat is needed to heat the steel to a required temperature. For this reason, in electron beam surfacing of steel parts using powdered materials, the range of optimal values of the power is narrower than in surfacing of copper plates. For example, in surfacing over the powder layer 1-mm thick, the power values ensuring good formation of the deposited metal may vary only in the range, $10.5 \cdot 10^{10}$–$11.0 \cdot 10^{10}$ J/m^3. For those with a powder layer 4–4.5-mm thick, it is in the range $4.0 \cdot 10^{10}$–$6.5 \cdot 10^{10}$ J/m^3.

The studies performed were used to develop technologies for surfacing of copper plates of moulds for continuous casting of steel and knives for cutting chemical fibers.

9.2.1.5   Important Conclusions

It was suggested, that PR-N80Kh13S2R powder should be used as a surfacing consumable for surfacing copper plates of the moulds (Figure 9.13), as this powder provides a deposited layer of high microhardness (250–500 kg/mm^2), with good formation, and no pores and cracks contained in the metal. Surfacing was performed into slots 10×10 mm^2 and 300-mm long in three passes. For the first and second

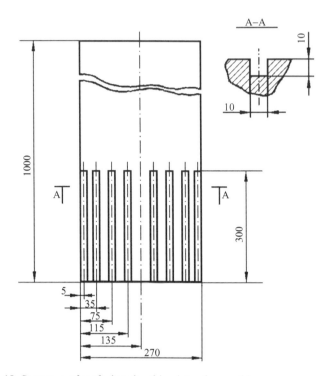

FIGURE 9.13 Sequence of surfacing the side plate of a mould.

passes, the powder layer should be 4–5-mm thick, while in the third pass, the powder layer should be 1–2-mm thick. Surfacing parameters for making the first and second passes are as follows: $I_s = 18$ mA, $V_s = 0.001$ m/s, and $v = 3$ Hz. The following parameters are recommended for deposition of the third layer: $I_s = 6$ mA and $V_s = 0.0008$ m/s. Use of the electron beam for surfacing allowed all the passes to be made without preheating of a workpiece, which would have been impossible with using other welding heat sources at the given surfacing speed.

Billets of St.3 steel and PR-N70Kh17S4R4 powder were used for surfacing the cutting edges (Figure 9.14) of knives for cutting chemical fibers. To raise the process productivity, surfacing was performed on four billets simultaneously. Surfacing parameters were as follows: $I_s = 10–12$ mA, $V_s = 0.002–0.0024$ m/s, and $v = 6–7$ Hz.

A more detailed description of the electron beam surfacing process is given in our study.[521]

9.3 WELDING AND SURFACING OF COPPER–STEEL PARTS

The metal fabrication industry is often faced with the necessity to manufacture the most diverse copper–steel parts. This relates to chemical equipment, moulds, blast furnace and conveyer tuyeres, refrigerating equipment, electrical engineering parts, etc.

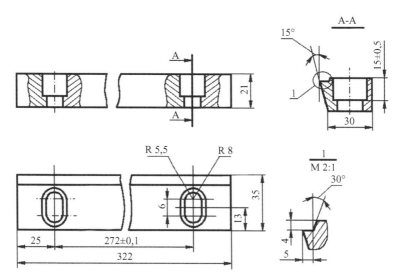

FIGURE 9.14 Blade blank for surfacing.

The parts, the manufacture of which requires a large scope of operations associated with joining copper–steel pipes are air separation units. Such parts are commercial cryogenic units intended for the production of liquid and gaseous air components. These units are subjected to substantial thermo-mechanical loads during the operation. This imposes increased requirements on the quality of the copper–steel joints.

Considerable experience has been accumulated in the field of producing such joints by flame brazing. However, the joints used in this case are of the bell-and-spigot type, which leads to a considerable increase of material consumption, and the filler metal is a silver-containing alloy of the PSr-45 (30% Cu; 45% Ag; 25% Zn) grade, which is rather expensive. In addition, the method of flame brazing has low productivity and requires highly skilled workers.

Decrease of labor consumption, saving of expensive scarce materials, and improvement of quality of welded joints are the main lines of upgrading of the technology of joining copper–steel pipes used in the air separation units. Application of the welding process for joining the copper–steel pipes instead of brazing is a promising way of achieving the above purposes. The tungsten argon-arc welding method can be used for this purpose, as it is characterized by versatility and sufficiently high productivity.

9.3.1 COPPER–STEEL PIPES

The results of studies given in Section 3.5 made it possible to start development of the technology for welding the copper–steel pipes.

As was shown earlier, contact of copper melts with steel is accompanied by a substantial decrease (Figure 9.8) in ductility of the solid metal, which may lead to formation of solidification cracks in the weld. Considering the results of the studies performed, brass LK62-0.5 can be recommended for use as the filler metal for welding the copper–steel pipes. When in contact with St.3 and 12Kh18N10T steels, the melt of this brass decreases their ductility to a lesser degree. However, the use of a brass wire under the arc welding conditions tends to cause intensive burning out of Zn, which leads to a decrease of the mechanical properties of the weld metal and possibly, to formation of pores in the weld. Therefore, in this case, it is advisable to use wires of MNZhKT5-1-0.2-0.2 and BrKMts3-1 grades.

In welding using a filler, the quality of the weld and its formation depend in many respects on the flow of the molten metal into the gap between the parts joined. Results of earlier experiments (Figure 9.8) demonstrate that MNZhKT5-1-0.2-0.2 wire provides a better flow of the melt under the effect of the arc discharge. This is particularly the case, when using plates of St.3 steel.

The flow of the melt into the narrow gap depends (see Equation (9.6)) also on the size of the gap between the parts being joined. The effect of size of the gap on the flow of the filler metal melt between the plates in shown in Figure 9.9. As is seen, under the arc welding conditions, in the case of using plates of St.3 steel, the best flow of the melt for any gap size was provided by alloy MNZhKT5-1-0.2-0.2. For steel 12Kh18N10T, the best flow of the melt was provided by the copper melt and a slightly worse flow by the melt of alloy MNZhKT5-1-0.2-0.2. For the latter, the optimal gap size ranges from 0.6 to 0.8 mm.

9.3.1.1   Crack Resistance

Experiments were conducted to determine operational strength of copper–steel, steel–steel, and copper–copper welded joints in order to evaluate their crack resistance. This was done using the Houldcroft-type test on plates 3.0-mm thick. Bead-on-plate welding was done by the argon-arc method under the following conditions: $I_w = 120$ A, $U_{arc} = 14$ V, and $V_w = 0.0011$ m/s. These conditions provided the required penetration and shape of the deposited bead. The experimental results are given in Table 9.3. As is seen, the largest cracks in the deposition on the plates of St.3 steel resulted from using wire of copper M3r, that is, the material,

TABLE 9.3
Length of Cracks in Deposition of Copper Materials on the Houldcroft-Type Test Specimens

Base Metal	Length of Cracks in Using Filler Material, mm		
	M3r	BrKMts3-1	MNZhKT5-1-0.2-0.2
St.3	25	15	20
12Kh18N10T	55	70	65
M3r + 12Kh18N10T	No	No	No
M3r	No	No	No

the melt of which decreases the ductility of this steel to the greatest degree. For steel 12Kh18N10T, the largest cracks resulted from using wires of copper alloys BrKMts3-1 and MNZhKT5-1-0.2-0.2. The results obtained are in good agreement with the available data on the wettability of steels by copper melts and decrease of ductility of steels in contact with the copper melts.

As seen from Table 9.3, in argon-arc welding of steel–steel joints, the cracks are formed in all the cases. Moreover, the better the copper melt wets the given metal, the larger the cracks. Welding of copper and copper–steel joints did not result in cracking. The absence of cracks in these joints can be explained as follows. First, copper retains its high ductility up to a high temperature (800 K). Second, liquid metal in contact with solid steel has a composition differing from that of the filler wire, as it contains components of solid steel dissolved in it. For example, in argon-arc welding of copper M3r to steel 12Kh18N10T using filler wire MNZhKT5-1-0.2-0.2, the weld metal has the following composition (wt%): 0.02–0.03 C, 58–63 Cu, 16–21 Fe, 5.7 Cr, 4.3 Ni, 0.37–0.47 Mn, 0.1–0.2 Si, and 0.05–0.12 Ti. Melting point of the metal of this composition is approximately 1470–1480 K. This is higher than the melting point of the MNZhKT5-1-0.2-0.2 alloy, and higher than the temperature at which the effect of decrease in the ductility of solid steel in contact with the copper melt is manifested.

Strength of the weld metal in the case of using the filler wire MNZhKT5-1-0.2-0.2 was 280–399 MPa, and elongation was 30–34%, which is higher than the corresponding values for copper of M3r grade.

The results obtained allowed the following technologies to be offered for welding copper–steel pipes of the air separation units. Welding is performed in two passes. First a copper edge is preheated by the welding arc at a current of 130–150 A, until it begins to melt. Welding of the root weld is performed at a current of 120–130 A, arc voltage is 14 V, at a gap between the weld edges equal to 0.6–0.8 mm. Alloy MNZhKT5-1-0.2-0.2 is used as a filler wire, the wire diameter being 2.0 mm. Diameter of the tungsten electrode is 3.0 mm, the argon flow rate is $(13.3–17) \cdot 10^{-5} \text{ m}^3/\text{s}$, and the welding speed is 0.0011–0.00167 m/s. Welding of the outside weld is performed under the same conditions as welding of the root one. Only the welding current is decreased to 100–120 A.

Inspection of the quality of the welded joints conducted by the x-ray methods, study of mechanical properties, metallography, and field tests demonstrated the high quality of welded joints between the copper and steel pipes of the air separation units made by the offered technology.

9.3.2 WETTING AND SPREADING

Wetting and spreading of the metal melt over the surface of the solid metal should be allowed for also in the development of the technology for deposition of copper and its alloys on steel parts. In this case, the general trend is to decrease the depth of penetration of base metal, as this minimizes transfer of the components to the deposited metal and increases the adhesion values. The value of the penetration depth and, therefore, transfer of components of the base metal to the deposited one depend first of all on the methods and parameters used for the deposition.

The depth of penetration of the base metal can be decreased by using vibration arc surfacing. This method provides decrease of transfer of iron to the deposited metal (Table 9.4) which, as is seen, depends to a certain degree on the welding consumables used.

However, with this surfacing method, the base metal is heated to a much lesser extent than in surfacing without electrode vibration. Therefore, the spread of the copper melt over the surface of the solid metal is worse in this case.

One of the first studies on wettability, which served as the basis for the development surfacing technology of steel with copper was completed in 1962. It was performed to investigate[522] the wettability of steel by different copper-based alloys. The study noted the relationship between the thickness of the plate treated and the area of spreading of the copper melt. Decrease of thickness of the plate was accompanied by increase of the area of spreading of the melt, which is associated with better heating of a thinner plate. In all the cases, spreading occurred rapidly and was limited by the isothermal line of steel at a temperature of about 1180 K. A part was heated mostly due to the heat of the arc. The better spreading of the copper melt took place in the case of using helium as a shielding gas. Surfacing in a gas mixture of $Ar + 0.25\%$ O_2 or in the atmosphere of pure Ar failed to provide a good formation of the deposited metal. This is likely to be associated with a lower thermal power of the arc discharge and, hence, insufficient heating of a steel part.

9.3.2.1 Important Parameters

The spreading area can be increased, either through increasing power of the arc or through weaving of the electrode (arc). The first way leads to increase in the depth of penetration of the base metal, as well as increase of transfer of iron to the deposited metal, which is undesirable. The second approach is more efficient, as weaving of the electrode or arc leads to increase in the area of spreading of the melt over the surface of the base metal and reduction in its penetration depth, thus decreasing transfer of iron to the deposited metal.

The investigations conducted allowed the development and successful application of the technology for surfacing of leading bands of the artillery shells.

The investigation results given in the study of Sivkov et al.[254] and presented in Section 3.3 allowed bronze BrOS8-12 to be used as a filler metal for surfacing large-sized friction pairs (sliding supports of walking excavator ESh65.100). This choice of the material was based on its better wettability and spreading over the steel surface. Also, it provided high adhesion between the phases.

Bronzes are used as inserts to be joined to pipes of commercial iron in electron beam welding of cylindrical bodies of electromagnetic valves (Figure 9.15). In this case, the sag of the metal in the weld root is permitted, because the welded billets of the bodies are subjected further on to machining on the inside and outside to a wall thickness of 1.5 mm.

TABLE 9.4

Dependence of Iron Content of the Deposited Metal on the Welding Consumables Used in the Case of Vibration Arc Surfacing

#	Base Metal	Electrode Wire	Shielding Gas	Thickness of Deposited Layer, mm	Iron Content of Deposited Metal, wt%
1	Steel 45	M1	Argon	2	2.38
				4	1.24
				6	0.40
		BrKMts3-1		2	2.30
				4	1.01
				6	0.40
2	Steel 45	M1	Nitrogen	2	2.32
				4	1.01
				6	0.42
		BrKMts3-1		2	2.30
				4	1.07
				6	0.4
3	Steel 45	M1	Air	2	2.98
				4	1.42
				6	0.51
		BrKMts3-1		2	2.63
				4	1.18
				6	0.43
4	Steel 20	M1	Argon	2	2.41
				4	1.28
				6	0.41
		BrKMts3-1		2	2.40
				4	1.16
				6	0.40
5	Steel 20	M1	Nitrogen	2	2.59
				4	1.16
				6	0.46
		BrKMts3-1		2	2.37
				4	1.06
				6	0.44
6	Steel 20	M1	Air	2	3.16
				4	1.58
				6	0.50
		BrKMts3-1		2	2.84
				4	1.22
				6	0.46

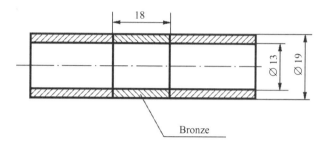

FIGURE 9.15 Sketch of a sample for optimization of the technology for electron beam welding of commercial iron to bronze.

One of the problems which had to be solved to develop the technology for welding of the given part consisted in finding conditions to ensure normal weld formation. As welding of iron to bronze BrAZhMts10-3-1.5 resulted in crack-free welded joints, and the depth of penetration of the low-melting melt into iron was small in this case, the investigations were conducted particularly for this pair of metals. Weld formation was considered satisfactory, if the edges were welded through the entire thickness (due to melting and mutual solidification of metals), and one-sided undercuts (resulting from spreading of bronze over the outside surface of the body) were absent.

9.3.2.2 Deflection of Electron Beam

The experiments showed,[523] that a satisfactory weld formation could be achieved due to deflection of the electron beam from the joint to the bronze or iron surface. However, in this case, the range of permissible deflection of the beam is very small. For example, when the beam is deflected to bronze, the optimal deflection range is from 0 to 0.25 mm. At a larger deflection of the beam, no melting of iron in the weld root takes place (Figure 9.16).

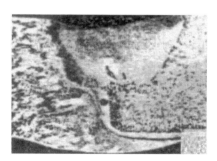

FIGURE 9.16 Macrosection of a joint between bronze and commercial iron with the electron beam shifted toward bronze to 0.5 mm.

Similar picture is observed in the deflection of the electron beam to iron. The sag of the weld surface decreases with increase of the welding speed from 0.0028 m/s to 0.0166 m/s due to deterioration of spreading of bronze over the iron surface. However, increase of the welding speed makes it more difficult to maintain the preset deflection of the electron beam. The task can be made simpler, if welding of the bodies is performed by oscillating the electron beam along and across the joint. In this case, a better heating of the parts provides good wetting of solid iron with the bronze melt, which leads to elimination of the undercuts. An excessive sag of the weld metal was avoided owing to increase of the welding speed to 0.014–0.0166 m/s.

9.4 MANUFACTURE OF METAL CUTTING TOOLS WITH A DEPOSITED CUTTING EDGE

Extension of life of cutting tools used for the machining of metals is still a challenge. An insufficient life of metal cutting tools is often a factor, which makes it impossible to fully realize the technological capabilities the machine tools are designed for, as well as to achieve high technical–economic indices of machining of the parts. One of the ways of handling this challenge is to develop a composite tool by depositing high-alloy tool steel on a billet of the cutting edge made from the conventional structural steel. In this case, it is important, that the metal has no pores and cracks, in addition to having a high wear resistance. Electrodes OZI-3 and OZI-5, providing the deposited metal of high hardness, were used as surfacing consumables, and the body material was 40X steel.

Chemical composition of the deposited metal resulting from using electrodes OZI-3 was as follows (wt%): $C = 0.9$–1.1; $Mn = 0.7$; $Mo = 2.5$–4.0; $Cr = 3.0$–4.0; $Si = 0.8$; $W = 0.9$–1.7; $V = 0.6$–1.3. In surfacing using electrodes OZI-5 the deposited metal had a different chemical composition (wt%), such as: $C = 0.1$; $Mn = 0.3$–0.5; $Si = 0.5$–1.0; $Cr = 2.0$–3.0; $Mo = 8.0$–11.0; $W = 9.0$–12.0; $Co = 16.0$–19.0; $V = 0.5$–0.8.

9.4.1 HARDNESS

Hardness of the deposited metal greatly depends on the share of the base metal in the deposited one. This is proved by the experimental results shown in Figure 9.17. As can be seen, in indirect arc surfacing of steel 40X samples, where there is a marked transfer of the base metal to the deposited one, hardness of the deposited metal is lower than 55 HRC_e in the case of using OZI-3 and OZI-5 electrodes. Hardness of the deposited metal increases almost to 60 HRC_e in three-layer surfacing using OZI-5 electrodes and to 62 HRC_e in surfacing with OZI-3 electrodes. The indirect arc used for surfacing allows hardness at a level of 63–65 HRC_e to be achieved already in the first layer, which is caused by a decrease of the share of the base metal in the deposited one.

In the manufacture of cutting tools, the cutting edges of which are made by surfacing with OZI-3 and OZI-5 electrodes, the hot cracks, the presence of which is

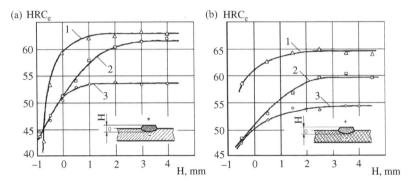

FIGURE 9.17 Variation in the hardness of the deposited metal through thickness in the case of electrodes: OZI- 3 (a) and OZI-5 (b). Arc surfacing mode: [1] one-layer non-transferred; [2] three-layer transferred; [3] one-layer transferred.

inadmissible, are often formed in the deposited metal. Their formation may be associated with the Rebinder effect, which can take place in the presence of tensile stresses in the solidifying metal and a low-melting medium, which readily wets the metal crystals.

In collaboration with E.N. Zubkova and E.P. Safonov,[524] we studied the formation of temporary stresses on samples of R18 and 08Kh17T steels. The studies were conducted using the testing machine IMASh-20-75 by simulating a welding thermal cycle. Analysis of the data obtained (Figure 9.18) showed that cooling of the rigidly fixed samples of R18 and 08Kh17T steels led to increase of tensile stresses. For steel 08Kh17T, the increase of tensile stresses occurs with cooling to room temperatures. In this case, the intensity of growth increases with decrease of temperature. This is likely to be associated with increase of strength and decrease of ductile properties of the metal. For R18 steel, the increase of tensile stresses occurs with cooling of the metal to a temperature close to that of the beginning of martensitic transformation. Further, cooling of the metal leads to a drastic decrease of the level of tensile stresses,

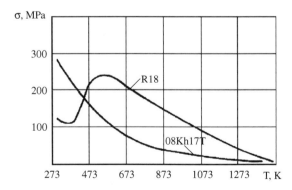

FIGURE 9.18 Variation in temporary stresses in specimens during cooling.

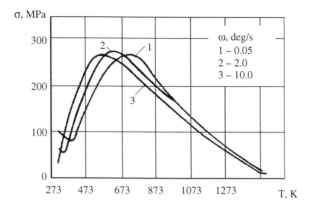

FIGURE 9.19 Effect of cooling rate on variations in tensile stresses in R18 steel.

which is proved by the effect of superplasticity taking place in heat-resistant steels of high hardness, which was first detected and investigated by B.B. Gulyaev.[383]

When a small amount of the metal is deposited on a comparatively massive body of a billet, the rate of cooling of the deposited metal may achieve substantial values. However, as is seen from Figure 9.19, this does not have a marked effect on the value of the tensile stresses in a temperature range of 1273–1473 K, where solidification cracks are most probably formed. The value of tensile stresses in the deposited metal can be influenced by preheating (Figure 9.20).

9.4.2 Cracks

Investigation results given in Chapter 8 are indicative of the fact that the probability of formation of solidification cracks rises with improvement of wettability of the solid metal by a low-melting melt. Results of investigations on the wettability of

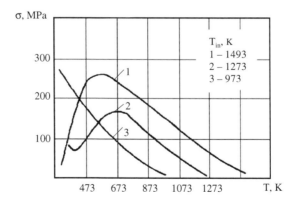

FIGURE 9.20 Effect of preheating on variations in tensile stress in R18 steel during cooling.

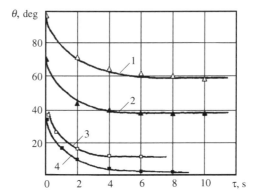

FIGURE 9.21 Variation in the contact angle of wetting of the deposited metal with the FeS melt using electrodes OZI-5 and OZI-3. (1,4) transferred arc; (2,3) non-transferred arc.

the metal deposited with OZI-3 and OZI-5 electrodes are shown in Figure 9.21. It turned out that the metal deposited by using OZI-5 electrodes was wetted worse, and it had less cracks. It was noted that wettability of the deposited metal varied from layer to layer, and that the first metal layers deposited at multilayer surfacing were better wetted. In addition, the first layers contained more cracks, compared with the last ones. This seems to be caused by the fact, that the first layers of the deposited metal contained a large amount of the base metal (steel 40Kh), which is well wetted with FeS melt (Figure 3.36). This is proved by the data shown in Figure 9.21, according to which the metal deposited by the indirect arc method has worse wettability than metal deposited by the direct arc method.

9.4.3 DUCTILITY

Investigations of the effect of the iron sulfide melt on the ductility of metal deposited using OZI-3 and OZI-5 electrodes showed (Figure 9.22) that in the case of OZI-5 electrodes, the deposited metal was characterized also by a smaller decrease of ductility in contact with FeS melt. Greater decrease of ductility taking place at the decrease of the rate of straining of a sample is indicative of a pronounced effect of corrosion embrittlement of the metal deposited with OZI-5 electrodes. Nevertheless, the use of these electrodes is more expedient for surfacing of edges of metal-cutting tools.

Based on the investigations conducted, OZI-5 electrode was selected for the manufacture of cutting tools with deposited edges, and surfacing was performed by the indirect arc method under the following conditions: $I_s = 120-130$ A; $U_{arc} = 24-26$ V; $V_s = 0.001$ m/s. Prior to surfacing, the billets were preheated to a temperature of $1073-1123$ K. This provided good formation of the deposited metal owing to improvement of wettability of the base metal by the deposited one. Further, it enabled avoidance of formation of rolls in addition to decreasing the magnitude of tensile stresses formed in the deposited metal.

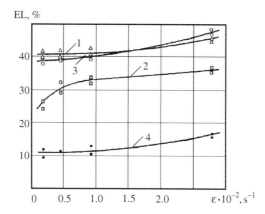

FIGURE 9.22 Variation in the ductility of the metal deposited with electrodes OZI-5 and OZI-3 depending upon the strain rate ($T = 1373$ K); (1,3) without and (2,4) with the contact of FeS melt.

9.5 THE ROLE OF SURFACE PHENOMENA IN THE SUBMERGED-ARC WELDING AND SURFACING PROCESSES

Submerged-arc welding is one of the processes, which is most extensively used for the fabrication of the most diverse structures from different materials. Therefore, increase of technical–economic indices of this process and improvement of quality of the resulting welded joints are the important tasks, the researchers and practitioners are faced with. The role of surface phenomena in processes of electrode-metal transfer, formation of various defects (pores, non-metallic inclusions, undercuts, lacks-of-penetration, etc.) was described above. In this section, we will consider the effect of surface phenomena on two aspects of the submerged-arc welding process, (i) productivity of the process; (ii) removal of the slag crust from surface of the weld or deposited metal.

Different techniques are employed in practice, allowing the productivity of the submerged-arc welding and surfacing processes to be raised. For example, metal fillers in the form of powder, wire, metal grit, or other materials are introduced into the groove in one-arc multilayer welding. To increase the melting coefficient of an electrode, use is made of automatic or semi-automatic submerged-arc welding with an increased electrode extension, which leads to preheating of electrode wire in the extension region due to Joule heat. The productivity of submerged-arc welding process can be increased also by using two or more electrodes, that is, by employing multi-arc or multi-electrode welding.

9.5.1 METHOD OF INCREASING THE PRODUCTIVITY

German scientists suggested the method for increasing the productivity of submerged-arc welding through improving wettability of the electrode wire by molten flux.

The mechanism of this method of increasing the welding-process productivity, the more detailed description of which is given in the study of Deyev,[525] is as follows. In submerged-arc welding, a part of the current is shunted by the molten flux, and the value of this current is usually equal to 1.5–2.0% of the welding current. It is likely that increase of the shunting current leads to extra heating of the electrode due to heat released in the slag pool, as well as to increase of the melting coefficient, which characterizes the welding-process productivity. According to the Joule's law, the amount of heat released in a conductor due to the electric current flowing through it is directly proportional to the square of the current, electric resistance of the conductor, and the time during which the current flows. Therefore, it is important to know, which factors affect the shunting current and how, as well as the value of the current density during the submerged-arc welding process. Calculations made to determine the current density at the slag–electrode boundary showed (Figure 3.41a), that the distribution of the current at the given boundary depends on the size of the cell, as well as on the specific electric conductivity of the arc column, χ_1.

As is seen (Figure 3.41a), increase of the length of contact of the electrode wire with the molten flux leads to growth of the shunting current flowing through the slag. According to calculations made, an increase of value H_2 from 5 to 10 mm leads to growth of the shunting current by more than a factor of 6. Undoubtedly, this growth of the shunting current should lead to extra heat released in the slag pool and, hence, to extra preheating of the electrode wire metal. Therefore, one of the ways of increasing the submerged-arc welding process productivity is to improve wettability of the electrode wire metal by the molten flux. This can be achieved by adding to the slag, the components which improve wettability of solid metal by molten slags, or by applying coatings, which can be well wetted with the slag melts, to the welding wire surface.

The relationship between the value of the shunting current and that of the melting coefficient of the electrode wire in submerged-arc welding is confirmed by the results of investigations of the welding process by adding an extra tungsten electrode to the slag.

Methods of adding the extra electrode to the molten slag pool can be different. In our experiments, we studied two methods: with (Figure 9.23a) and without

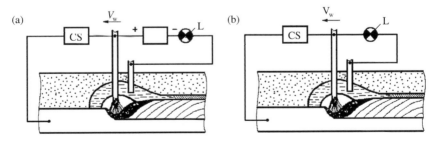

FIGURE 9.23 Submerged-arc welding using an additional tungsten electrode immersed into slag: (a) with and (b) without the auxiliary welding current source.

(Figure 9.23b) an extra current source. Lamp (L) served to monitor the closed loop "electrode–slag–extra electrode." The ABS head, plates of St.3 steel, welding wire Sv-08, and flux AN-348A were used in the experiments. The extra tungsten electrode was a ring enclosing the main metal electrode.

As shown by oscillography of the process, adding the tungsten electrode to the molten flux caused the shunting current to increase to a value of 5–10% of the welding current. This was enough to increase the value of the melting coefficient by 11–14%. It should be noted, that increase of the shunting current leads to decrease of the welding current and increase of the arc voltage (by 3–4 V), which causes a decrease of the penetration depth and some increase of the weld width.

Release of an extra amount of heat in the slag pool due to rise of the current flowing through the slag also leads to extension of the time of existence of the weld pool, which should reduce the probability of formation of undercuts. This is proved also by the experimental results (Figure 9.24). Thus, with a conventional welding method, the undercuts appeared at $V_w = 0.0197$ m/s, whereas in welding using the extra electrode introduced into the slag, welding parameters being the same, the undercuts were formed at $V_w = 0.0266$ m/s.

In submerged-arc welding of a thick metal, or in multilayer surfacing, the slag crust should be removed from the previous layer of the deposited metal prior to welding or deposition of each next layer. Detachability of the slag crust from the surface of the weld or deposited metal affects the productivity of the welding operations, as in the case of poor detachability of the slag, it is necessary to perform the operations of cleaning of the weld surface, which are time- and labor-consuming. In addition, in the case of poor detachability of the slag crust, the pieces of slag remaining on the surface of the weld or deposited metal may cause formation of non-metallic inclusions in the weld metal, or lacks-of-fusion and lacks-of-penetration in the deposition of the next metal layers.

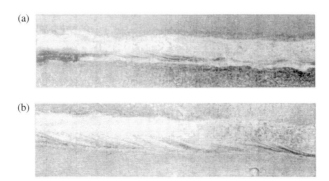

FIGURE 9.24 Appearance of the weld made by welding performed according to a typical flow diagram (a) and by using an additional electrode (b). $I_w = 300$ A; $U_a = 30$ V; $v_w = 0.0197$ m/s.

9.5.2 SLAG ADHESION

Adhesion of the slag to the metal surface is usually attributed to two causes: (1) mechanical sticking of the slag crust; (2) chemical bond of the slag to the metal. In the case of mechanical sticking of the slag crust, the lower the strength of the slag crust and the higher the thermal expansion coefficient of the slag, or, to be more exact, the higher the value of difference in thermal expansion coefficients between the slag and deposited metal, the better the detachability of the slag crust from the surface of the weld (deposited) metal. As shown by practice, other conditions being equal, detachability of the slag crust deteriorates with decrease of the groove angle, as this leads to formation of additional forces which keep the slag crust on the surface of the deposited metal.

Chemical bond of the slag to the metal is determined by the processes, which occur at the interface between the contacting phases (metal–slag). The value of the adhesion energy can serve as the criterion of interaction of particles at the interface. The adhesion energy can be determined from the Dupre equation with the known values of surface tension of the slag and the metal, and interphase tension at the metal–slag boundary.

Surface tension of slags, even at an insignificant variation in their compositions, changes to a comparatively low degree. Therefore, the highest effect on the value of adhesion is exerted by the surface tension of metal and the interphase tension at the metal–slag boundary. Values of the surface tension of the slags and the metals, and values of the interphase tension at the metal–slag boundary, taken at a temperature close to the melting point of the metal, as well as values of adhesion between metal and slag determined from Equation (2.6) are given in Table 9.5.

TABLE 9.5
Effect of Composition of Metal and Slag in Contact with It on the Value of Adhesion between Them

Welding Wire Grade	Flux Grade or Composition, wt%	σ_{sl-g}, MPa	σ_{m-g}, MPa	σ_{m-s}, MPa	W_A, mJ/m^2
Sv-08	AN-348A	480	1270	882	877
Sv-08	AN-26	492	1279	1180	591
Sv-08	SiO$_2$–38 FeO–62	460	1279	414	1325
Sv-08	SiO$_2$–60 MgO–40	410	1279	1198	491
Np-4Kh13	AN-26	492	1157	1064	585
Np-4Kh13	OF-6	306	1157	1137	326
Np-60Kh3V10F	AN-26	492	1304	1163	633
Np-60Kh3V10F	OF-6	306	1304	1209	401

As is seen from Table 9.5, the value of the adhesion energy at the metal–slag interface greatly depends on the composition of the welding consumables used. To establish a relationship between the adhesion value and detachability of the slag crust, additional experiments were conducted to determine the energy required to remove the slag from the deposited metal surface. This was done by using a procedure based on the measurements of the value of the energy consumed for the removal of the slag from the groove, described in the book, "Arc Welding Consumerables" by Potapov.[304] Specimens and a pendulum impact testing machine used for the tests are shown in Figure 9.25 and Figure 9.26.

An indicator of detachability of the slag crust from the metal surface is a value determined from the following relationship:

$$A_s = Pl(1 - \cos \varphi)/S_w,$$

where P is the weight, kg; l is the length of pendulum of the machine, m; φ is the weight drop angle; S_w is the surface of the weld exposed from the slag after impact on the specimen of the weight, cm^2.

The tests were conducted as follows. The specimen with a single-layer weld deposited by the submerged-arc method was placed on support 2 of the pendulum impact testing machine. Then, a free-falling weight 3 impacted the specimen on the reverse side of the weld. The impact energy can be regulated by varying the angle of lifting of the weight.

The A_s value was determined as an arithmetic mean of the data of three tests. The investigation results given in Table 9.6 are indicative of the fact that detachability of the slag crust from the deposited metal surface decreases with increase of adhesion between the metal and the slag. The slag crust can be removed comparatively easily from the surface of welds deposited using wire Np-4Kh13 by the submerged-arc method under the layer of flux AN-26 and OF-6. The slag crust detaches slightly less easily from the surface of the welds deposited using wire Np-60Kh3V10F. Surfacing with this wire results in the slag particles which remain on the surface of the welds after impact tests of a specimen. Detachability of the slag crust is still worse in the case of the welds deposited with wire Sv-08 by the submerged-arc method using a flux consisting of SiO_2 and FeO. In this case, oxidation of the surface of the

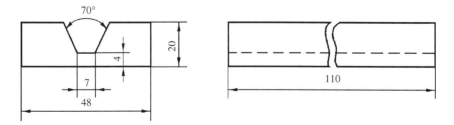

FIGURE 9.25 Sketch of a specimen used to study detachability of the slag crust from the weld surface.

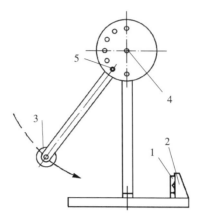

FIGURE 9.26 Schematic of an impact testing machine used to study detachability of the slag crust from the weld surface: (1) test specimen; (2) pendulum impact testing machine base; (3) weight; (4) pendulum axis; (5) fixation pin.

solidified weld metal with liquid slag and presence of iron and silicon oxides in the slag create conditions leading to formation of strong bonds between the metal and the slag crust, which is proved by a high value of adhesion.

As the adhesion value at the metal–slag interface greatly depends on the interphase tension at this interface, for systems in which polarization is of a concentration nature, variations in polarity of the current during d.c. welding should affect detachability of the slag crust. This fact should be allowed for in the development of the submerged-arc welding and surfacing technologies, especially, if the weld or deposited metal has several layers.

As is seen from the established relationships between detachability of the slag crust and the value of adhesion between the metal and the slag, as early as at the stage of development of welding consumables (electrodes for manual arc welding, welding wires, fluxes) it is possible to predict how strong the adhesion will be between the slag crust and the deposited metal on the basis of the results of measurement of surface tension of the metal and the slag, as well as tension between the phases at the metal–slag interface.

TABLE 9.6
Variation in Value A_s Depending on the Adhesion Energy at the Metal–Slag Interface

W_A, mJ/m^2	326	401	491	585	591	633	877	1325
A_s, kgm/cm^2	12.7	14.6	20.6	23.4	22.8	24.2	36.1	45.3

9.6 SURFACE PHENOMENA OCCURRING IN WELDING AND SURFACING IN SPACE

Practical implementation of many ideas in the field of cosmonautics is impossible[526] without a wide application of welding of the metals directly in space. Repair and assembly operations, which are unavoidable in long-term operation of space stations, will require efforts on the welding and cutting of materials. It is most probable, that this will involve fusion welding methods, as in this case, the formation of a welded joint occurs without the application of an external pressure. As a rule, devices necessary to create external pressure are cumbersome, have a large weight, and thus, require considerable expenditures to launch them into space.

However, conditions of occurrence of fusion welding processes in space differ substantially from the earth ones. The presence of vacuum, microgravity, temperature gradients, etc. will exert a marked effect on physical–chemical processes occurring during welding. Thus, under the earth conditions, the gravity forces play an important and sometimes even dominating role in many processes taking place during fusion welding. Under the space conditions, the gravity forces having a substantial radius of action do not fully disappear. At a space station, there occurs the so-called self-gravity, which consists of the forces of gravity interaction of structural elements of a spacecraft and its content.

9.6.1 MAJOR COMPARISONS WITH EARTH CONDITIONS

To establish the effect of these forces, compared with the surface forces, it is proper to equate the accelerations $a_G = a_\sigma$ generated in the metal melt formed during the fusion welding process under the effect of the corresponding forces. Then $Gm_1/R^2 = \sigma/4\rho R^2$, or

$$4\pi\rho GR/3 = \sigma/4\rho R^2, \tag{9.14}$$

where ρ and m_1 are the density and weight of the melt, and G is the gravity constant. From Equation (9.14), $R \approx \sqrt[3]{\sigma/16G\pi\rho^2}$. For molten steel, it can be assumed that $\rho = 7000 \text{ kg/m}^3$ and $\sigma = 1200 \text{ mN/m}$. In this case, the self-gravity forces are comparable with the surface forces only for a sphere of molten metal having a radius of $R = 2.83$ m. As the sizes of the weld pool and electrode drops are much smaller under the space conditions, the effect of the self-gravity forces on welding processes will be manifested to a much lower degree, compared with the effect of the surface forces.

9.6.2 THE FORCE OF ACCELERATION

According to Steg,[527] the melts in space are subjected to low accelerations generated due to a simultaneous effect exerted by the following main forces: gravity gradient of the earth; rarefaction of the atmosphere; centripetal force formed in rotation of a spacecraft; pressure of the solar wind; gravity gradients generated due to motion of the moon; acceleration formed at switching on of engines and due to movements of cosmonauts; vibrations on board caused by working of equipment; elastic deformations caused by non-stationary heat flows; electromagnetic forces.

Values of some types of the accelerations are given in Table 9.7, which was compiled on the basis of data given in the study of Peinter.[528] The values of the accelerations given in Table 9.7 were obtained for the following conditions: a flying vehicle has a spherical shape, weight of the vehicle is 45.36 ton, its diameter is 19.2 m, mid section area is 290 m^2, resistance coefficient is 2, electrostatic charge is 1600 V, and distance of the elementary volume of liquid from the center of gravity (or rotation) is 3.05 m.

Accelerations, which can be formed in the melt under the effect of surface forces can be approximately estimated as follows. The magnitude of the surface forces is $F \approx \sigma L$, where L is the characteristic linear size. In this case, the weight of the liquid is $m \approx \gamma L^3$, and acceleration of liquid particles is $n \approx a/g_0 = \sigma/\rho g_0 L^2$. Assuming, as before, that for a molten metal $\rho = 7000\ kg/m^{-3}$ and $\sigma = 1200\ mN/m$, we find for a weld pool, 0.01-m long and of a volume of $0.25 \cdot 10^{-6}\ m^3$, that $n = 1.75 \cdot 10^{-1}$. Therefore, accelerations formed in the melt under the effect of surface forces dominate over the rest of the accelerations.

Variation of the free fall acceleration leads to variation of the relationship between the mass and surface forces, which can be seen from the determination of the Bond number:

$$Bo = \rho dg^2/\sigma, \tag{9.15}$$

where d is the characteristic size. It follows from this formula, that under conventional earth conditions, a pronounced effect of the surface forces is seen only at small values of d.

According to the data of the study of Avduevsky et al.,[529] the gravity forces will dominate over the surface tension forces at $Bo \geq 100$. At $Bo \leq 1$, the surface

TABLE 9.7
Variations of Relative Accelerations Depending on the Altitude of Circular Orbit of a Flying Vehicle

Cause	Altitude of Circular Orbit, km	
	240	1610
Aerodynamic (maximum solar activity)	$7.0 \cdot 10^{-6}$	$4.6 \cdot 10^{-12}$
Geomagnetism	$9.5 \cdot 10^{-12}$	$5.1 \cdot 10^{-13}$
Light pressure	$3.1 \cdot 10^{-9}$	$3.1 \cdot 10^{-9}$
Internal gravity	$3.3 \cdot 10^{-8}$	$3.3 \cdot 10^{-8}$
Control of orientation in flight	$4.3 \cdot 10^{-7}$	$2.4 \cdot 10^{-7}$
External gravity (heterogeneity of gravity field of the earth)	$4.3 \cdot 10^{-7}$	$2.4 \cdot 10^{-7}$

tension forces will be dominant. Al low values of the free fall acceleration, the Bond number becomes less than 1. Under the space conditions, the gravity generated due to the above causes does not fully disappear. However, the free fall acceleration in this case is low and amounts to not more than $10^{-3}-10^{-5}$ of that observed under the earth conditions.[530,531] Therefore, there is no doubt that the surface tension forces under the space conditions will dominate over the mass forces and will play the primary role. In this connection, it is important to know how the above processes will proceed during welding operations under the microgravity conditions. At the same time, it should be noted that values of surface tension of materials and contact angles of wetting of solid bodies with the melts under the earth conditions are the same as in space.

9.6.3 MICROGRAVITY

First of all, let us analyze how the presence of microgravity will affect the metal-arc welding process, and the electrode-metal transfer in particular. As follows from Equation (2.23), a decrease of mass forces should lead to an increase of weight and size of the electrode-metal drop. This is proved also by the data of reported studies[526,532] conducted to investigate the melting and transfer of the electrode metal under the microgravity conditions. As shown by these investigations, in this case, the time of existence of the drop at the electrode tip increased by a factor of 15–20 and amounted to 4.5–6.0 s at small values of the welding current (40–50 A). Size of the drop was several times larger than the diameter of the electrode, while transfer of the electrode metal to the weld pool took place only at contact of the drop with the weld-pool surface as a result of the effect exerted by the wetting forces.

According to Paton et al.,[532] the main forces affecting the drop of the electrode metal under microgravity are the surface tension, reactive forces, and electromagnetic forces. Moreover, only the last can lead to detachment of the electrode metal from the electrode tip. The reactive forces seem to have no marked effect on the electrode-metal transfer due to decrease of the current density in the active spot on the electrode and stability of the arc discharge observed with increase of size of the drop.[526] Therefore, the main role in this case will be played by the surface tension forces. In particular, this is evidenced also by a shape of the drop formed at the electrode tip, which as a rule is spherical or close to spherical.

In this connection, it can be assumed that $z = 2R$ and $R_1 = R_2 = R$ in welding without short circuiting at the stage of a stable growth of the drop having a spherical shape. Then Equation (4.5) describing the process of a stable growth of the drop at the electrode tip will have the following form:

$$\sigma_{m-g}/\rho g = R^2, \tag{9.16}$$

that is radius of the drop becomes equal to the capillary constant for a given material. Therefore, in the case of a metal electrode used under the microgravity conditions, the electrode-metal drops may remain stable, reaching very large sizes.

It should be noted that the capillary pressure, which under conventional earth conditions favors detachment of the drop from the electrode tip, will have no such effect under microgravity, because no necking between the electrode and the drop will take place under microgravity. Therefore, the drop can be detached from the electrode under the space conditions either due to the electrodynamic forces (welding at a modulated current) or by performing the process with short-circuiting between the drop and the weld pool.

The presence of microgravity also affects the formation, weld or the deposited metal. Decrease of the arc stability, increase of the role of surface tension forces, and the impossibility of providing a fine-drop or spray transfer of the electrode metal lead to decrease of the depth of penetration of the metal in metal–electrode welding in space.[526] As noted in Chapter 5 (see Section 5.3), formation of the weld bead and root depends on the value of the capillary constant, which is equal to $a_c = \sqrt{\sigma/g\Delta\rho}$. Decrease of the value of free fall acceleration, g observed under the space conditions leads to growth of the a_c value and increase of the role of capillary forces in the processes of formation of the weld bead and root. Apparently, at small values of g, the radius of curvature of the bead surface depends only on the surface tension of the weld-pool metal. Increase of the role played by the surface tension forces in the process of formation of the weld bead under microgravity can be seen also from the system of Equations (5.14). In addition, decrease of the gravity forces should lead to increase of height of the weld bead in welding of similar and dissimilar materials. This is confirmed also by the experimental data[526] on metal-arc welding under microgravity (Figure 9.27). So, metal-arc welding performed under the microgravity conditions is accompanied by formation

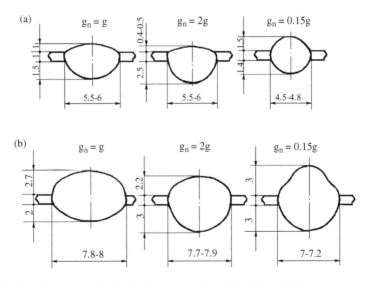

FIGURE 9.27 Shape and size (mm) of a cross-section of welds made by metal-arc welding at different values of acceleration due to gravity g_{Π}:[526] (a) 1Kh18N9T steel; (b) AMg6 alloy.

of large drops at the electrode tip, the transfer of which to the weld pool can take place either due to short circuiting or due to feeding the current pulses. In addition, this is also accompanied by lower arc stability and greater height of the weld bead. Therefore, this process can hardly be regarded as the optimal one for application under the space conditions.

In tungsten electrode welding, the weld formation occurs under the effect of two forces: surface tension and arc pressure. Here, there are no forces of hydrostatic pressure; then, if follows from Equation (5.1), that the depth of penetration of the base metal should be larger than in welding under the earth conditions. In particular, this was noted in the study of Nogi, Aoki, and Fuzi,[533] who investigated peculiarities of tungsten electrode welding under microgravity.

With thorough penetration of the metal under the space conditions, the weld formation will not depend on the mass forces, in contrast to the earth conditions. Therefore, the condition of equilibrium of the weld pool in the case of an isothermal system will have the following form:

$$P_{arc} - P_s^{out} - P_s^r = 0 \tag{9.17}$$

Here, the designations are the same as in Equation (5.7). The absence of mass forces should lead to the formation of a flatter weld, which is proved by the experimental data of Nogi, Aoki, and Fuzi.[533] The sag of the weld and values of parameters h and h' (Figure 5.10) in the welding of a given material will depend only on the difference $P_{arc} - (P_s^{out} + P_s^r)$ and hence, on the arc pressure P_{arc}.

Conditions for the formation of welds in horizontal and overhead positions are substantially changed in welding in space. It can be seen from Equation (5.16) that decrease of the gravity force acceleration leads to an increase of weight of the drop, which is contained on an inclined surface, because values of adhesion between solid and molten metals are the same under both earth and space conditions. Therefore, the smaller the gravity force acceleration, the larger the volume of the weld-pool metal and the higher the value of the critical angle of inclination of the part surface. Decrease of the effect of gravity forces on weld formation also allows rolls in welding in the horizontal position to be avoided. Good weld formation in the horizontal position was noted[533] in tungsten arc welding in the helium atmosphere and in electron beam welding under microgravity.

Under the earth conditions, it is very difficult to perform overhead welding, as there is a constant risk of metal flowing out of the weld pool. Under microgravity conditions, the normal weld formation in welding in the overhead position is possible at much larger sizes of the weld pool, which follows from Equation (5.18) and Equation (5.19).

Under microgravity, the formation of the deposited layer should also occur in a different way. As is seen from Equation (5.29), decrease of value of the gravity force acceleration leads to increase of the equilibrium thickness of the deposited layer, and its size will depend only on the wettability of the surface treated with the metal melt, as well as on the surface properties of the metal being deposited.

Decrease of the gravity forces leads to decrease of the probability of burns-through and enables welding to be performed at large gaps without a risk of metal flowing out of the weld-pool. In addition, welding can be carried out without backing, which is usually used to contain the weld-pool metal. For this reason, the condition of formation of craters in the weld can be given as following: $P_{arc} > 2\sigma_{m-g}/r_c$. Therefore, in welding in space, the probability of formation of craters is higher than in welding under the earth conditions, because of a lower hydrostatic pressure.

9.6.4 THE QUESTION OF CONVECTION

As was noted above, penetration of the metal and formation of undercuts, pores, and non-metallic inclusions are related to a considerable degree to the hydrodynamic processes occurring in the weld pool. Therefore, it is of interest to consider peculiarities of displacement of the metal melt under the microgravity conditions, as well as the role of the Marangoni effect in this case.

First of all, it should be noted that under microgravity, there is almost no natural convection.[534] Hence, it has almost no effect on stirring of the weld-pool metal. Displacement of liquid layers and particles can be of several types under the space conditions.[535] It includes Marangoni flow, Brownian motion, motion driven by Lorentz forces, motion caused by temperature, pressure and concentration gradients, motion caused by phase transformations, etc.

According to the experimental data obtained under the space conditions, two basic types of convection are distinguished: thermocapillary and diffusion. The role of this or that type of convection in the entire stirring process depends on many factors: temperature gradient, configuration of the channel via which the metal melt flows, concentration of surface-active materials, etc.

As has been discussed in Chapter 5 (see Section 5.4), one of the components of the velocity of motion of surface layers of the metal melt, G_r, depends on the value of the free fall acceleration and decreases with decrease of the value of g. This creates more favorable conditions for an inequality (Equation (5.38)) to be met, which is indicative of greater probability of a flow of the metal surface layers of Marangoni-type during welding in space.

Let us consider now how the conditions of formation, growth, and removal of gas bubbles and non-metallic inclusions change in welding in space. Apparently, formation of nuclei of pores and inclusions in a homogeneous medium, as well as in a heterogeneous one (in the absence of chemical interaction between the contacting phases), will be the same as on the earth. This is attributable to the fact, that in these cases, the nucleation depends only on the surface (interface) tension, wettability of the contacting phases, and the degree of oversaturation of the melt with a material of the precipitating phase. However, if chemical reactions leading to variation of the degree of oversaturation occur at the interface, the presence of vacuum may lead to increase of the intensity of these reactions. For example, it is reported[25] that creation of vacuum over the surface of the metal pool leads to acceleration of the de-oxidizing ability of carbon. Therefore, in the presence of oxide inclusions in the weld-pool metal, occurrence of the reaction, $y[C] + (Me_xO_y) = x[Me] + y\{CO\}$ may lead to local oversaturation of the melt with CO, which favors nucleation of the gas bubbles at the interface between the metal melt and non-metallic inclusion.

9.6.5 Growth of Gas Bubbles and Non-Metallic Inclusions

Conditions affecting the growth of the gas bubbles and non-metallic inclusions will greatly change in welding in space. As noted in Section 7.4, growth of the gas bubbles occurs mostly as a result of diffusion of gas from the metal melt into a bubble and due to coalescence of the bubbles. In the first case, the process of growth of the gas bubbles depends in many respects on the value of the atmospheric pressure. Therefore, in welding outside the orbital chamber, a decrease of P_{atm} (see Figure 7.16) leads to acceleration of growth of the gas bubble in the weld pool. Efficiency of the process of coalescence of the bubbles is determined by the intensity of stirring of the weld-pool metal, which under the microgravity conditions is identified with the Marangoni number.[535]

$$Mn = [r^2(\partial\sigma/\partial T)\text{grad } T]/\mu\chi.$$

In this case, the velocity of movement of the bubble can be described by the following equation[536]:

$$V = -(r/2\mu)(\partial\sigma/\partial T)\text{grad } T, \qquad (9.18)$$

where r is the bubble radius.

It follows from Equation (9.18), that the velocity of the bubbles depends on the presence of surface-active materials in the melt and the temperature gradient in the weld pool. Therefore, adding surface-active materials to the metal melt should have a substantial effect on the motion of metal and the gas bubbles contained in it.

It is likely, that motion of non-metallic inclusions is also determined mostly by the thermocapillary convection. No motion of the gas bubbles and non-metallic inclusions should occur under microgravity because of difference in densities of the metal melt and gas (non-metallic inclusion), which follows from the Stokes formula. This is proved by the experimental data generated by B.V. Volynov and V.M. Zholobov[537] on the behavior of the gas bubbles in water. The cosmonauts watched the behavior of the bubbles in liquid under the microgravity conditions. Approximately, 100 bubbles with an initial diameter of 0.1–1.0 mm were present in water in a spherical flask 30 mm in diameter. After about 100 h there occurred coalescence of the bubbles into one spherical bubble located almost in the center of the flask. If coalescence occurred in a pure form, this would take time by an order of magnitude longer.

As is shown in Chapter 7 (see Section 7.5), under welding conditions, because the process of weld formation is fast, the removal of the gas bubbles from the weld pool depends to a considerable degree on the rate of destruction of the metal film raised with a gas bubble by the metal that approached the surface. According to

Equation (7.37) and Equation (7.38), the smaller the thickness of the metal film, the higher the rate of its destruction. Thickness of the film depends on the velocity at which the metal located between the outer and inner surfaces of the film flows out to the weld pool. As a decrease of the gravity force acceleration leads to a decrease of the velocity of metal flowing out, which can be seen from Equation (7.36), under the space conditions, the film becomes less stable, thus hindering the removal of the gas bubbles from the weld pool, even if they come to the weld-pool surface. Therefore, under the space conditions, an increased porosity of the weld metal may be expected, which is proved also by the welding experiments conducted in space.[538]

Apparently, the content of oxide non-metallic inclusions in the welds produced in space should be lower than in welding performed under the earth conditions. This is associated with the presence of vacuum, which leads to decrease of the values of over-saturation depending (see Section 6.2) on the actual concentrations of de-oxidizing elements and oxygen dissolved in metal. Decrease of the degree of the deoxidation of the metal, taking place in vacuum, leads to decrease of the C/C_S value, as well as lowering of the probability of formation of oxide non-metallic inclusions. The presence of vacuum may also lead to dissolution[26] of oxide non-metallic inclusions contained in the metal produced under the earth conditions.

Decrease of the role of the gravity forces increases to a still higher degree the importance of the surface tension forces in the process of removal of non-metallic inclusions from the weld pool in welding in space even for the case of coarse inclusions (see Section 6.4).

The necessity to allow for the surface phenomena in performing welding operations under microgravity is emphasized also in a study.[539] It is noted in this study that the electron beam process, which can be used both for welding and cutting of materials, holds the highest promise for the space conditions.

Therefore, it follows from the above-said points, that surface properties of welding consumables and surface phenomena occurring at the interfaces between the contacting phases have a substantial effect on the most diverse processes taking place in fusion welding. Their role is changed depending on the welding method employed, type of a weldment, and conditions under which the welding process is performed. Nevertheless, they should be taken into account in the development of the welding and surfacing technologies, as well as the related consumables, as this determines in many respects the quality of the welded joints and deposited metal. This is especially important nowadays, because welders have to solve increasingly difficult problems associated with welding of new materials and fabrication of more sophisticated structures intended for operation under severe conditions.

It should be noted that the relationships derived in this study are of a general nature and do not depend on the type of materials used. It is important to know just the surface properties of these materials and the surface phenomena occurring at the interfaces between the contacting phases. Moreover, the surface phenomena should be allowed for, in the performance of welding operations both in space, where they become of a primary importance, and on the earth.

Appendix

TABLE A1
Composition of Metals Mentioned in the Book

Metal	Base	Chemical Condition, (wt%)						
		C	Mn	Si	Ni	Cr	S	Other
Armco-iron	Fe	0.05	0.07	0.03	—	—	0.018	
Sv-08	Fe	0.09	0.33	0.03	—	—	0.021	
Sv-08G2S	Fe	0.08	1.67	0.82	—	—	0.03	
Sv-10GS	Fe	0.15	0.99	1.17	—	—	0.032	
U8A	Fe	0.83	0.21	0.28	—	—	0.01	
08Kh20N10G6	Fe	0.13	6.10	0.40	9.64	18.62	—	
St 3	Fe	0.18	0.42	0.07	—	—		
Steel 20	Fe	0.17–0.24	0.35–0.65	0.17–0.37	—	—	0.04	
Steel 50	Fe	0.47–0.55	0.5–0.8	0.17–0.37	—	—	0.04	
Steel 45	Fe	0.42–0.50	0.5–0.8	0.17–0.37	—	—	0.04	
U10	Fe	0.95–1.04	0.17–0.28	0.17–0.33	—	—	0.018	
ShKh 15	Fe	0.95–1.05	0.2–0.4	0.17–0.37	—	1.3–1.65	0.02	Al–0.05
08Yu	Fe	0.05	0.27	<0.03	—	—	0.025	Cu < 0.2
10KhSND	Fe	0.12	0.5–0.8	0.8–1.1	0.5–0.8	0.6–0.9	0.04	
40Kh	Fe	0.36–0.44	0.5–0.8	0.17–0.37	—	0.8–1.1	0.035	
30KhGSA	Fe	0.28–0.34	0.8–1.1	0.9–1.2	—	0.8–1.1	0.025	
9KhF	Fe	0.8–0.9	0.3–0.6	0.15–0.35	—	0.4–0.7	0.03	Mo-0.15–0.25
12Kh18N9T	Fe	0.12	2.0	0.8	8–9.5	17–19	0.02	Ti 0.6–0.8
12Kh18N10T	Fe	0.12	2.0	0.8	9–11	17–19	0.02	Ti 0.6–0.8
10Kh17N13M3T	Fe	0.1	2.0	0.8	12–14	16–18	0.02	Ti 0.5–0.7; Mo 3–4
E3A	Fe	0.078	0.15	2.88	—	—	0.016	
07Kh25N13	Fe	0.09	1.0–2.0	0.5–1.0	12–14	24–25	0.028	

	Base							
R18	Fe	—	—	—	—	3.8–4.4	0.02	W 17–18.5; V 1–1.4; Mo < 1.0; Co < 0.5
G13L	Fe	1.38	11.6	0.5	—	0.32	0.009	—
G13FL	Fe	1.12	12.45	0.71	—	—	0.007	—
12KhMNF	Fe	0.08–0.15	0.4–0.7	0.17–0.37	0.7–1.0	0.9–1.2	0.025	Mo 0.25–0.35
16GNMA	Fe	0.12–0.18	0.9–1.2	0.15–0.35	0.9–1.2	—	0.025	—
40KhNMA	Fe	0.37–0.44	0.5–0.8	0.17–0.37	1.0–1.1	—	0/025	Mo 0.2–0.3
10Kh18N10T	Fe	0.08	1.0–2.0	—	9–11	17–19	0.02	Ti 0.7
10Kh18N9T	Fe	0.07–0.12	—	—	8.5–10.0	17–19	0.02	Ti 1.1–1.6
65G	Fe	0.67–0.7	0.9–1.2	0.17–0.37	—	—	0.03	—
09G2S	Fe	0.12	1.3–1.7	0.5–0.8	—	—	—	—
08KhGSMF	Fe	0.06–0.1	1.2–1.5	0.45–0.7	<0.3	0.95–1.25	0.03	Mo 0.5–0.7; V 0.2–0.35
15KhMNF	Fe	0.12–0.18	0.35–0.6	<0.35	0.8–1.2	0.7–0.9	0.025	Mo 0.4–0.6; V 0.2–0.35
18Kh5N5AM3 (VNS-9)	Fe	0.14–0.20	0.3–0.5	0.17–0.37	4–7	4–6	0.03	Mo 2.5–3.0
08Kh17T	Fe	0.06	0.8	0.03	0.6	16–18	0.025	Ti 0.8
Np-4Kh13	Fe	0.4	—	—	—	13	—	—
Np-60Kh3V10F	Fe	0.6	—	—	—	3	—	V 1.0
M1	Cu-99.9	—	—	—	—	—	0.004	O_2 0.05
M3r	Cu 99.5	—	—	—	—	—	0.01	O_2 0.01
L-63	Cu	—	—	—	—	—	—	Zn 38–35
LK-62-0.5	Cu	—	—	0.5	—	—	—	Zn 37.5–34.5
L-90	Cu	—	—	—	—	—	—	Zn 9–12
BrKMts 3-1	Cu	—	1–1.5	2.75–3.5	—	—	—	Fe < 1.1
BrAMts 9-2	Cu	—	1.5–2.5	—	—	—	—	Al 8–10

(Table continued)

TABLE A1 *Continued*

Metal	Base	Chemical Condition, (wt%)						
		C	Mn	Si	Ni	Cr	S	Other
BrAZhNMts 8.5-4-5-1,5	Cu	—	1.7	—	5.1	—	—	Al 8.7; Fe 4.2
BrON8.5-3	Cu	—	—	—	3–4	—	—	Sn 9–10
BrOS8-12	Cu	—	—	—	—	—	—	Sn 7–9; Pb 11–13
BrAZhNMts 10-3-1.5	Cu	—	—	—	—	—	—	Al 9–11; Fe 2–4
BrMNZhKT 5-1-0.2-0.2	Cu	—	0.3–0.8	0.15–0.3	5.0–6.5	—	—	Fe 1.0–1.4
A99	Al	—	—	—	—	—	—	0.01
AK5	Al	—	0.15–0.35	0.5–1.2	—	—	—	—
AMg6	Al	—	0.5–0.8	0.4	—	—	—	Mg 5.6–5.8
AD33	Al	—	—	0.4–0.8	—	—	—	Mg 0.4–0.8
VT-1	Ti	—	—	—	—	—	—	0.2
OT4-1	Ti	—	1.0	—	—	—	—	Al 1.5
VT-20	Ti	—	—	—	—	—	—	Al 6.0; Zr 2.0; V 1.0
VT-14	Ti	—	—	—	—	—	—	Al 4.5; Mo 3.0; V 1.0
VT6	Ti	—	—	—	—	—	—	Al 6.0; V 4.5
VT6C	Ti	—	—	—	—	—	—	Al 5.0; V 4.0
NO	Ni	—	—	—	—	—	—	<0.07

TABLE A2
Composition of Fluxes Mentioned in the Book

Flux	Chemical Composition, (wt%)					
	SiO_2	MnO	CaO	CaF_2	Al_2O_3	MgO
OSZ-45	38–44	38–44	<6.5	6–9	<5	
AN-5	50–52	—	26–30	5–6	—	10–14
AN-26	29–33	2.5–4	4–8	20–24	19–23	15–18
AN-18[a]	17–21	2.5–5	14–18	19–23	14.5–18.5	—
ANF-14	14–16	—	<8	60–65	10–12	4–8
AN-42	30–34	14–19	12–16	14–20	13–18	—
OF-6	3–6	<0.3	16–20	50–60	20–24	<3.0
AN-20	19–24	<0.5	3–9	25–33	27–32	9–13
AF-4a[c]	—	—	—	—	—	—
AN-348A	41–44	34–38	<6.5	4–5.5	<4.5	6.0
AN-15M[b]	6–10	<1.0	29–33	16–20	36–40	<2.0

[a]Fe_2O_3; 13.5–16.5%.

[b]NaF; 2–6%.

[c]Flux AF-4a; KCl – 50%; LiCl – 14%; NaF – 8%; NaCl – 28%.

TABLE A3
Composition of Surfacing Powders Mentioned in the Book

Powder Grade	Composition, (wt%)							
	Ni	C	Cr	Si	B	Fe	Mo	V
PR-N80 Kh13S2R	Base	0.2–0.4	12–14	2.8–3.5	1.8–2.3	Upto 5	—	—
PR-N70 Kh17S4R4	Base	0.8–1.2	16–18	3.8–4.5	3.1–4.0	Upto 5	—	—
PR-Kh N80SR2	Base	0.3–0.6	12–15	1.5–3.0	1.5–2.5	Upto 5	—	—
M6F3	—	1.15	4.2	—	—	Base	6.2	2.6

TABLE A4
Composition of Slags Formed in Coated-Electrode Manual Arc Welding

Electrode	Chemical Composition, (wt%)					
	SiO_2	MnO	TiO_2	$FeO + Fe_2O_3$	CaO	CaF_2
OMM-5	27.3	28.9	15.2	13.2	3.6	—
UONI 13/55	31.0	4.6	2.2	7.9	42.0	—
ANO-4	27.0	14.7	43.2	3.7	0.3	—
UONI 13/45	15.0	5.0	9.0	4.0	27.0	34.0
VSTs-4A	18.8	15.0	29.0	8.2	2.0	—

References

1. Paton, B.E., Ed., *Technology of Electric Fusion Welding of Metals and Alloys*, Mashinostroenie, Moscow, 1974, p. 768.
2. Rykalin, N.N. et al., *Laser and Electron Beam Teratment of Materials*, Reference book, Mashinostroenie, Moskow, 1985, p. 496.
3. Seferian, D., *Welding Metallurgy*, Mashgiz, Moscow, 1963, p. 347.
4. Morozov, A.N., *Hydrogen and Nitrogen in Steel*, Metallurgiya, Moscow, 1968, p. 281.
5. Kozlov, R.A., *Hydrogen in Welding of Hull Steels*, Sudostroenie, Leningrad, 1965, p. 175.
6. Ram, V.A., *Adsorption of Gases*, Khimiya, Moscow, 1966, p. 767.
7. Shimmyo, K. and Takami, T., Kinetics of reaction between levitated iron droplet and nitrogen, *Procedings of the 1st Confrence on Science and Technology of Iron and Steelmeking*, Tokyo, 1971, p. 543.
8. Barr, G., *Viscosimetry*, GONTI, Moscow, 1938, p. 274.
9. Darken, L.S. and Gurri, G.V., *Physical Chemistry of Metals*, Metallurgizdat, Moscow, 1960, p. 582.
10. Dzhoshi, V.B., Visharev, A.F., and Yavoiskii, V.I., Role of surface phenomena in the processes of nitrogen distribution between metallic and gas phases, *Izv. Vuzov. Chyorn. Metall.*, 11, 36, 1960.
11. Lakomsky, V.I., *Plasma-Arc Remelting*, Teknnika, Kyiv, 1974, p. 335.
12. Fedorchenko, V.I. and Averin, V.V., Investigation of solubility and kinetics of nitrogen dissolution in iron–oxygen and iron–sulphur melts, using the method of crucibleless melting, *Interaction between Metals and Gases in Steel-Making Processes*, Metallurgiya, Moscow, 1973, p. 26.
13. Fedorchenko, V.I. and Averin, V.V., Influence of surface-active elements on kinetics of interaction between nitrogen and iron-base melts, *Kinetics and Thermodynamics of Interaction between Gases and Liquid Metals*, Nauka, Moscow, 1974, p. 49.
14. Fisher, W.A. and Hoffman, A., Aufnahmegeschwindket und loslichkeit von stickstoff in flussigen eisen in abhangigkeit vom gelosten sauerstoff, *Arch.Eisenhut.*, 4, 215, 1960.
15. Arsentiev, P.P., Filippov, S.I., and Burkin, S.I., Viscosity of melts of iron–oxygen and iron–sulphur system, *Izv. Vuzov. Chyorn. Metallurgiya*, 9, 132, 1968.
16. Svyazhin, A.G., El Gammal, T., and Dal, V., Diffusion of nitrogen in liquid iron and iron–oxygen melts, *Interaction between Metals and Gases During Steel Making Processes*, Metallurgiya, Moscow, 1973, p. 51.
17. Fedorchenko, V.I., Averin, V.V., and Samarin, A.M., Kinetics of nitrogen dissolution in nickel–base melts, *Izv. AN SSSR. Metally*, 1, 102, 1969.
18. Okorokov, G.N., Interaction of nitrogen with liquid iron in plasma-arc discharge conditions, *Kinetics and Thermodynamics of Interaction between Gases and Liquid Metals*, Nauka, Moscow, 1974, p. 54.
19. Emee, T., Theotherical calculation of solubility of nitrogen and carbon in liquid iron, cobalt and nickel, *Interaction of Gases with Metals*, Nauka, Moscow, 1973, p. 109.

20. Erokhin, A.A., *Plasma-Arc Melting of Metals and Alloys*, Nauka, Moscow, 1975, p. 188.

21. Popel', S.I. et al., Temperature dependence of surface tension of iron-oxygen melts, *Physico-Chemical Investigations of Metallurgical Processes*, UPI, Sverdlovsk, 1974, p. 54.

22. Pokhodnya, I.K., Principles of hydrogen adsorption and desorption in arc welding, *Scientific Problems of Welding and Special Electrometallurgy*, Vol. 3, Naukova Dumka, Kyiv, 1970, p. 142.

23. Novozhilov, N.M., *Fundamentals of Metallurgy in Gas-Shielded Welding*, Mashinostroenie, Moscow, 1979, p. 231.

24. Erokhin, A.A., *Fundamentals of Fusion Welding*, Nauka, Moscow, 1975, p. 188.

25. Fischer, W.A. and Hoffman, A., Einflus von kohlenstoff und sauerstoff auf das verhalten einidger begleitelemente in eisenschmelzen under hochvakuum, *Arch. Eisenhut.*, 7, 411, 1960.

26. Oeters, F., Uber die auflosing von suspendierten oxydeteilchen in flussingem eisen, *Arch. Eisenhut.*, 2, 83, 1967.

27. Potapon, N.N., *Fundamentals of Selecting Fluxes for Welding Steels*, Mashinostroenie, Moscow, 1979, p. 168.

28. Sapiro, L.S., On nitrogen and hydrogen evolution during metal solidification, *Svar. Proizvod.*, 7, 41, 1965.

29. Novokhatsky, I.A., *Gases in Oxide Melts*, Metallurgiya, Moscow, 1975, p. 216.

30. Holzgruber, V., Schneidhofer, A., and Eger, H., *Hydrogen in Electroslag Remelting, Special Electrometallurgy*, Part 2, Naukova Dumka, Kyiv, 1972, p. 122.

31. Lubavsky, K.V., Metallurgy of automatic submerged-arc welding of low-carbon steel, *Problems of Theory of Welding Processe*, Mashgiz, Moscow, 1948, p. 86.

32. Lubavsky, K.V. and Timofeev, M.M., Influence of composition variations of high-manganese flux on its properties, *Avtogen. Delo*, 6, 5, 1951.

33. Chalmers, B., *Solidification Theory*, Metallurgiya, Moscow, 1968, p. 88.

34. Efimov, V.A., *Pouring and Solidification of Steel*, Metallurgiya, Moscow, 1966, p. 552.

35. Braun, M.N. and Skok, Yu.V., Influence of rare-earth metals on steel solidification, *Litejnoe Proizvod.*, 9, 27, 1966.

36. Kreshchanovski, N.S. and Sidorenko, M.F., *Modification of Steel*, Metallurgiya, Moscow, 1970, p. 296.

37. Danilov, V.I. and Pomogajlo, A.G., On solidification of sodium and potassium, *Dokl. AN SSSR*, 68(5), 843, 1949.

38. Ovsienko, D.E. and Kostyuchenko, V.I., Influence of oxides on nucleation of solidification centers in overcooled melt of iron and some of its alloys, *Growing of Crystals*, Vol. 3, AN SSSR Publ. House, Moskow, 1961, p. 104.

39. Rusanov, A.I., *Phase Equilibria and Surface Phenomena*, Khimiya, Leningrad, 1967, p. 388.

40. Teumin, I.I., Ultrasonic treatment of metals during solidification, *Problems of Physical Metallurgy and Physics of Metals*, No. 7, Metallurgiya, Moscow, 1962, p. 375.

41. Ubbelode, A., *Fusion and Crystalline Structure*, Mir, Moscow, 1969, p. 365.

42. Khollomon, D.N. and Turnbull, D., Nuclei formation in phase transformations, *Progress of Physics of Metals*, Metallurgizdat, Moscow, 1956, p. 304.

43. Turnbull, D. and Cech, R.E., *J. Appl. Phys.*, 21, 804, 1950.

44. Dokhov, M.P. and Zadumkin, S.N., Surface energy on melt-crystal boundary, *Wettability and Surface Properties of Melts and Solids*, Naukova Dumka, Kyiv, 1972, p. 13.

45. Taova, T.M. and Khokonov M.Kh., Calculation of interphase energy at the boundary of crystal-melting of metals and inorganic compounda, *Physics of Interphase Phenomena*, KBGU, Nalchik, 1984, p. 88.

46. Przhibyl, I., *Theory of Casting Processes*, Mir, Moscow, 1967, p. 328.

47. Meyer, K., *Physical–Chemical Crystallography*, Metallurgiya, Moscow, 1972, p. 479.

48. Kunin, L.L., *Surface Phenomena in Metals*, Metallurgiya, Moscow, 1955, p. 304.

49. Zaletaeva, R.I., Kreshchanovski, N.S., and Kunin, L.L., Influence of calcium on crystallization and surface tension austenitic chrome-nickel steel, *Litejnoe Proizvod.*, 2, 15, 1951.

50. Skvortsov, V.N., The role adsorbtion components during arc electrowelding of metals, *Applied Chemistry in Mashine Building*, Mashgiz, Moskow, 1955, p. 15.

51. Gasyuk, G.N. and Savchuk, V.P., On influence of ultrasound on the process of substance solidification, *Acoustics and Ultrasonic Technique*, Tekhnika, Kyiv, 1978, p. 19.

52. Dyatlov, V.I., Chernysh, V.P., and Samotryasov, M.S., *Electromagnetic Stirring and Solidification of Welding Pool*, UkrNIINTI, Kyiv, 1970, p. 19.

53. Balandin, G.F., *Formation of Crystalline Structure of Castings*, Mashinostroenie, Moscow, 1965, p. 255.

54. Gruzin, V.G., Structure formation depending on liquid phase temperature field, *Ingot and Properties of Steel*, AN SSSR Publ. House, Moscow, 1961, p. 32.

55. Pokhodnya, I.K. et al., Increase in resistance of weld metal to crystalline cracking in welding with rutile electrodes, *Avtom. Svarka.*, 10, 23, 1980.

56. Bruk, B.I., *Radioactive Isotopes in Metallurgy and Physical Metallurgy of Welding*, Sudpromgiz, Leningrad, 1959, p. 232.

57. Movchan, B.A. and Poznyak, L.A., Radiographic examination of intragranular non-homogeneity of sulphur and phosphorus in welds, *Avtom. Svarka.*, 4, 76, 1956.

58. Tiller, W.A., Solute segregation during ingot solidification, *J. Iron and Steel Inst.*, 192(4), 338, 1959.

59. Landau, L.D. and Lifshits, E.M., *Statistical Physics*, Nauka, Moscow, 1976, p. 584.

60. Orlov, A.N., Perevezentsev, V.N., and Rybin, V.V., *Grain Boundaries in Metals*, Metallurgiya, Moscow, 1980, p. 156.

61. Hondros, E.D. and Seah, M.P., The theory of grain boundary segregation in terms of surface adsorption analogues, *Met. Trans.*, 9, 1363, 1977.

62. Geguzin, Ya.E., *Macroscopic Defects in Crystals*, Metallurgizdat, Moscow, 1962, p. 342.

63. Semenchenko, V.K., *Surface Phenomena in Metals and Alloys*, Gosteoretizdat, Moscow, 1957, p. 491.

64. Esin, O.A., *Electrolytic Nature of Liquid Slags*, UPI Publ. House, Sverdlovsk, 1946, p. 41.

65. Deryabin, A.A., Esin, O.A., and Popel', S.I., Features of electrocapillary curves in oxide melts, *J. Phys. Chimi.*, 4, 966, 1965.

66. Adam, N.K., *Physics and Chemistry of Surfaces*, OGIZ, Moscow, 1947, p. 342.

67. Najdich, Yu.V., *Contact Phenomena in Metal Melts*, Naukova Dumka, Kyiv, 1972, p. 196.

68. Kindgery, W.D., *High-Temperature Measurements*, Metallurgizdat, Moscow, 1963, p. 465.

69. White, D., Theory and experiment in methods for the preosion measurment of surface tension, *Trans. Amer. Soc. for Met.*, 5 (March–June–September), 757, 1962.

70. Maurakh, M.A., Mitin, B.S., and Rojtberg, M.B., Contactless measurement of density and surface tension of liquid oxides at high temperatures, *Zavod. Lab.*, 8, 14, 1967.

71. Yakobashvili, S.B., Influence of chromium and vanadium on surface and interphase tension of liquid steel and flux, *Avtom. Svarka.*, 8, 38, 1983.

72. Pekarev, A.I., Determination of surface tension of tungsten, molybdenum and rhenium, *Izv. Vuzov. Chyorn. Metall.*, 6, 111, 1963.

73. Carverley, A.A., Determination of the surface tension of liquid tungsten by the drop weight method, *Prog. Phys. Soc.*, 11, 215, 1957.

74. Namba, S., Measurment of surface tension of molten metals by electron bombardment heating, *Procedings of the 4th Sympium an Electron Beam Technol.*, Alloyed Electron. Corp., 1962, p. 95.

75. Muu, Bui Van., Fenzke, H.W., and Kraus, S., Erfahrungen bei der experimentellen Bestimmung der Oberflachenspannung von Eisen- und Schlackenmelzen, *Neue Hutte.*, 29(3), 113, 1984.

76. Baum, B.A. et al., Surface tension of some commercial grade steels, *Physical Chemistry of Interphases*, Naukova Dumka, Kyiv, 1976, p. 80.

77. Pugachevich, P.P., Some aspects of surface tension measurement in a gas bubble, *Surface Phenomena in Metallurgical Processes*, Metallurgizdat, Moscow, 1963, p. 127.

78. Glebovskii, V.G. and Burtsev, V.T., *Melting of Metals and Alloys in Suspended State*, Metallurgia, Moscow, 1974, p. 176.

79. Ei-Sharont, M. and Lange Klaus, W., Optimierungsverfazhen zur berechnung der grenzflaschen spannung nach der methode des hangenden tropfens, *Z. Metallk.*, 78(3), 184, 1987.

80. Taran, V.D. and Chudinov, M.S., Determination of surface tension of molten metal pool in the conditions of welding, *Svar. Proizvod.*, 1, 7, 1972.

81. Emelyanov, I.L., Estimation of interphase properties by deposited bead shape, *Svar. Proizvod.*, 10, 56, 1976.

82. Tyulkov, M.D., Influence of surface tension on butt weld root formation in shielded-gas electric arc welding, *Aspects of Gas-Shielded Arc Welding*, Mashgiz, Moscow, 1957, p. 55.

83. Skorov, D.M. et al., *Surface Energy of Solid Metallic Phases*, Atomizdat, Moscow, 1973, p. 172.

84. Udin, H., Shaler, A.J., and Wulfs, J., The surface tension of solid copper, *J. Metals*, 1, 186, 1949.

85. Kopan', V.S., The variant of the metod of zero creep, *Ukrain. Phys. J.*, 2, 14, 1965.

86. Khokonov, Kh.B., Shebzukhova, I.G., and Kokov, Kh.N., Measurement of surface tension of tin, indium and lead in solid state, *Wettability and Surface Properties of Melts and Solids*, Naukova Dumka, Kyiv, 1972, p. 156.

87. Missol, V., *Interfacial Energy in Metals*, Metallurgia, Moscow, 1978, p. 176.

88. Popel', S.I., Esin, O.A., and Gel'd, N.V., On the procedure of interphase tension measurement at high temperatures, *Dokl. AN SSSR.*, 74(6), 1097, 1950.

89. Mikiashvili, Sh.M., Tsylev, L.M., and Samarin, A.M., Properties of melts of MnO, $SiO_2-Al_2O_3$ system, *Physical–Chemical Principles of Steel Production*–Metallurgizdat, Moscow, 1964, p. 423.

90. Yakobashvili, S.B., *Surface Properties of Welding Fluxes and Slags*, Tekhnika, Kyiv, 1970, p. 208.

91. Patrov, B.V., On the double layer charge and capacitance in cast iron-slag system, *Izv. vuzov. Chern. Metall.*, 7, 33, 1961.

92. Patrov, B.V., Electrocapillary phenomena in cast iron–slag system, *Izv. Vuzov. Chern. Metall.*, 6, 3, 1958.

93. Minaev, Yu.A., *Surface Phenomena in Metallurgical Processes*, Metallurgia, Moscow, 1984, p. 152.

94. Patskevich, I.R., Ryabov, V.R., and Deyev, G.F., *Surface Phenomena in Welding of Metals*, Naukova dumka, Kyiv, 1991, p. 240.

95. Esin, O.A., Nikitin, Yu.P., and Popel', S.I., Electrocapillary phenomena at high temperatures, *Dokl. AN SSSR*, 83(3), 341, 1952.

96. Nicitin, Yu.P. and Esin, O.A., Electrocapillary phenomena in pirometallurgical systema, *Dokl. AN SSSR*, 107(6), 847, 1956.

97. Deryabin, A.A. and Popel', S.I., Shape of electrocappillary curves for cast irons and steels, contacting oxide melts, *Elektrokhimiya*, 2(3), 295, 1966.

98. Deryabin, A.A., Popel', S.I., and Saburov, L.N., Change of interphase tension in polarization of metal-oxide phase boundary by direct and alternating current, *Izv. Vuzov. Chyorn. Metall.*, 6, 10, 1969.

99. Saurwald, F. and Drath, G., The surface tension of molten metals and alloys, *Z. Anorg. Chemie*, 181, 353, 1929.

100. Sapiro, S.I., Thermodynamic analysis of coalescence and coagulation of nonmetallic inclusions *Steel*, 7(6), 449, 1946.

101. Andreev, I.A., Process of boiling of open-hearth bath in terms of physics of surface phenomena, *Trans. of TsNII NKTP*, 2–3, 17, 1944.

102. Filippov, S.I., Menshikov, M.R., and Yakovlev, V.V., Relationship between decarbonization rate and melt properties, *Izv. Vuzov. Chyorn. Metall.*, 5, 9, 1971.

103. Esche, W. and Peter, O., Bestimmung der oberflaschenspannung an reinem und legirtem eisen, *Arch. Eisenhutt.*, 6, 355, 1956.

104. Van Zin-tan, Karasev, R.A., and Samarin, A.M., Influence of carbon and oxygen on surface tension of liquid iron, *Izv. AN SSSR. OTN. Metall. Toplivo*, 1, 30, 1960.

105. Eremenko, V.P., Ivashchenko, Yu.N., and Bogatyrenko, B.B., Surface tension of pure iron and Fe–C system alloys, *Surface Phenomena in Metals and Alloys and their Role in Powder Metallurgy Processes*, AN Ukr.SSR Publ. House, Kyiv, 1962, p. 56.

106. Kashik, I. and Skala, I., Measurement of surface tension of iron alloys in liquid state, *Physical–Chemical Principles of Steel Production*, AN SSSR Publ. House, Moscow, 1961, p. 133.

107. Tsarevsky, B.V. and Popel', S.I., Influence of alloying elements on surface properties of iron, *Izv. Vuzov. Chyorn. Metall.*, 12, 12, 1960.

108. Halden, F.A. and Kindgery, W.D., Surface tension of elevated temperatures 11. Effect of C, N, O, and S on liquid iron surface tension and interfacial energy with Al_2O_3, *J. Phys. Chem.*, 51(6), 557, 1955.

109. Van Zin-tan, Karasev, R.A., and Samarin, A.M., Surface tension of iron–manganese and iron–sulphur melts, *Izv. AN SSSR. OTN. Metall. Toplivo*, 2, 49, 1960.

110. Popel', S.I. et al., Simultaneous influence of oxygen and sulphur on surface tension of iron, *Izv. AN SSSR. Metally*, 4, 54, 1975.

111. Dragomir, I., Vishkarev, A.F., and Yavojsky, V.I., Investigation of the properties of iron–phosphorus melts. Surface tension and density, *Izv. Vuzov. Chyorn. Metall.*, 11, 50, 1963.

112. Grigoriev, V.P. et al., Influence of phosphorus and manganese on surface tension of iron–carbon alloys, *Izv. Vuzov. Chyorn. Metall.*, 4, 4, 1960.

113. Volkov, S.E., Levets, N.P., and Samarin, A.M., Surface tension of iron–phosphorus–oxygen melts, *Surface Phenomena in Melts and Solid Phases Formed from them*, Kabardino–Balkarskoe Knizhnoe Izdatelstvo, Nalchik, 1965, p. 411.

114. Gel'd, P.V. and Petrushevsky, M.S., Isotherm of surface energy of silicon–iron liquid alloys, *Izv. AN SSSR. OTN. Metall. Toplivo*, 3, 160, 1961.

115. Kawai, Y., Viscosity and surface tension of liquid iron alloys, *Proceeding of the International Conference on the Science and Technology of Iron and Steel Tokyo, 1970*, Part 1, Tokyo, 1971, p. 2949.

116. Gel'd, P.V., Baum, B.A., and Akshentsev, Yu.P., Surface properties of liquid alloys on manganese and iron base, *Physicochemical Principles of Steel Production*, Nauka, Moscow, 1971, p. 165.

117. Van Zin-tan, Karasev, R.A., and Samarin, A.M., Surface tension of iron–sulphur–carbon, iron–manganese–sulphur, and iron–manganese–carbon melts, *Izv. AN SSSR. OTN. Metall. Toplivo*, 1, 45, 1961.

118. Popel', S.I., Dzhemilev, N.K., and Tsarevsky, B.V., Density and surface tension of Fe–Mn–Si melts at 1550°C, *Zh. Fizi. Khim.*, 7, 1545, 1966.

119. Levin, A.M., Measurement of liquid steel surface tension, *Trans. Dnieprop. Metall.*, Metallurgizdat, Moscow, 28, 105, 1952.

120. Mori, M. et al., Surface tension of liquid iron–nickel–chromium alloys system, *J. Jpn. Inst. Metals*, 12, 1301, 1975.

121. Kaufman, S.M. and Whalen, T.J., Surface tension and surface adsorption in liquid iron–carbon alloys: the systems Fe–C–Ni and Fe–C–Co, *Trans. Metal. Soc. AIME*, 230(4), 263, 1964.

122. Dzhemilev, N.K., Popel', S.I., and Tsarevsky, B.V., Density and surface properties of iron–cobalt–nickel melts at 1500°C, *Zh. Fiz. Khim.*, 41(1), 47, 1967.

123. Kolesnikova, T.P. and Samarin, A.M., Influence of manganese, chromium and vanadium on surface tension of liquid metal, *Izv. AN SSSR*, 5, 63, 1956.

124. Smirnov, A.A., Popel', S.I., and Pastukhov, A.I., Influence of vanadium on density and surface properties of iron–carbon alloys, *Izv. Vuzov. Chyorn. Metall.*, 4, 13, 1965.

125. Smirnov, A.A., Popel', S.I., and Tsarevsky, B.V., Influence of titanium on surface properties of iron and iron–carbon alloys, *Izv. Vuzov. Chyorn. Metall.*, 3, 10, 1965.

126. Tsarevsky, B.V., Popel', S.I., and Babkin, V.G., Influence of tungsten on surface properties and density of iron, *Surface Phenomena in Melts*, Naukova Dumka, Kyiv, 1968, p. 176.

127. Tsarevsky, B.V., Popel', S.I., and Domozhirov, B.F., Influence of molybdenum and chromium on surface properties and density of iron, *Surface Phenomena in Melts and Solid Phases Formed from Them*, Kabardino-Balkarskoe Knizhnoe Izdatelstvo, Nalchik, 1965, p. 316.

128. Baum, B.A., Gel'd, P.V., and Akshentsev, Yu.N., Density and surface energy of Fe–C–Cr system liquid alloys, *Surface Phenomena in Melts*, Naukova Dumka, Kyiv, 1968, p. 202.

129. Ayushina, G.G., Levin, E.S., and Gel'd, P.V., Influence of temperature and composition on density and surface energies of iron–aluminium melts, *Zh. Fiz. Khim.*, 42(11), 2799, 1968.

130. Bliznyukov, S.A., Pirogov, N.A., and Kryakovsky, Yu.V., Influence of boron and cerium on surface tension of liquid iron and steel, *Surface Phenomena in Melts and Solid Phases Formed from Them*, Kishinev, 1968, p. 86.

131. Vashchenko, K.I., Rostovtsev, D.I., Larin, V.K., Influence of lanthanum and cerium master alloys on surface tension of Kh21L steel, *Surface Phenomena in Melts*, Naukova Dumka, Kyiv, 1968, pp. 143–147.
132. Vishkarev, A.F. et al., Influence of rare-earth elements on surface tension of liquid iron, *Izv. Vuzov. Chyorn. Metall.*, 3, 60, 1962.
133. Bobkova, O.S. et al., Influence of boron additions on surface tension of 12KhMF steel and identification of traps by nonisothermal hydrogen diffusion method, *Surface Phenomena in Melts and Solid Phases Formed from Them*, Kabardino-Balkarskoe Knizhnoe Izdatelstvo, Nalchik, 1965, p. 338.
134. Pirogov, N.A., Bliznyukov, S.A., and Kryakovsky, Yu.V., Influence of boron and cerium on surface tension of liquid iron and steel, *Physical Chemistry of Surface Phenomena at High Temperatures*, Naukova Dumka, Kyiv, 1971, p. 106.
135. Tavadze, F.N., Bairamshvili, I.A., and Khantadze, D.V., Surface tension and structure of melted borides of iron, cobalt and nickel, *Dokl. AN SSSR*, 162(1), 62, 1965.
136. Sveshkov, Yu.V. et al., Adsorption and surface activity of selenium solution in liquid iron, *Izv. AN SSSR. Metally*, 8, 74, 1973.
137. Kindgery, W.D., Surface tension at elevated temperatures. 1V. Surface tension of Fe–Se and Fe–Te alloys, *J. Phys. Chem.*, 62(7), 878, 1958.
138. Borodulin, E.K., Kurochkin, K.T., and Umrikhin, P.V., Influence of nitrogen on surface tension of iron and its alloys, *Physicochemical Principles of Steel Production*, Nauka, Moscow, 1968, p. 21.
139. Borodulin, E.K. et al., Examination of surface activity of iron base triple systems, *Surface Phenomena in Melts*, Naukova Dumka, Kyiv, 1968, p. 225.
140. Zhu Iun and Mukai Kusihiro., The surface tension of liquid iron containing nitrogen and oxygen, *ISIJ Inst.*, 38(10), 1039, 1998.
141. Ershov, G.S. and Byshev, V.M., Influence of gases on surface tension of liquid iron and alloyed steel, *Izv. AN SSSR, Metally*, 4, 59, 1975.
142. Kreshchanovski, N.S., Prosvirin, V.I., and Zaletaeva, R.P., Nitrogen influence on surface tension and solidification of austenitic steel, *Liteinoe proizvodstvo*, 1, 23, 1954.
143. Antipin, V.G., Chebotarev, V.V., and Korotkikh, V.F., Surface properties of the melts of some grades of steel, *Izv. AN SSSR*, 1, 62, 1978.
144. Baum, B.A., Kurochkin, K.T., and Umrichin, P.V., Hydrogen influence of surface tension of iron and its alloys, *Izv. AN SSSR, Metall. i Toplivo*, 3, 82, 1961.
145. Dzhoshi, V.B., Vishkarev, A.F., and Yavoiskii, V.I., Role of surface phenomena in the processes of hydrogen redistribution between the metal and the gas phase, *Izv. Vuzov. Chern. Metall.*, 3, 23, 1961.
146. Kirdo, I.V., On composition of gases in the welding arc atmosphere, *Avtomaticheskaya Svarka*, 1(10), 50, 1950.
147. Frolov, V.V., *Physicochemical Processes in Welding Arc*, Mashgiz, Moscow, 1954, p. 132.
148. Smittels, K., *Gases and Metals*, Metallurgizdat, Moscow-Leningrad, 1940, p. 228.
149. Mazanek, T., Navellanie zeiaza za pomoca tienku welga, *Arch. Hutn.*, 4, 186, 1956.
150. Kulikov, G.D., *Current Methods of Reconditioning of Parts by Surfacing*, Yuzhno Uralskoe Knizhn. Publ. House, Cheliabinsk, 1974, p. 182.
151. Esche, W. and Peter, O., Bestimmung der oberflachenspanning an reinen und legirtem eisen, *Arch. Eisenhut.*, 6, 355, 1956.
152. Rykov, O.A., Investigation of Some Peculiarities of Arc Surfacing in Air Flow, Ph.D. thesis, Cheliabinsk, 1973.

153. Medzhibozhsky, M.Ya., On mechanism of oxidizing of iron additions in metal purging by oxygen and air, *Izv. Vuzov. Chyorn. Metall.*, 11, 37, 1967.
154. Streltsov, F.N., Perevalov, N.N., and Travin, O.V., Some aspects of kinetics of iron–carbon alloys decarborinizing, *Theory of Metallurgical Processes*, Metallurgiya, Moscow, 1967, p. 45.
155. Popel', S.I., Influence of oxide melt components on interphase tension between the melt and iron, *Zh. Fizi. Khim.*, 32(10), 2398, 1958.
156. Levin, E.S. and Gel'd, P.V., Polyterms of density and surface energy of liquid aluminium, *Therm. Phy. High Temp.*, 6(3), 432, 1968.
157. Lang, G., Einflus von zustzelement auf die oberflachenspannung von flussigem rienstaluminium, *Aluminium*, 50(11), 731, 1974.
158. Yatsenko, S.P., Kononenko, V.I., and Sukhman, A.L., Experimental investigation of temperature dependence between surface tension and density of tin, indium, aluminium and gallium, *Therm. Phy. High Temp.*, 10(1), 66, 1972.
159. Korolkov, A.M., Surface tension of aluminium and its alloys, *Izv. AN SSSR. OTN*, 2, 35, 1956.
160. Lang, G., Glesseigenschaften und oberflachenspannung von aluminium und binaren aluminium legierungen. Teil 111. Oberflachenspannung, *Aluminium*, 49(3), 231, 1973.
161. Kubichek, L., Influence of some elements on surface tension of aluminium alloys. *Izv. AN SSSR. Metall. i Toplivo*, 2, 96, 1959.
162. Pogadaev, A.M. and Lukashenko, E.E., Surface tension and density of aluminium-magnesium liquid alloys, *Advances in Theory and Technology of Metallurgical Processes*, Krasnoyarskoe Publ. House, Krasnoyarsk, 1973, p. 83.
163. Najdich, Yu.V. et al., Temperature dependence of liquid copper surface tension, *Zh. Fiz. Khim.*, 35(3), 694, 1961.
164. Bykova, N.A. and Shevchenko, V.G., Density and surface tension of copper, aluminium, gallium, indium and tin, *Physicochemical Investigations of Liquid Metals and Alloys*, No 29, Sverdlovsk, 1974, p. 42.
165. Khilya, G.P., Free Surface Energy and Molar Volumes of Some Systems with a Maximum on Liquidus Line, Ph.D. thesis, Kyiv, 1975.
166. Norin, P.A., Osipov, A.M., and Deyev, G.F., Influence of silicon, manganese and aluminium on surface tension of copper, *Abstracts of Papers, 11th All-Union Conference on Welding of Non-Ferrous Metals*, E.O. Paton Welding Institute, Kyiv, 1982, p. 122.
167. Najdich, Yu.V., Eremenko, V.N., and Kirichenko, L.F., Surface tension and density of liquid alloys of copper-aluminium system, *Zh. Fiz. Khim.*, 7(2), 333, 1962.
168. Chursin, V.M. and Gerasimov, S.P., Influence of chemical composition on surface tension of tin bronzes, *Trans. Mosc. Vechern. Metall. Inst.*, 12, 121, 1972.
169. Wigbert, G., Pawlek, F., and Ropenack, A., Einflus von verunreinigyngen und beimengyngen auf viskositat und oberflachenspannungen geschmolzenen kupfers, *Z. Metallk.*, 54(3), 147, 1963.
170. Eremenko, V.N., Najdich, Yu.V., and Nosonovich, A.A., Surface activity of oxygen in copper–oxygen liquid alloys, *Zh. Fiz. Khim.*, 34(5), 1018, 1960.
171. Joud, J.C. et al., Determinasion de la tension superficielle des alliages Ag–Pb et Cu–Pb par la methode de la goutle posse, *J. Chim. Phys. Et phys—Chim. Biol.*, 70(3), 1290, 1973.
172. Tille, J. and Kelly, J.C., The surface tension of liquid titanium, *Brit. J. Appl. Phys.*, 14(10), 717, 1963.

173. Elyutin, V.P., Kostikov, V.I., and Maurakh, M.A., Physical properties of liquid refractory metals and oxides, *High-Temperature Materials*, No. 49, Metallurgiya, Moscow, 1968, p. 106.

174. Eremenko, V.N. et al., Procedure and results of measurement of free surface energy of refractory and active metals, *Surface Phenomena in Melts*, Naukova Dumka, Kyiv, 1968, p. 148.

175. Elyutin, V.P., Maurakh, M.A., and Pugin, V.S., Surface tension of titanium alloys with tin, aluminium and iron, *Izv. Vuzov. Chyorn. Metall.*, 5, 117, 1964.

176. Elyutin, V.P., Kostikov, V.I., and Maurakh M.A., Investigation of contact interaction between liquid titanium and graphite, *Surface Phenomena in Melts and Solid Phases Occurring from Them*, Kabardino-Balkarskoe Publ. House, Nalchik, 1964, p. 345.

177. Ahmad, U.M. and Murr, L.E., Surface free energy of nickel and stainless steel at temperatures above the melting point, *J. Mater. Sci.*, 11(2), 224, 1976.

178. Popel', S.I., Shergin, L.M., and Tsarevsky, B.V., Temperature dependence of densities and surface tension of iron-nickel melts, *Zh. Fiz. Khim.*, 40(9), 2365, 1969.

179. Eremenko, V.N. et. al., Surface tension of liquid alloys of binary metallic systems with a maximum on liquidus curves, *Ukr. Khim. Zh.*, 28(4), 500, 1962.

180. Shergin, L.M., Popel', S.I., and Tsarevsky, B.V., Temperature dependence between densities and surface tension of cobalt–silicon and nickel–silicon melts, *Physical Chemistry of Metallurgical Melts*, No.25, Sverdlovsk, 1971, p. 50.

181. Whalen, T.J. and Humenik, M., Surface tension and contact angles of copper-nickel alloys on titanium carbide, *Trans. Metal. Soc. AIME*, 218(5), 952, 1960.

182. Kurkjian, C.R. and Kingery, W.D., Surface tension at elevated temperatures. III. Effect of Cr, In, Sn and Ti on liquid nickel surface tension and interfacial energy with Al_2O_3, *J. Phys. Chem.*, 60(7), 961, 1956.

183. Eremenko, V.N., Nizhenko, V.I., and Sklyarenko, L.I., Surface properties of chrome–nickel melts, *Surface Phenomena in Melts and Solid Phases, Formed from Them*, Kabardino-Balkarskoe Knizhnoe Izdatelstvo, Nalchik, 1965, p. 287.

184. Eremenko, V.N., Nizhenko, V.I., and Sklyarenko, L.I., Density and free surface energy of liquid alloys of nickel–chrome system, *Poroshk. Metall.*, 5, 24, 1965.

185. Mori, M. et. al., Surface tension of liquid iron–nickel–chromium alloys system, *J. Jp. Inst. Metals*, 39(12), 1301, 1975.

186. Sveshkov, Yu.V., Kalmykov, V.A., and Mironov, V.A., Surface tension of Fe–Se, Ni–Se and Ni–Te melts, *Izv. AN SSSR. Metally*, 3, 84, 1975.

187. Sveshkov, Yu.V., Kalmykov, V.A., and Alferov, V.I., Free surface energy and adsorption of selenium and tellurium solutions in metals of iron group, *Physical Chemistry of Boundaries of Contacting Phases*, Naukova Dumka, Kyiv, 1976, p. 71.

188. Vaisburd, S.E., Surface properties of binary sulfide-metallic Fe–S, Co–S, Ni–S melts, *Surface Phenomena in Melts and Solid Phases, Formed from Them*, Kabardino-Balkarskoe Knizhnoe Izdatelstvo, Nalchik, 1965, p. 333.

189. Eremenko, V.N., Nizhenko, V.I., and Sklyarenko, L.I., On temperature dependence of free surface energy of liquid iron, *Surface Phenomena in Melts and Solid Phases, Formed from Them*, Kabardino-Balkarskoe Knizhnoe Izdatelstro, Nalchik, 1965, p. 287.

190. Kupriyanov, A.A. and Filippov, S.I., Surface tension and structural transformations in iron–carbon melts, *Izv. Vuzov. Chyorn. Metall.*, 11, 16, 1968.

191. Deyev, G.F. and Patskevich, I.R., Determination of surface tension of metals in different gaseous media, *Investigation and Application of Dip-Transfer Surfacing*, Yuzhno-Uralskoe Publ. House, Chelyabinsk, 1968, p. 16.

192. Deyev, G.F. and Patskevich, I.R., Determination of surface tension of some welded steels in nitrogen atmosphere, *Theory and Practice of Welding Production*, No. 82, Chelyabinsk, 1969, p. 9.

193. Deyev, G.F. and Sergeyev, I.V., Determination of surface tension of some steels in hydrogen atmosphere, *Advanced Technology in Welding Production*, Voronezh, 1975, p. 62.

194. Patskevich, I.R., Bojko, V.P., and Deyev, G.F., Surface tension of molten titanium and its alloys, *Avtom. Svarka*, 8, 73, 1981.

195. Leontieva, A.A., Influence of free carbon on interphase surface tension in silicate-iron sulphide system, *Kolloidn J.*, 11(3), 176, 1949.

196. Popel', S.I., Esin, O.A., and Nikitin, Yu.P., About surface activity of carbon and phosphorus at the metal–slag boundary, *Surface Phenomena in Pyrometallurgy*, Sverdlovsk, 1954, p. 82.

197. Mikiashvili, Sh.M. and Samarin, A.M., Surface properties of the interphase between iron–carbon melts and aluminium–manganese silicates, *Physicochemical Principles of Steel Production*, Nauka, Moscow, 1968, p. 29.

198. Popel', S.I. et al., Influence of sulphur on interphase tension at the metal–slag boundary, *Dokl. AN SSSR*, 112(1), 104, 1957.

199. Mikiashvili, Sh.M. and Samarin, A.M., Interphase tension at the boundary of liquid iron with oxysulphide melts, *Physicochemical Principles of Steel Production*, Nauka, Moscow, 1964, p. 42.

200. Bobkova, O.S. and Petukhov, V.S., Influence of sulphur on interphase interaction of steel with slags of $CaO–Al_2O_3$ system, *Surface Phenomena in Melts and Solid Phases, Formed from Them*, Kabardino-Balkarskoe Knizhnoe Izdatelstvo, Nalchik, 1965, p. 532.

201. Kozacevitsh, P., Urbain, G., and Sage, M., Activite super ficielle et activite thermo-dynamique du sourte dans les alliages liquides fercarbone-sourte, *Rev. de Metal*, 52(2), 161, 1955.

202. Konovalov, G.F. and Popel', S.I., Interphase tension at the boundary of steel with slags and deoxidation products, *Phys. Chem. of Metallurgical Proc.*, 93, 73, 1959.

203. Krinochkin, E.V., Kurochkin, K.T., and Umrikhin, P.V., Influence of some components on interphase properties of iron, *Wettability and Surface Properties of Melts and Solids*, Naukova Dumka, Kyiv, 1972, p. 83.

204. Mazanek, T., Wlasnosci powierzchniowe metalu I zuzla w procesach stalowniezych, *Arch. Hutn.*, 2, 83, 1963.

205. Bobkova, O.S., Interphase tension of iron–chromium alloys at the boundary with multicomponent slags, *Surface Phenomena in Melts*, Naukova Dumka, Kyiv, 1968, p. 321.

206. Deryabin, A.A., Popel', S.I., and Sajdunin, R.A., Interphase tension and adhesion of ferroalloys to oxides, *Surface Phenomena in Melts and Solid Phases, Formed from Them*, Kishinev, 1968, p. 110.

207. Gel'd, P.V., Popel', S.I., and Nikitin, Yu.P., On liquid silicon oxide, *Zh. Prikl. Khim.*, 25(6), 592, 1952.

208. Mikiashvili, Sh.M. and Gogiberidze, Yu.M., Interphase tension and adhesion at the interface of iron–silicon alloys with melts of manganese oxide–alumina–silica system, *Inform. AN GSSR*, 38(3), 28, 1965.

209. Mikiashvili, Sh.M., Gogiberidze, Yu.M., and Kekelidze, A.M., Interphase tension at the interface of silicon–manganese–iron alloys with oxide melts, *Surface Phenomena in Melts*, Naukova Dumka, Kyiv, 1968, p. 316.

210. Popel', S.I., Esin, O.A., and Gel'd, P.V., On interphase tension of iron alloys at the boundary with slags, *Dokl. AN SSSR*, 74(6), 1097, 1950.

211. Bobkova, O.S. and Petukhov, V.S., Role of surface phenomena in mixing of steel with synthetic slags, *Surface Phenomena in Melts and Processes of Powder Metallurgy*, AN Ukr. SSR Publ. House, Kyiv, 1963, p. 212.

212. Deyev, G.F. and Patskevich, I.R., Influence of slag composition on interphase tension, *Avtom. Svarka*, 2, 5, 1971.

213. Popel', S.I. and Konovalov, G.F., Interphase tension of low-carbon steel at the boundary with deoxidation products, *Izv. Vuzov. Chyorn. Metall.*, 8, 3, 1959.

214. Deryabin, A.A. and Popel', S.I., Adhesion of ShKh15 steel to slags, containing sodium oxide, *Izv. Vuzov. Chyorn. Metall.*, 5, 18, 1964.

215. Bobkova, O.S. and Petukhov, V.S., Influence of fluor-spar and sodium oxide on slag surface tension and on interphase tension at the boundary with ferrochrome, *Theory of Metallurgical Processes*, No. 50, Metallurgiya, Moscow, 1967, p. 30.

216. Kazumi, M. and Macatito, P., Surface Tension between molyen Iron and melted Slag, *J. Iron Steel Inst. (Jpn)*, 4, 274, 1954.

217. Georgiev, A.I., Yavojsky, V.I., and Bobkova, O.S., Interphase phenomena at the interface of manganese low-phosphorous slag with accompanying metal, *Theory of Metallurgical Processes*, Metallurgiya, Moscow, 1967, p. 28.

218. Pupynina, S.M., Volkov, S.E., and Bobkova, O.S., Interphase tension of stainless steel at the interface with fluorine slags, *Physical Chemistry of Surface Phenomena at High Temperatures*, Naukova Dumka, Kyiv, 1971, p. 246.

219. Pokhodnya, I.K. and Garpenyuk, V.N., Temperature of electrode metal drops in heavy-coated electrode arc welding, *Avtom. Svarka*, 12, 1, 1967.

220. Frumin, I.I. and Pokhodnya, I.K., Investigation of average temperature of weld pool, *Avtom. Svarka*, 4, 13, 1955.

221. Podgaetsky, V.V., Galinich, V.I., and Goloshubov, V.I., Investigation of the properties of some commercial fluxes, *Tekhnol. Organizat. Proizvod.*, 12, 48, 1976.

222. Sergienko, A.I., Interphase tension at the liquid phases boundary, *Avtom. Svarka*, 6, 26, 1965.

223. Tarlinskii, V.D. and Yatsenko, V.P., Determination of interphase tension of stick electrode slags, *Avtom. Svarka*, 7, 77, 1980.

224. Slivinsky, A.M., Kopersak, V.N., and Solokha, A.M., Physicochemical and technological properties of fluxes of $CaF_2-SiO_2-Al_2O_3-MgO$ system, *Avtom. Svarka*, 7, 31, 1981.

225. Safonnikov, A.N. and Nikitin, Yu.P., About lacks-of-fusion in electroslag welding of chrome-nickel austenitic steels and alloys, *Avtom. Svarka*, 9, 27, 1962.

226. Deyev, G.F. and Patskevich, I.R., Investigation of interphase tension in metal–slag system as applied to welding processes, *Advanced Technology in Welding Production*, Voronezh, 1969, p. 102.

227. Najdich, Yu.V. and Kolesnichenko, G.A., Investigation of liquid metal wettability and adhesion to graphite and diamond, *Surface Phenomena in Metallurgical Processes*, Metallurgizdat, Moscow, 1963, p. 255.

228. Lesnik, N.D., Pestun, T.S., and Eremenko, V.N., Kinetics of spreading of liquid metals over the surface of solids, *Poroshk. Metall.*, 10, 83, 1970.

229. Summ, B.D. and Goryunov, Yu.V., *Physico-Chemical Principles of Wetting and Spreading*, Chemistry, Moscow, 1976, p. 232.

230. Eremenko, V.N., Lesnik, N.D., and Ivanova, T.S., Kinetics of spreading and contact interaction in metallic systems with intermediate phases, *Methods of Investigation*

and Properties of Interphases of Contacting Phases, Naukova Dumka, Kyiv, 1977, p. 51.

231. Zhukhovitsky, A.A., Grigoryan, V.A., and Mikhalik, E., Action of chemical process on surface properties, *Zhurnal. Fizicheskoj. Khimii.*, 39(5), 1179, 1964.

232. Nicholas, M. and Poole, D.M., The kinetics of sessile drop spreading in reaction metal- metal systems, *Trans. Met. Soc. AIME*, 236(11), 1515, 1966.

233. Najdich, Yu.V. and Perevertajlo, V.M., Investigation of wettability of solids by metal melts as a result of deviation of the system from the equilibrium position, *Wettability and Surface Properties of Melts and Solids*, Naukova Dumka, Kyiv, 1972, p. 32.

234. Natapova, R.I. et al., Wettability of refractory metals by melts of tellurium, antimony and their alloys. *Physical Chemistry of Interphases of Contacting Phases*, Naukova Dumka, Kyiv, 1976, p. 112.

235. Opalovsky, A.A. and Fedorov, V.B., New data in the field of investigation of molybdenum chalcogenides, *Chalcogenides*, 2, 77, 1970.

236. Opalovsky, A.A. et al., Investigation of the processes of interaction of tungsten and rhenium with chalcogenides, *Chalcogenides*, 2, 86, 1970.

237. Tarasova, A.L. and Kirdyashkina, L.I., Influence of temperature on zinc spreading over steel surface, *Physical Chemistry of Boundaries of Contacting Phases*, Naukova Dumka, Kyiv, 1976, p. 117.

238. Ukhov, V.F. et al., Investigation of wettability of nonmetallic solids by liquid palladium-base alloys, *Physical Chemistry of Surface Phenomena at High Temperatures*, Naukova Dumka, Kyiv, 1971, p. 139.

239. Patskevich, I.R. and Deyev, G.F., Study of solid oxides wettability by molten metal, *Avtom. Svarka*, 12, 60, 1979.

240. Novosadov, V.S. et al., Kinetics of spreading of metals over iron, copper, nickel depending on degassing degree, *Wettability and Surface Properties of Melts and Solids*, Naukova Dumka, Kyiv, 1972, p. 53.

241. Esin, O.A. and Gel'd, P.V., *Physical Chemistry of Pyrometallurgical Processes*, Vol. 2, Metallurgiya, Moscow, 1966, p. 703.

242. Deryagin, B.V. On the dependence of contact angle on the microrelief or roughness of wetted surface, *Dokl. AN SSSR*, 51, 357, 1946.

243. Deryagin, B.V. and Sherbakov, L.M., On influence of surface forces on phase equilibrium of polymollecular layers and of the angle of wetting, *Colloid. Zh.*, 23, 40, 1961.

244. Budurov, S.I. et al., Determination of wettability angles at microgravity, *"Salyut-6" – "Soyuz"*. *Science of Materials and Technological Materials, International meeting, Riga, 18-23 May, 1983*, Moscow, 1985, p. 64.

245. Bykhovsky, A.I., *Spreading*, Naukova Dumka, Kyiv, 1983, p. 192.

246. Bykhovsky, A.I. and Glushchenko, A.A., Influence of polarization on spreading of cadmium in salt melt at different temperatures, *Surface Physics and Chemistry*, Nalchik, 1985, p. 31.

247. Bykhovsky, A.I. and Pashchenko, A.V., Kinetics of mercury spreading over metals in the presence of a temperature gradient, *Metallofizika*, No. 67, 1977, p. 80.

248. Bykhovsky, A.I. et al., Influence of polarisation and temperature gradient on spreading of aluminium alloy over stainless steel in chloride melts, *Capillary and Adhesion Properties of Melts*, Naukova Dumka, Kyiv, 1987, p. 49.

249. Turygin, V.N., Myalin, M.I., and Sagalevich, V.M., Interaction of cast iron with copper-base alloys in surfacing, *Svar. Proizvod.*, 6, 3, 1988.

250. Summm, B.D. and Goryunov, Yu.V., *Physico-Chemical Principles of Wetting and Spreading*, Chemistry, Moscow, 1976, p. 232.

251. Bogatyrenko, B.B. et al., Surface phenomena in the processes of steel surfacing, using copper-base alloys, *Adhesion of Melts and Brazing of Materials*, No. 14, 1985, p. 45.

252. Nikitin, I.I., *Physical–Chemical Phenomena in the Case of Action of Liquid Metals on Solid Metals*, Atomizdat, Moscow, 1976, p. 423.

253. Popel', S.I., Kinetics of melt spreading over solid surfaces and kinetics of wetting, *Adhesion of Melts and Brazing of Materials*, No. 1, 1976, p. 3.

254. Sivkov, M.N. et al., Spreading of bronzes over steel surface, *Theoriya i Praktika Svar. Proizvod.*, 5, 133, 1986.

255. Eremenko, V.N., Lesnik, N.D., and Ryabov, V.R., Investigation of the kinetics of aluminium spreading over iron, *Physical Chemistry of Surface Phenomena in Melts*, Naukova Dumka, Kyiv, 1971, p. 203.

256. Eremenko, V.N., Pestun, T.S., and Ryabov, V.R., Kinetics of spreading of aluminium and iron–aluminium melts over iron, *Porosh Metall.*, 7, 58, 1973.

257. Ryabov, V.R., *Fusion Welding of Aluminium to Steel*, Naukova Dumka, Kyiv, 1969, p. 232.

258. Eremenko, V.N., Natanzon, Ya.V., Ryabov, V.R., On interaction of aluminium with iron in the case of fusion welding, *Avtom. Svarka*, 4, 14, 1974.

259. Ryabov, V.R., *Welding of Aluminium and its Alloys to Other Metals*, Naukova Dumka, Kyiv, 1983, p. 264.

260. Eremenko, V.N., Lesnik, N.D., and Ivanova, T.S., Peculiarities of spreading of aluminium over nickel, *Adhesion of Metals and Alloys*, Naukova Dumka, Kyiv, 1977, p. 12.

261. Sebo, P. and Havalda, A., Zmacanie molibdenum kvapalmim hlinikom, *Kovove Materially*, 13(5), 654, 1975.

262. Ko, S.H., Choi, S.K., and Yoo, C.D., Effect of surface depression on pool convection and geometry in station GTAW, *Weld. J.*, 80(2), 39, 2001.

263. Berezovsky, B.M. and Stikhin, V.A., Influence of surface tension forces on formation of butt weld reinforcement, *Svaro. Proizvod.*, 1, 51, 1977.

264. Emelyanov, I.L., Influence of surface tension forces and external pressure on the shape of the deposited bead surface, *Trans. Lenig. In-ta Ing. Vodn. Transporta*, 135, 135, 1972.

265. Popel', S.I., *Theory of Metallurgical Processes. Results of Science and Technology. 1969*, VINITI, Moscow, 1971, p. 132.

266. Lakedemonsky, A.V., *Bimetallic Castings*, Mashinostroenie, Moscow, 1964, p. 180.

267. Khansen, M. and Anderko, K., *Structures of Binary Alloys*, Mashinostroenie, Moscow, 1962, p. 608.

268. Najdich, Yu.V. and Kolesnichenko, G.A., *Interaction of Metal Melts with Surface of Diamond and Graphite*, Naukova Dumka, Kyiv, 1967, p. 89.

269. Gevlich, S.O., Tylkina, M.I., and Chernyshova, T.A., Peculiarities of interaction between boron and liquid aluminium, *Metall. Termich. Obrab. Metallov*, 8, 21, 1984.

270. Akopyants, K.S. et al., Influence of accelerating voltage on penetration parameters in electron beam welding, *Avtom. Svarka*, 11, 11, 1972.

271. Borland, I.C., Fundamentals of solidification cracking in welds, *Weld. Metal Fabr.*, 17(1, 2), 19–21, 23–26, 28–29, 99–101, 103–105, 107, 1979.

272. Kujanpaa, V. et al., Collerations between Solidification cracking and microstructure in austenitic and austenitic-ferritic stainless steel welds, *Weld Res. Inst.*, 2, 55, 1979.

273. Lipetsky, I.A., Crack formation during cooling of welds on structural carbon steels, *Vest. Metalloprom-st*, 18(23), 32, 1938.

274. Deyev, G.F. and Patskevich, I.R., *Defects in Welding Joints*, Naukova dumka, Kyiv, 1984, p. 208.

275. Patskevich, I.R. and Deyev, G.F., *Surface Phenomena in Weld Processes*, Metallurgia, Moscow, 1974, p. 120.

276. Deyev, G.F., Sharapov, V.N., and Karikh, V.V., Wetting of metals and oxides by iron sulphide, *Avtom. Svarka*, 11, 74, 1983.

277. Patskevich, I.R. et al., Investigation of wettability of titanium by fluoride fluxes melts, *Current Problems of Welding of Welding of Non-Ferrous Metals*, 1985, p. 132.

278. Gurevich, S.M., Ed., *Metallurgy and Technogy of Welding Titanium and its Alloys*, Naukova Dumka, Kyiv, 1979, p. 300.

279. Samarsky, A.A., *Introduction to Theory of Difference Schemes*, Nauka, Moscow, 1971, p. 650.

280. Samarsky, A.A. and Andreev, B.V., *Difference Methods for Elliptical Equations*, Nauka, Moscow, 1976, p. 352.

281. Korn, G. and Korn, T., *Reference Book on Mathematics*, Nauka, Moscow, 1968, p. 720.

282. Tikhodeev, G.M., *Energety Properties of Electric Welding Arc*, AN SSSR Publ. House, Moscow-Leningrad, 1961, p. 251.

283. Deyev, G.F. and Patskevich, I.P., Influence of external electric field on interphase tension in low-carbon steel–slag system, *Avtom. Svarka*, 1, 12, 1973.

284. Ludemann, K.P. and Fenzke, H.W., Elektrische potenzial-differenzen zwischen eisen- und stahlschmelzen und flussigen schlaken, *Neue Hutte*, 11, 651, 1962.

285. Deryabin, A.A., Popel', S.I., and Kuznetsov, V.A., Estimation of the fraction of surplus charges in the double electric layer in adhesion of metals to salt and oxide melts, *Elektrokhimiya*, 4(8), 955, 1968.

286. Nikitin, Yu.P., Ion exchange and interphase tension at the boundary of metal with slag, *Surface Phenomena in Metallurgical Processes*, Metallurgiya, Moscow, 1963, p. 147.

287. Esin, O.A. and Chechulin, V.A., Cathode polarization in iron and sodium segregation from oxide melts, *Zh. Fiz. Khim.*, 32(2), 355, 1958.

288. Nikitin, Yu.P., Esin, O.A., and Popel', S.I., Interphase tension and peculiarities of metal–slag boundary structure, *Surface Phenomena in Melts and Processes of Powder Metallurgy*, Naukova Dumka, Kyiv, 1963, p. 208.

289. Ershler, B.V., Investigation of the kinetics of electrode reactions, using alternating currents, *Zh. Fiz. Khim.*, 22(6), 683, 1948.

290. Gorodynsky, A.V. and Panov, E.V., Diffusion measurements in melts, *Physical Chemistry of Molten Salts*, Metallurgiya, Moscow, 1965, p. 193.

291. Levich, V.G., Theory of diffusion kinetics of heterogeneous chemical processes, *Zh. Fiz. Khim.*, 22(5), 575, 1948.

292. Elliot, D.F., Gleiser, M., and Ramakrishna, V., *Thermochemistry of Steelmaking Processes*, Metallurgiya, Moscow, 1969, p. 252.

293. Pinchuk, S.I., Postaushkin, V.F., and Kulikov, G.D., Nature of action of surface tension forces in fracture of the bridge in the drop, *Avtom. Svarka*, 11, 24, 1974.

294. Beckert, M. et al., Neues in der Schweisstechnik 1985. Werkstoffgrundlagen und Zusatzwerkstoffentwicklung, *Schweisstechnik (DDR)*, 36(5), 214, 1986.

295. Akulov, A.I. and Kopaev, B.V., On effect of plasma flow on the drop in argon-shielded welding, *Svar. Proizvod.*, 7, 47, 1972.

296. Ivashchenko, Yu.N. and Brodsky, V.P., On formation of concurrent drops during drop detachment from vertical nozzle, *Physical Chemistry of Boundaries of Contacting Phases*, Naukova Dumka, Kyiv, 1974, p. 47.

297. Erokhin, A.A., *Kinetics of Metallurgical Processes in Arc Welding*, Mashinostroenie, Moscow, 1964, p. 256.

298. Pokhodnya, I.K., *Gases in Welded Joints*, Mashinostroenie, Moscow, 1972, p. 266.

299. Deyev, G.F., Semykina, V.A., and Patskevich, I.R., Metal spattering during gas bubble escaping from the weld pool, *Avtom. Svarka*, 9, 34, 1987.

300. Fedko, V.T. and Sapozhnikov, S.B., Investigation of thermal fields in the zone of contact of molten metal drops (spatter) with the surface of the metal being welded, *Svar. Proizvod.*, 10, 12, 1998.

301. Fedko, V.T., Thermal interaction between molten metal spatter (drops) and surface of items in CO_2 welding, *Svar. Proizvod.*, 11-12, 23, 1993.

302. Sapozhnikov, S.B., Fedko, V.T., and Bubenshchikov, Yu.M., Adhesion of molten metal spatter in CO_2 welding with welded item surface, *Svar. Proizvod.*, 6, 23, 1999.

303. Petrov, A.V., Metal transfer in the arc in consumable electrode gas-shielded welding, *Avtom. Svarka*, 2, 26, 1955.

304. Potapov, N.N. Ed., *Arc Welding Consumables*, Mashinostroenie, Moscow, 1989, p. 544.

305. Dyatlov, V.I., Elements of theory of electrode metal transfer in electric arc welding, *New Problems of Welding Technique*, Kyiv, 1964, p. 167.

306. Yavojsky, V.I., Role of surface phenomena in the ferrous metallurgy, *Surface Phenomena in Metallurgical Processes*, Metallurgizdat, Moscow, 1963, p. 23.

307. Ibatullin, B.L. and Mukhin, V.F., Conditions of spray transfer of electrode metal in CO_2 welding, *Avtom. Svarka*, 7, 25, 1980.

308. Voropay, N.M. and Lavrishchev, V.E., Conditions of electrode metal transfer in welding, *Avtom. Svarka*, 5, 8, 1976.

309. Bukarov, V.A. and Ermakov, S.S., Mechanism of drop formation and its transfer into the pool in arc welding, *Svar. Proizvod.*, 11-12, 20, 1993.

310. Protsenko, P.P. and Privalov, N.T., Influence of alloying elements on electrode metal transfer in gas-shielded arc welding, *Avtom. Svarka*, 12, 29, 1999.

311. Yuzvenko, Yu.A., Kirilyuk, G.A., and Krivchikov, S.V., Model of fusion of self-shielded flux-cored wire, *Avtom. Svarka*, 1, 26, 1983.

312. Karpenko, V.M. et al., Influence of alloying elements of flux-cored wire on characteristics of electrode metal transfer, *Svar. Proizvod.*, 7, 18, 1981.

313. Mazel, A.G., *Technological Properties of Welding Electric Arc*, Mashinostroenie, Moscow, 1972, p. 178.

314. Essers, W.G., Gelmozoni, G., and Tichelanz, G.W., The transfer of metal from coated electrodes, *Metal Constr. Brit. Weld. J.*, 4, 151, 1971.

315. Pokhodnya, I.K., Gorpenyuk, V.N., and Milichenko, S.S., Some methods of improvement of the mode of metal transfer in welding with basic-coated electrodes, *Avtom. Svarka*, 1, 30, 1985.

316. Latash, Yu.V. and Medovar, B.I., *Electroslag Remelting*, Metallurgiya, Moscow, 1970, p. 239.

317. Finkel', V.M., *Physical Principles of Fracture Retardation*, Metallurgiya, Moscow, 1977, p. 360.

318. Trufiakov, V.I. et al., *Load-Carrying Capacity of Welded Joints Having Technological Defects*, Mashinostroenie, Moscow, 1988, p. 48.

319. Iida, K., Sato, M., and Nagai, M., Fatigue strength of butt-welded joint with undercut, *Weld. World*, 39(5), 262, 1997.
320. Bashenko, V.V. and Vainshtein, V.I., Analysis of forces, acting on the weld pool in electron beam welding, *Svar. Proizvod.*, 8, 1, 1970.
321. Russo, V.L., Kudoyarov, B.V., and Suzdalev, N.V., About penetration process in welding of thick items, *Svar. Proizvod.*, 11, 1, 1971.
322. Makara, A.M., Savitsky, M.M., and Varenko, N.I., Influence of refining on metal penetration in arc welding, *Avtom. Svarka*, 9, 7, 1977.
323. Paton, B.E., Makara, A.M., and Medovar, B.I., Weldability of structural steels, subjected to refining remelting, *Avtom. Svarka*, 6, 1, 1974.
324. Deyev, G.F., Dimensions of penetration zone and surface properties of metal, *Advanced Technology in Weld. Production*, Boronezh, 1985, p. 3.
325. Eager, T.W., Physics of arc welding, *Proceedings of the Physics of Steel, Industrial Conference. APS/AISI, Bethlemen, PA. October 5-7, 1981*, New York, 1982, p. 277.
326. Heiple, C.R. and Roper, J.R., Mechanism for minor element on GTA fusion zone geometry, *Weld. J.*, 1, 97, 1982.
327. Friedman, B.E., Analysis of weld puddle distortion and its effect on penetration, *Weld. J.*, 6, 161, 1978.
328. Petrov, G.L. and Tumarev, A.S., *Theory of Welding Processes*, Vysshaya Shkola, Moscow, 1967, p. 508.
329. Arsentiev, P.P., Filippov, S.I., and Burkin, V.P., Viscosity of melts of iron−oxygen and iron−sulphur system, *Izv. Vuzov. Chyorn. Metall.*, 9, 132, 1968.
330. Kosoj, L.L., Shalimov, A.G., and Ludkovsky, V.M., Surface phenomena and their role in welding of superpure high-strength steel, *Izv. Vuzov. Chyorn. Metall.*, 6, 73, 1971.
331. Killing, U., Korkhaus, J., and Niederhoff, K., On the problem of the dependence od penetration behavior on the weld pool in fully mechanized inert-gas tungsten-frc welding, *Schweis. Schneid*, 1(E4-E8), 13, 1988.
332. Patskevich, I.R., Palash, V.N., and Bashur, M., Influence of composition of $CO_2 + O_2$ mixture on weld formation in welding of cast iron with steel wire, *Avtom. Svarka*, 11, 29, 1979.
333. Chernyshov, G.G., Rybachuk, A.M., and Kubarev, V.F., On the movement of metal in the weld pool, *Izv. Vuzov. Mashinostr.*, 3, 134, 1979.
334. Olshansky, N.A., Gutkin, A.M., and Girimadzhi, G.D., Displacement of molten metal during electron beam welding, *Svar. Proizvod.*, 9, 12, 1974.
335. Gladush, G.G. et al., Thermocapillary convection in liquid under the action of powerful laser radiation, *Kvantovaya Elektron.*, 9(4), 660, 1982.
336. Anthony, T.R. and Cline, H.E., Surface rippling induced by surface tension dradients during laser surface melting and alloing, *J. Appl. Phys.*, 48(9), 3895, 1977.
337. Heiple, C.R. and Burgard P., Effects of SO_2 shielding gas additions on GTA weld shape, *Weld. J.*, 64(6), 159, 1985.
338. Petrov, A.V., Metal transfer in the arc and penetration of base metal during gas-shielded welding, *Avtom. Svarka*, 4, 19, 1957.
339. Essers, W.G. and Walter, R., Heat transfer and penetration mechanisms with GMA and plasma-GMA welding, *Weld. J.*, 60(2), 37, 1981.
340. Kuzmenko, V.G., Determination of the velocity of liquid metal movement in submerged-arc welding, *Avtom. Svarka*, 10, 21, 1996.
341. Voloshkevich, G.Z., Welding of vertical welds by forced formation method, *Jubilee Collection of Papers, Dedicated to E.O. Paton*, Kyiv, 1951, p. 371.

342. Stikhin, V.A., Berezovsky, B.M., and Krylov, V.G., Peculiarities of weld formation in narrow-gap welding of martensite-ageing steels, *Problems of Welding Production*, Cheliabinsk, 1979, p. 111.

343. Mojsov, L.P., Chernyshov, G.G., and Khokhlov, V.G., Influence of wetting of slag on weld formation in the downhand position, *Trans. of VNII on Mounting and Spec. Construc. Works*, 28, 35, 1978.

344. Yakobashvili, S.B., Influence of surface tension forces on the shape of the deposited bead, *Svarochnye Processy Metallurgii*, No. 1, Tbilisi, 1974, p. 89.

345. Andrews, I.G., Atthey, D.R., and Byatt-Smith, I.C., Weldpool sag, *J. Fluid Mech.*, 100(4), 785, 1980.

346. Komarov, A.I., Khodakov, V.D., and Starchenko, E.G., Influence of surface tension of steels and fluxes on deposited metal formation, *Avtom. Svarka*, 3, 25, 1983.

347. Frenkel', Ya.I., *Collected Works*, Vol. 3, AN SSSR Publ. House, Moscow—Leningrad, 1957, p. 460.

348. Ryabtsev, I.A., Kuskov, Yu.M., and Chernyak, Ya.P., Deposited metal formation on inclined and vertical surfaces, *Avtom. Svarka*, 4, 23, 2000.

349. Berezovsky, B.M. and Stikhin, V.A., Calculated definition of weld reinforcement shape and critical sizes of weld pool in overhead welding, *Problems of Welding Production*, Chelyabinsk, 1979, p. 103.

350. Chudinov, M.S. and Taran, I.D., Weld formation in position welding of pipes with forced containment of the weld pool, *Svar. Proizvod.*, 10, 6, 1970.

351. Tyulkov, M.D. and Turbin, V.V., On the problem of circumferential joint welding, *Advanced Methods of Welding in Repair and Manufacturing of Petrochemical and Oil-refining Equipment*, Angarsk, 1984, p. 95.

352. Akulov, A.I., Doronin, Yu.V., and Chernyshov, G.G., Physico-chemical properties of molten slag of the melt backing and their influence on back bead formation in consumable electrode welding, *Svar. Proizvod.*, 3, 18, 1981.

353. Akulov, A.I., Chernyshov, G.G., and Doronin, Yu.V., Influence of hydrodynamic phenomena in the weld pool on butt weld formation in consumable electrode welding, *Izv. Vuzov. Mashinostr.*, 8, 135, 1978.

354. Asnis, A.E., *Dynamic Strength of Welded Joints of Low-Carbon and Low-Alloyed Steels*, Mashgiz, Moscow — Kyiv, 1962, p. 173.

355. Trufiakov, V.I., *Fatigue of Welded Joints*, Naukova Dumka, Kyiv, 1973, p. 216.

356. Belchuk, G.A. and Naletov, V.S., On some principles of weld formation in weld-to-base metal transition zone, *Welding in Shipbuilding*, No. 79, Leningrad, 1972, p. 15.

357. Berezovsky, B.M. and Stikhin, V.A., Peculiarities of formation of weld reinforcement–base metal transition zone, *Problems of Welding Production*, Chelyabinsk, 1981, p. 99.

358. Fedorov, S.A. and Ovchinnikov, V.V., Influence of the difference of metal surface tension in the pool on weld formation in light beam welding of sheet materials, *Svar. Proizvod.*, 11, 33, 1986.

359. Heiple, C.R. et al., Surface active element effects on the shape of GTA, laser and electron beam welds, *Weld. J.*, 2, 124, 1983.

360. Deyev, G.F., Gerasomenko, T.A., and Pevzner, E.P., Thermo-capillary phenomena in surfacing with an electron beam, *Svar. proizvod.*, 2, 31, 1993.

361. Najdich, Yu.V., Zabuga, V.V., and Perevertajlo, V.M., Study of Marangoni effect in metal melts Information 1, Concentrated-capillary flows, *Adhesion of Melts and Brazing of Materials*, No. 26, 1, 1991, p. 1.

362. Smitls, C. J., *Metals*, Metallurgiya, Moscow, 1980, p. 447.

363. Camel, D., Tison, P., and Favier, J., Marangoni flow regimes in liquid metals, *Acta Astronaut.*, 13(11–12), 723, 1985.

364. Lin, M.I. and Eagar, T.W., Influence of arc pressure on weld pool geometry, *Weld. J.*, 64(6), 163, 1985.

365. Potekhin, V.P., Development of models of the thermal and force effect of electric arc on metal in non-consumable electrode welding, Ph.D. thesis, Volgograd, 1987.

366. Suzdalev, I.V., Yavno, E.I., and Russo, V.L., Calculation of the shape and size of overlap in horizontal welding on a vertical plane, *Svar. Proizvod.*, 9, 44, 1977.

367. Berezovsky, B.M., Stikhin, V.A., and Bakshi, O.A., Mathematical model of formation of horizontal welds on a vertical plane, *Theory and Practice of Welding Production*, Sverdlovsk, 1980, p. 28.

368. Alov, A.A., Slag inclusions in welds during arc welding of low-carbon steels, *Avtogen. Delo*, 4, 4, 1945.

369. Alov, A.A., Principles of theory of arc welding metallurgy of low-carbon steels, *Problems of Theory of Welding Processes*, Mashgiz, Moscow, 1948, p. 5.

370. Mazel', A.G., Tarlinsky, V.D., and Sbarskaya, N.P., Non-metallic inclusions in gas-shielded electrode welding, *Svar. Proizvod.*, 9, 39, 1967.

371. Podgaetsky, V.V., *Non-Metallic Inclusions in Welds*, Mashgiz, Moscow, 1962, p. 84.

372. *Steel Production in Basic Open-Hearth Furnace*, Metallurgizdat, Moscow, 1969, p. 708.

373. Yavojsky, V.I., Bliznyukov, S.A., and Vishkarev, A.F., *Inclusions and Gases in Steels*, Metallurgiya, Moscow, 1979, p. 272.

374. Kuslitsky, A.B., Mizetsky, V.L., and Karpenko, G.V., On the influence of non-metallic inclusions on the mechanism of fatigue crack initiation, *Dokl. AN SSSR*, 187(1), 79, 1969.

375. Brooksbank, D. and Andrews, K.W., Thermal expansion of some inclusions found in steel and relation to tessellated stresses, *J. Iron Steel Inst.*, 206(6), 595, 1968.

376. Brooksbank, D. and Andrews, K.W., Tessellated stresses associated with some inclusions and steel, *J. Iron Steel Inst.*, 207(4), 474, 1969.

377. Goodier, J.N., Concentration of stress around spherical and cylindrical inclusions and flaws, *Appl. Mech. ASME*, 1(1), 157, 1933.

378. Melan, E. and Parkus, G., *Thermoelastic Stresses, Induced by Stationary Temperature Fields*, Fizmatgiz, Moscow, 1958, p. 167.

379. Ishii, Y., Kihara, H., and Tada, Y., On the relation between the nondelstructive testing information of steel welds an their mechanical strength, *J. Nondestruct. Test. (Jp.)*, 16(8), 319, 1967.

380. Frenkel', Ya.I., *Kinetic Theory of Liquids*, AN SSSR Publ. House, Moskow, 1959, p. 592.

381. Turpin, M.L. and Elliott, J.F., Nucleation of oxide in iron melts, *J. Iron and Steel Inst.*, 204(3), 217, 1966.

382. Sigworth, G.K. and Elliott, J.F., Conditions for nucleation of oxides in Fe–Si–O alloys, *Can. Met. Quart.*, 2, 337, 1972.

383. Gulyaev, B.B., *Solidification and Inhomogeneity of a Steel Ingot*, Metallurgizdat, Moscow, 1950, p. 228.

384. Podgaetsky, V.V., Parfesso, G.I., and Menzhelej, G.I., Investigation of composition and shape of sulphides in welds, *Avtom. Svarka*, 8, 34, 1963.

385. Laptev, D.M., Thermodynamics of formation of two-component nuclei in presence of chemical reactions, *Izv. Vuzov. Chyorn. Metall.*, 2(5), 4(5), 1968.

386. Drozdin, A.D., Povolotsky, D.Ya., and Roshchin, V.E., Investigation of thermodynamic principles of deoxidation product nucleation in liquid iron, *Izv. Vuzov. Chyorn. Metall.*, 12, 35, 1976.

387. Roshchin, V.E., Nucleation, formation and behavior of deoxidation products in metal melt and during its solidification process, Ph.D. thesis, Moscow, 1979.

388. Levin, A.M., Investigation of wetting at high temperatures, *Trans. of Dnieprop. Metall. Inst.*, 28, 89, 1952.

389. Popel', S.I., Wetting of refractory materials with molten metal and slag, *Theory and Practice of Foundry*, Mashgiz, Sverdlovsk, 1959, p. 162.

390. Popel', S.I., Influence of surface properties of steel and its oxidation on sand blend sticking to castings, *Theory and Practice of Foundry*, Mashgiz, Sverdlovsk., 1959, p. 173.

391. Zelbet, B.M., Laposhko, L.D., and Kapyrin, A.A., Investigation of qualitative composition of non-metallic inclusions in ShKh-15 steel, using the x-ray spectrum microanalysis, *Trans. of VNIIPP*, Moscow, 1972, p. 70.

392. Bogdandy, L., Meyer, W., and Stranski, I., Beitrage zur kinetik der desoxydation flussigen eisen, *Arch. Eisenhut.*, 13, 451, 1961.

393. Sapiro, S.I., Thermodynamic analysis of coalescence and coagulation of nonmetallic inclusions, *Steel*, 7(8), 449, 1945.

394. Babkin, V.G., Tsarevsky, B.V., and Popel', S.I., Spreading of oxide melts over the surface of molding materials, *Wettability and Surface Properties of Melts and Solids*, Naukova Dumka, Kyiv, 1972, p. 91.

395. Baptizmansky, V.I., Bakhman, N., and Dmitriev, Yu.V., Investigation of principles of non-metallic inclusion coagulation in liquid steel, *Izv. Vuzov. Chyorn. Metallurgiya*, 3, 42, 1969.

396. Khlynov, V.V. and Ishimov, V.I., On the influence of surface properties of iron at the boundary with oxides on the floating rate of nonmetallic inclusions, *Izv. Vuzov. Chyorn. Metallurgiya*, 10, 10, 1970.

397. Patskevich, I.R. and Deyev, G.F., On the role of individual gases in pore formation, *Advanced Methods of Welding and Spraying of Metals*, Kursk, 1972, p. 112.

398. Muller, G., Coagulation of colloids with particles, having the shape of "sticks" and "leaves". Theory of any polydisperse systems and coagulation in flowing, *Coagulation of colloids*, ONTI NKTB SSSR, Moscow, 1936, p. 74.

399. Fortier, A., *Suspension Mechanics*, Mir, Moscow, 1971, p. 264.

400. Deryagin, B.V. and Kusakov, M.M., Experimental investigations of surfaces solvation as applied to construction of mathematical theory of lyophobic colloids stability, *Izv. AN SSSR*, Ser. of Chem. Scs., 5, 48, 1937.

401. Parsons, P., Equilibrium properties of charged interphases, *Some Problems of Current Electrochemistry*, Inostr. Liter. Publ. House, Moscow, 1958, p. 125.

402. Markhasev, B.I. and Pereverzev, D.D., On some peculiarities of formation of double electric layer at the liquid metal–solid oxide boundary, *Surface Phenomena in Melts*, Naukova Dumka, Kyiv, 1968, p. 287.

403. Batsanov, S.S., *Electronegativity of Elements and Chemical Bond*, AN SSSR Publ. House, Novosibirsk, 1962, p. 196.

404. Khlynov, V.V., Gornovoj, V.A., and Sorokin, Yu.V., Kinetics of slag drops coalescence, *Izv. Vuzov. Chyorn. Metallurgiya*, 6, 6, 1969.

405. Knyuppel, G., Brotsman, K., and Ferster, N., Investigation of oxide inclusions in mild aluminium-killed steels, *Chyorn. Metally*, 11, 32, 1965.

406. Geguzin, Ya.E., *Sintering Physics*, Nauka, Moscow, 1967, p. 360.

407. Deryabin, V.A., Popel', S.I., and Deryabin, Yu.A., Coalescence of spherical particles by fringes of $CaO-SiO_2-Al_2O_3$ melts, *Izv. Vuzov. Chyorn. Metallurgiya*, 7, 15, 1974.
408. Eremenko, V.N., Najdich, Yu.V., and Lavrinenko, I.A., *Sintering in the Presence of Liquid Metallic Phase*, Naukova Dumka, Kyiv, 1968, p. 122.
409. Popel', S.I., Deryabin, V.A., and Deryabin, Yu.A., Forces, contracting the particles by fringes of melts, *Methods of Investigation and Properties of Boundaries of Contacting Phases*, Naukova Dumka, Kyiv, 1977, p. 3.
410. Popel', S.I. and Deryabin, A.A., Factors, influencing the rate of inclusion floating in steel, *Izv. Vuzov. Chyorn. Metallurgiya*, 4, 25, 1965.
411. Khlynov, V.V., Sorokin, Yu.V., and Stratonovich, V.N., Factors, influencing the elimination of liquid inclusions from the metal, *Physicochemical Investigations of Metallurgical Processes*, 1973, No 1, 114.
412. Nelson, R.S., *J. Nucl. Mater.* 2, 149, 1966.
413. Grigoryan, V.A., Shvindlerman, L.S., and Aleev, R.A., Transition of gas bubbles through the interface, *Izv. Vuzov. Chyorn. Metall.*, 3(5), 5(11) 1968.
414. Khlynov, V.V., Gornovoj, V.A., and Stratonovich, V.N., Some factors, influencing the clustering and elimination of non-metallic inclusions from steel, *Izv. AN SSSR. Metally*, 5, 47, 1970.
415. Popel', S.I., Physicochemical peculiarities of clustering of non-metallic inclusions and their elimination from steel, *Steel and Non-Metallic Inclusions*, Metallurgiya, Moscow, 1976, p. 56.
416. Gurevich, Yu.G., Precipitation of solid non-metallic inclusions into slag on a flat interface, *Izv. Vuzov. Chyorn. Metall.*, 8, 5, 1968.
417. Frenkel', Ya.I., Viscous flow in crystalline bodies, *ZhETF*, 16(1), 29, 1946.
418. Moldavsky, O.D., Features of oxide inclusion assimilation by slag during film flow over the cone of consumable electrode, *Izv. AN SSSR. Metally*, 6, 36, 1970.
419. Povolotsky, D.Ya., Roshchin, V.E., and Katkov, A.V., About a limiting element of the process of metal purification from deoxidation products, *Izv. Vuzov. Chyorn. Metall.*, 4, 68, 1972.
420. Masi, O., Erra, A., Lesame radiografico delle saldatura, una completa valutazione dei diffeti in termine di resistenza statica, *La Metall. Ital.*, 45(8), 273, 1953.
421. Norrish, J. and Moore, D., Porosity in are welds and its effect on mechanical properties, *Proceedings of the Second Conference on the Significance of Defect in Welds, London*, Welding Institute, 1968, p. 86.
422. Potthoff, F., Enfluss von poren auf technologische usutewerte, *Praktiker*, 8, 140, 1975.
423. Ashton, R.F., Wesley, R.P., and Duxon, C.R., The effect of porosity on 5086-H116 aluminum alloy welds, *Weld. J.*, 3, 95, 1975.
424. Makarov, I.I., Volynsky, V.N., and Prokhorov, N.N., Influence of pores and oxide inclusions on the strength of AMg6 alloy welded joints, *Avtom. Svarka*, 4, 27, 1976.
425. Khakymyanov, R., Influence of the technological defects of welding on the load-carrying capacity of circumferential welds of the main pipelines, *Advanced Methods of Welding of Pipelines*, Moscow, 1976, p. 66.
426. Makarov, I.I., Prokhorov, N.N., and Zavalishin, G.I., Stress concentration near spherical and cylindrical pores in butt welded joints, *Svar. Proizvod.*, 5, 25, 1976.
427. Goodier, J.N., Concentration of stress around spherical and cylindrical inclusions and flaws, *Appl. Mech. ASME*, 1(1), 157, 1933.
428. Babaev, A.V., Influence of pores on fatigue resistance of welded joints, *Avtom. Svarka*, 10, 6, 1980.

429. Homes, G.A. and Rosenthal, D., Beziehungen zwischen ursprungsfestigkeit und dichtigkeit von flusstahlschweissungen, *Areos*, 89, 1951, 1938.
430. Hufnagel, W., Einfluss von porositat auf die statische und dynamische festigkeit von aluminium, *Schweisverbindungen, Aluminium*, 53(10), 613, 1977.
431. Sirgo, A., On the influence of defects on the carrying capacity of welded joints, *Avtom. Svarka*, 5, 13, 1981.
432. Vegener, P.P. and Mak, L.M., Condensation in supersonic and hypersonic wind tunnels, *Problems of Mechanics*, No. 3, Inostr. Liter. Publ. House, Moscow, 1961, p. 68.
433. Zel'dovich, Ya.B., Theory of formation of new cavitation phase, *ZhETF*, 12, 525, 1942.
434. Arakelov, A.G., Bernova, L.A., and Solodkov, A.I., Study of the dynamics of pore formation in welds on OT-4 titanium alloy with the help of x-ray TV defectoscopy, *Avtom. Svarka*, 5, 34, 1976.
435. Nesis, E.I. and Frenkel', Ya.I., Boiling of gassy liquid, *Zh. Techn. Fiziki*, 22(9), 1500, 1952.
436. Neumann, K. and Doring, W., Tropfchenbilding in ubersattigkeiten dampigemischen zweier vollstandig mischbarer flussgkeiten, *Z. Phys. Chem.*, 186(4), 203, 1940.
437. Knyuppel, G., *Deoxidation and Vacuum Treatment of Steel*, Metallurgiya, Moscow, 1973, p. 312.
438. Bondarev, A.M. and Dorofeev, E.B., About some causes of pore formation in argon-arc welding of OT4-1 alloy, *Svar. Proizvod.*, 9, 48, 1971.
439. Tyvonchuk, P.A., Smiyan, O.D., and Deyev, G.F., Influence of non-metallic inclusions on weld porosity, *Advanced Technology of Welding Production*, Voronezh, 1989, p. 23.
440. Rabkin, D.M., Dovbishchenko, I.V., and Psaris, G.G., Metallurgical features of welding of commercial aluminium in an argon–oxygen mixture, *Avtom. Svarka*, 4, 25, 1974.
441. Karasev, R.A., Chemisorption of oxygen on liquid iron surface, *Physicochemical Principles of Interaction of Liquid Metal and Gases and Slags*, Nauka, Moscow, 1976, p. 43.
442. Levine, H.S., Vormation of vapor nuclei in high temperature melts, *J. Phys. Chem.*, 76(18), 2609, 1972.
443. Wei, Q. et al., A study of weld pore sensitivity of self-shilded, flux coren electrodes, *Weld. J.*, 6, 90, 2002.
444. Prandtl, L., *Fluid Dynamics*, Inostr. Liter. Publ. House, Moscow, 1961, p. 575.
445. Patskevich, I.R. and Deyev, G.F., Influence of interphase boundary properties on pinholes formation in welding, *Svar. Proizvod.*, 9, 42, 1973.
446. Appen, A.A. and Kayalova, S.S., Some generalized data, concerning the surface tension of silicate melts, *Surface Phenomena in Melts and Processes of Powder Metallurgy*, AN Ukr.SSR Publ. House, Kyiv, 1963, p. 347.
447. Frumin, I.I., *Automatic Electruc-Arc Surfacing*, Metallurgizdat, Khar'kov, 1961, p. 421.
448. Gapchenko, M.N. and Futer, I.E., *Porosity of Welds and Measures of its Prevention*, Gostekhizdat, Kyiv, 1953, p. 76.
449. Rabkin, D.M., Energy study of near-electrode zones of a powerful welding arc, *Avtom. Svarka*, 2, 17, 1951.
450. Dudko, D.A., Rublevsky, I.N., and Chernega, D.F., Features of hydrogen behavior in the electroslag process, *Avtom. Svarka*, 6, 28, 1957.

451. Dudko, D.A. and Rublevsky, I.N., Influence of kind and polarity of current on metallurgical processes in electroslag welding, *Avtom. Svarka*, 1, 8, 1959.

452. Potapov, N.N. and Lyubavsky, K.V., Hydrogen in the deposited metal in submerged-arc welding, *Svar. Proizvod.*, 7, 4, 1970.

453. Holzgruber, V. and Peterson, K., Metallurgical processes in ac and dc electroslag remelting, *Scientific Problems of Welding and Special Electrometallurgy*, Naukova Dumka, Kyiv, 1970, p. 82.

454. Peover, M.E., Electroslag remelting: a review of electrical and electrochemical aspects, *J. Inst. Metals*, 100(4), 97, 1972.

455. Yavojsky, V.I., *Gases in Steel Furnace Baths*, Metallurgizdat, Moscow, 1952, p. 245.

456. Potapov, N.N. and Lyubavsky, K.V., On the problem of metal–slag interaction in the reaction zone in submerged-arc welding, *Svar. Proizvod.*, 7, 9, 1971.

457. Malholland, E.W., Hazeldian, G.S., and Davies, M.W., Vizualization of slag-metal reactions be x-ray fluoroscopy: decarburization in basic oxyten steelmaking, *J. Iron Steel Inst.*, 211(9), 632, 1973.

458. Romadin, K.P., Electrolytic transfer in metal liquid and solid solution, *Trans. of N.B. Zhukovsky VVIA.*, 1947, No. 167, p. 114.

459. Romadin, K.P., Electric transfer in metal solid solutions, *Investigations on Refractory Alloys*, AN SSSR Publ. House, Moscow, 1958, p. 292.

460. Yavojsky, V.I., Hydrogen behavior in iron alloys in the liquid state, as well as during their solidification and subsequent cooling, *Kinetics and Thermodynamics of Gas–Liquid Metal Interaction*, Nauka, Moscow, 1974, p. 5.

461. Shchepetkin, F.S. and Ageev, N.Ya., On the problem of metal dehydrogenation by solid particles, *Izv. Vuzov. Chyorn. Metall.*, 5, 57, 1971.

462. Mazurin, O.V., *Electric Properties of Glass*, Leningrad, 1962, p. 58.

463. Volmer, M., *Kinetik der Phasenbildung*, Dresden-Leipzig, 1938, p. 219.

464. Shcherbakov, A.M. and Bajbakov, V.S., Surface tension of metals at the boundary with gas and own melt, *Surface Phenomena in Metallurgical Processes*, Metallurgizdat, Moscow, 1963, p. 200.

465. Frank-Kamenetsky, D.A., *Diffusion and Heat Transfer in Chemical Kinetics*, Nauka, Moscow, 1967, p. 490.

466. Vesterhaid, D. and Westwater, D., Isothermic growth of hydrogen bubbles in electrolysis, *Problems of Boiling Physics*, Mir, Moscow, 1964, p. 354.

467. Klassen, V.I. and Mokrousov, V.A., *Introduction into Flotation Theory*, Gosgortekhizdat, Moscow, 1953, p. 636.

468. Pernik, A.D., *Problems of Cavitation*, Sudostroenie, Leningrad, 1966, p. 440.

469. Ginsburg, V.M. Ed., *Optical Holography*, Sov. Radio, Moscow, 1978, p. 238.

470. Gorshkov, A.I., Kinetics of gas bubble growth in the molten pool and pore formation in welding of titanium, *Svar. Proizvod.*, 6, 54, 1975.

471. Pokhodnya, I.K. and Demchenko, L.I., Numerical investigation of gas bubble growth in solidification of weld pool metal, *Avtom. Svarka*, 11, 8, 1977.

472. Szekely, J. and Martins, G.P., On spherical puase growth in multicomponent systems, *Trans. Metal. Soc. AIMD*, 245(8), 1741–1747, 1969.

473. Popel', S.I., Surface phenomena in steel-making processes, *Stal'*, 4, 260, 1979.

474. Kapustina, O.S., Degassing of liquid, *Physical Principles of Ultrasonic Technology*, Nauka, Moscow, 1970, p. 253.

475. Figurovsky, N.A., *Sedimentometric Analysis*, AN SSSR Publ. House, Moscow, 1948, p. 332.

476. Nemchenko, V.P., Koz'min, V.A., and Popel', S.I., Diffusion peculiarities of melt component adsorption on the surface of floating bubbles and desorption from it, *Izv. Vuzov. Chyorn. Metall.*, 2, 34, 1976.

477. Malenkov, I.G., On the motion of large gas bubbles, floating in liquid, *PMTF*, 6, 130, 1968.

478. Berdnikov, V.I. and Levin, A.M., On the rate of gas bubble floating in metal and slag melts, *Izv. Vuzov. Chyorn. Metall.*, 12, 24, 1977.

479. Frumkin, A.I., Physicochemical phenomena in motion at the boundary of two liquids, *Jubilee Collection of Papers, Dedicated to 30th Anniversary of the Great October Revolution*, AN SSSR Publ. House, Moscow, 1947, p. 512.

480. Smolukhovsky, M., Experiment with mathematical theory of kinetic coagulation of colloid solutions, *Coagulation of Colloids*, ONTI NKTB SSSR, Moskow, 1936, p. 7.

481. Shevtsov, E.K. et al., Mechanism of formation of slag-metal transition zone in a liquid open-hearth bath, *Izv. Vuzov. Chyorn. Metall.*, 1, 42, 1974.

482. Tikhomirov, V.K., *Foams. Theory and Practice of Producing and Destroying Them*, Khimiya, Moscow, 1975, p. 264.

483. Mc Entee, W. and Mysels, K.J., The bursting of soup films. 1. An experimental study, *J. Phys. Chem.*, 73(9), 3018, 1969.

484. Bashkirov, M.M. and Goncharov, G.F., On the time measurement of liquid film disintegration, *Kolloid. Zh.*, 34(5), 753, 1972.

485. Frankel, S. and Mysels, K.J., The bursting of soup films. 11. Theoretical considerations, *J. Phys. Chem.*, 73(9), 3028, 1969.

486. Talmud, D.L. and Bresler, S.E., *Surface Phenomena*, Gostechizdat, Moscow — Leningrad, 1934, p. 131.

487. Glinkov, G.M. and Shevtsov, E.K., Transition of gas bubbles through phase boundary surface, *Izv. Vuzov. Chyorn. Metall.*, 6, 168, 1971.

488. Orlov, Yu., Nemchenko, V.P., and Popel', S.I., Movement of liquid by floating bubbles, *Izv. Vuzov. Chyorn. Metall.*, 1,13, 1977.

489. Baptizmansky, V.I., *Mechanism and Kinetics of Processes in the Converter Bath*, Metallurgizdat, Moscow, 1960, p. 283.

490. Medzhibozhsky, M.Ya., *Intensification of Open-Hearth Melting by Air Injection into the Bath*, Mashgiz, Moscow, 1959, p. 174.

491. Shinoda, T., Hoshibo, K., and Yamashita, R., A new method for hot crack in welds, *Schweis. schneid*, 3, 17, 1992.

492. Podgaetsky, V.V. and Parfesso, G.I., *Sulphide Origin Cracks in Welding of Steel*, Naukova Dumka, Kyiv, 1977, p. 149.

493. Rabkin, D.M. and Frumin, I.I., Causes of hot cracks formation in welds, *Avtom. Svarka*, 2, 3, 1950.

494. Lejnachuk, E.I. and Parfesso, G.I., Influence of vanadium on metal susceptibility to hot cracking, *Avtom. Svarka*, 1, 14, 1969.

495. Lejnachuk, E.I., Podgaetsky, V.V., and Parfesso, G.I., Influence of niobium on weld susceptibility to solidification cracking in welding, *Avtom. Svarka*, 9, 10, 1974.

496. Lejnachuk, E.I., Podgaetsky, V.V., and Parfesso, G.I., Influence of niobium on weld metal resistance to solidification crackig, *Avtom. Svarka*, 1, 20, 1978.

497. Mchedlishvili, V.A., Lyubimov, G.A., and Samarin, A.M., *Role of Manganese in Elimination of Detrimental Effect of Sulphur on Hot-Brittleness of Steel*, Metallurgizdat, Moscow, 1960, p. 54.

498. Shorshorov, M.Kh. et al., *Hot Cracks in Welding of Heat-Resistant Alloys*, Mashinostroenie, Moscow, 1973, p. 224.

499. Kazenkov, Yu.I. et al., Influence of base metal grain size on hot cracking susceptibility of welds on austenitic steels and alloys, *Svar. Proizvod.*, 3, 6, 1977.
500. John, R. and Richards, W.G., Hot cracking in 5% nickel steel weld-metal, *Metal
Constr. and Brit. Weld. J.*, 4(4), 127, 1972.
501. Turi, A., and El-Hebeary, M.R., The interrelation among material, technology and
structure on cracking of welding joints, *Proceeding of the 7th Confence on
Welding Budapest*, 1976, p. 249.
502. Likhtman, V.I., Shuchkin, E.D., and Rebinder, P.A., *Phisico-Chemical Mechanic of
Metals*, AN SSSR Publ. House, Moskow, 1962, p. 304.
503. Wilken, K. and Kleistner, H., The classification and evaluation of hot cracking tests
for weldments, *Welding in the world*, 20(7/8), 126, 1990.
504. Rostoker, U., Mc-Cogi, D., and Marcus, G., *Brittleness Under the Action of Liquid
Metals*, Inostr. Liter Publ. House, Moscow, 1962, p. 176.
505. Novikov, I.I., *Hot Brittleness of Non-Ferrous Metals and Alloys*, Nauka, Moscow,
1966, p. 299.
506. Pertsov, N.V. et al., Influence of strain rate and temperature on the value of adsorption
effect of metal strength and plasticity lowering in low-melting metal melts, *Inzh. Fiz.
Zh.*, 2(12), 77, 1959.
507. Popovich, V.V. and Dmukhovskaya, I.G., *Liquid-Metal Embrittlement of Wrought
Metals*, Lvov, 1983, p. 68.
508. Shchukin, E.D., Criterion of crystal deformability and adsorption effect, *Dokl. AN
SSSR*, 68, 6, 1105.
509. Popovich, V.V., Mechanisms of liquid metal embrittlement, *Fiz.-Khim. Mekh.
Mater.*, 5, 11, 1979.
510. Mc Lin, D., *Grain Boundaries in Metals*, Metallurgizdat, Moscow, 1960, p. 322.
511. Prokhorov, N.N., *Hot Cracks in Welding*, Mashgiz, Moscow, 1952, p. 200.
512. Tsarevsky, B.V. and Popel', S.I., Surface properties of iron–carbon alloys, *Izv.
Vuzov. Chyorn. Metall.*, 8, 15, 1960.
513. Pozdnyak, L.A., On carbon influence on dendritic heterogeneity of sulphur
distribution in welds, *Avtom. Svarka*, 1, 3, 1957.
514. Maslov, B.G., *Liquid Penetrant Inspection of Items and Welded Joints*, Mashinostroenie,
Moscow, 1987, p. 52.
515. Ivanov, I.S., Sherstobitov, M.A., and Strelkov, L.L., Influence of refractory material
temperature field on the rate of capillary impregnation by slag, *Wettability and
Surface Properties of Melts and Solids*, Naukova Dumka, Kyiv, 1972, p. 229.
516. Frenkel', Ya.I., *Introduction into the Theory of Metals*, Nauka, Moscow, 1972, p. 424.
517. Arsentiev, P.P. and Polyakova, K.I., Viscosity of liquid aluminium, *Izv. AN SSSR
Metally*, 2, 65, 1977.
518. Fridlyander, E.I. and Kolpachev, A.K., On viscosity of extrapure aluminium, *Izv. AN
SSSR Metally*, 4, 38, 1980.
519. Koledov, L.A. and Lyubimov, A.I., Viscosity of diluted metal melts on the base of
aluminium, *Izv. Vuzov. Chyorn. Metall.*, 9, 136, 1963.
520. Skrinskii, A.N. et al., High-efficient surfacing and surface melting of powder coatings
with the beams of relativistic electrons, *Papers of USSR Academy of Scinces*, 4, 865,
1985.
521. Ryabov, V.R. and Deev, G.F., *Surfacing by an Electron Beam Released into the
Atmosphere*, Taylor and Francis Books, London, 2001, p. 83.
522. Antes, H.W., Edelman, R.E., and Rosenthal, H., Flow of arc welder copper base filler
alloys on steel, *Weld. J.*, 41(5), 207, 1962.

523. Bashkatov, A.V. et al., Electron beam welding of commercial iron to bronze, *Svar. Proizvod.*, 8, 11, 1977.

524. Deyev, G.F., Zubkova, E.N., and Safonov E.P., Technology of surfacing and thermal-treatment of deposited metal-cutting tool. *Trans. of LGTU-LEGI*, Lipetsk, 1998, p. 83.

525. Deyev, G.F., *Wettability and Spreading in Fusion Welding*, LEGI, Lipetsk, 2002, p. 155.

526. Paton, B.E., Welding in space, *Scientific Problems of Welding and Special Electrometallurgy*, Naukova Dumka, Kyiv, 1970, p. 123.

527. Steg, L., *Space Technology*, Mir, Moscow, 1980, p. 419.

528. Peinter, G., Liquid in zero gravity conditions, *Liquid Fuel Rocket Propulsion Units*, Nauka, Moscow, 1966, p. 149.

529. Avduevsky, V.S. et al., *Problems of Space Production*, Mashinostroenie, Moscow, 1980, p. 221.

530. Belyakov, I.T. and Borisov, Yu.D., *Technology in Space*, Mashinostroenie, Moscow, 1974, p. 290.

531. Grishin, S.D. et al., Measurement of low accelerations in scientific orbital station "Salyut-6", *Space Res.*, 20(3), 479, 1982.

532. Paton, B.E. et al., Examination of electrode metal melting and transfer in welding under varying gravity conditions, *Space Science of Materials and Technology*, Nauka, Moskow, 1977, p. 22.

533. Nogi, K., Aoki, Ya., and Fuzi, H., Peculiarities of EBW and nonconsumable electrode arc welding in microgravity conditions, *Avtom. Svarka*, 10, 39, 1999.

534. Ostrakh, S., Role of convection in technological processes carried out in microgravity conditions, *Space Technology*, Nauka, Moscow, 1980, p. 9.

535. Petrovsky, G.T. and Voronkov, G.L., *Optical Technology in Space*, Mashinostroenie, Leningrad, 1984, p. 158.

536. Clain, H. and Beversdorf, L., Dependence between velocity of bubble movement and Marangoni number, *Rocket Systems and Cosmonautics*, 19(7), 148, 1981.

537. Geguzin, Ya.E., Dzyuba, A.S., and Kachanovsky, Yu.S., Evolution of gas bubbles cluster in liquid in zero gravity conditions, *Dokl. AN SSSR*, 260(4), 876, 1981.

538. Bondarev, A.A. et al., Examination of the structure and distribution of elements in electron beam welded joints on 1201 and AMg6 alloys in zero gravity conditions, *Production and Behavior of Materials in Space*, Nauka, Moscow, 1978, p. 21.

539. Paton, B.E. and Lapchinsky, V.F., *Welding and Related Technologies in Space*, Naukova Dumka, Kyiv, 1998, p. 184.

Index

A

A.C. welding, *see* Alternating current
    (a.c.) welding
Acceleration force, 347–349, *348*
Acidic coating, *152*, 152–153
Adam studies, 43
Adhesion
  interphase tension, 37, 39
  non-metallic inclusions, *221*,
    225–226, *226*
  preheating, 142
Adsorption process
  cracks, 286, 288
  surface properties and phenomena,
    40–41
Advanced mathematical treatment,
    porosity, 269–271
Air
  behavior, 70, 73–75, *73–75*
  surface properties and phenomena, 70,
    73–75, *73–75*
Akshentsev, Baum, Gel'd and, studies,
    67, 70
Aleev, Grigoryan, Shvindlerman and,
    studies, 232
Allowances, 271–273, *272*, *see also*
    Technologies, development
Alloy composition influence, 65–67
Alternating current (a.c.) welding
  arc discharge action duration, 178
  comparison to d.c. welding, 4–7
  electric arc welding, 10
  electric field influence, 254
  electrochemical interactions, 125
  external electric field, 133
  slag, role, 22
  welding polarity, 153
Alumina, 209
Aluminum (Al) and aluminum alloys
  arc discharge, 105
  binary gas mixtures, behavior, 76
  composite materials, 112–113
  copper and copper alloys, 77–78

cracks, 281
interphase tension, 87
isothermal conditions, *98*, 98–99
nickel, 80
penetration depth, 166
physical interactions, 261
porosity, 237
substrate composition, 110
sulfur, 121
surface properties and phenomena,
    98, 99
ternary slags, 88
Angle of transition, 188–189
Anode region, 6
Antimony (Sb), 91
Aoki and Fuzi, Nogi, studies, 351
Arc active spot, 49–51, *51*
Arc discharge
  action duration, 178–179, *179*
  d.c. *vs.* a.c. welding, 7
  surface properties and phenomena,
    105–113, *106–107*
Arc parameters influence, 101–102, *102*
Arc welding, 172–173, *173*, *see also*
    Manual arc welding; *specific types*
Argon (Ar)
  arc discharge, 106
  chlorides/fluorides, 123
  current polarity, 174–176
  electrode welding, 172
  gases influence, 68–69
Armco-iron and steels, *see also* Steels
  binary gas mixtures, behavior, 76, 77
  comfort area, 295
  corrosion, 299
  cracks, 307
  grain boundaries, 305
  substrate composition, 109
  sulfur, 121
Austenitic steels, 95–96, 301, *302*
Autoelectronic emission method, 54
Automatic welding processes, 10–11
Avduevsky studies, 348
Avogadro constant, 41, 310

B

Backing, bead, 187–188
Bakhman and Dmitriy, Baptizmansky, studies, 221
Bakshi, Berezovsky, Stikhin and, studies, 204
Baptizmansky, Bakhman and Dmitriy studies, 221
Barium (Ba), 123
Base-metal surface properties, 166–168, *166–168*
Basic-type coating, 150–152, *151*
Baum, Gel'd and Akshentsev studies, 67, 70
Bcc, *see* Body-centered cubic (bcc) lattice
Beam power, surfacing, 327–329, *328–329*
Beckert studies, 140
Belchuk and Naletov studies, 188
Berezovsky, Stikhin and Bakshi studies, 204
Berezovsky and Stikhin studies, 183–184, 186, 189
Bernoulli's law, 141
Binary gas mixtures, behavior
  aluminum and aluminum alloys, 76
  armco-iron and steels, 76, *77*
  copper and copper alloys, 77–78, *78–79*
  metal temperature, *71–72*, 81–82, *83*
  nickel and nickel alloys, 80–81, *81–82*
  titanium and titanium alloys, *71–72*, 79–80, *79–80*
Bismuth (Bi), 102
Bobkova, Georgiev, Yavojsky and, studies, 86
Bobkova and Petukhov studies, 84
Bobkova studies, 67
Body-centered cubic (bcc) lattice, 35–36
Bogatyrenko studies, 95
Boltzmann constant
  groove smoothing, 54
  non-metallic inclusions, 259
  nucleation, 23
  reactive forces, 116
  stirring effect, 223
Bond number, 348
Borodulin, Kurochkin and Umrikhin studies, 67

Boron (B)
  alloy composition influence, 67
  arc discharge, 105
  composite materials, 110
  substrate composition, 109–110
  wettability, 110–111
Boundaries of contacting phases, 35–43, *see also* Grain boundaries
Braun and Stock studies, 24
Brittle temperature range (BTR), 279, 312
Brittleness, 158, 209, *see also* Embrittlement
Brodsky, Ivashchenko and, studies, 139–140
Brownian motion, 352
BTR, *see* Brittle temperature range (BTR)
Bubbles, *see* Gas bubbles
Bukarov and Ermakov studies, 147
Bulk self-diffusion, 55
Burgard, Heiple and, studies, 172
Burgers vector, 27
Burns-through formation, 200, *201*, 201–204
Butt welds
  downhand position, 183, 200
  formation issues, 182
  reinforcement, 184
  undercuts, 161
Bykhovsky studies, 93

C

Cadmium (Cd), 102
Calcium (Ca) and calcium alloys
  chlorides/fluorides, 123
  cracks, 281
  crystallization phenomena, 29
  interphase tension, 87
  non-metallic inclusions, 207
  ternary slags, 87–88
Calculations
  non-metallic inclusions, 233–236, *235*, 257–259, *258–259*
  penetration zone, experimental findings, *181*, 181–182
Camel, Tison and Favier studies, 196
Capillary forces, 316–319

Carbon (C)
  alloy composition influence, 66
  carbon dioxide/air behavior, 73
  composite materials, 113
  cracks, 280
  gas-adsorption ability, 18
  gases influence, 67
  interphase surface energies, 84
  material composition influence, 91
  metal structure, 282
  solidification cracking, 280–281
  substrate composition, 110
  sulfur, 121
  surface properties and phenomena, 84
  surface *vs.* volume concentration, 268
  titanium and titanium alloys, 79–80
Carbon dioxide (CO_2)
  behavior, 70, 73–75, *73–75*
  current polarity, 174–175
  electrode welding, 172–173
  surface properties and phenomena, 70,
      73–75, *73–75*
Carbon monoxide (CO)
  behavior, 69–70, *70–72*
  electric field influence, 252
  gas influence, 3
  physical interactions, 261
  surface properties and phenomena,
      69–70, *70–72*
Carbon-rich steels, 300–301, *300–301*
Cathode region, 6
Cech, Turnbull and, studies, 88
Cell dimension influence, 128–131,
      129–132
Chalmers and Fleming, Pfank, studies, 31
Chalmers studies, 23
Chemical inhomogeneity, 4
Chernyak, Ryabtsev, Kushkov and,
      studies, 186
Chlorides influence, *122*, 122–123
Chromium (Cr)
  alloy composition influence, 66–67
  cracks, 281
  gas-adsorption ability, 18
  interphase surface energies, 85
  nickel, 80
  sulfur, 121
  surface properties and phenomena, 85
Chudinov and Taran studies, 187

Coagulation and coalescence, 219–222
Coarsening, weld pool, 218–222
Coated-electrode manual arc welding,
      43, *359*
Coatings
  acidic, *152*, 152–153
  basic-type, 150–152, *151*
  single-component, 153–154, *154*
Cobalt (Co), 66
Coefficient of spreading, 39
Cohesion, 37, 39
Cold cracks, 279, *see also* Cracks and
      crack-like defects
Comfort area effect, 294–296, *294–296*
Composite fiber-reinforced materials,
      313–323
Composite materials, *100*, 110–113,
      111–114
Composition effect, 325, *325–326*
Composition of system, 83–86
Concentrated capillary flows, 194
Condensation, 55–56
Constricted-arc welding, 7, 11
Consumable electrode welding,
      172–173
Contacting phases, boundaries, 35–43
Convection, 352–353
Cooling, 3
Copper (Cu) and copper alloys
  arc discharge, 106
  binary gas mixtures, behavior, 77–78,
      78–79
  composite materials, 113
  corrosion, 291–292
  cracks, 312
  gas flame, 118
  gas-shielded welding, 102
  isothermal conditions, 95–96, *97–98*,
      98–99
  nickel, 80
  surface properties and phenomena,
      95–96, *97–98*, 98–99
Copper–steel parts, 330–337
Corrosion
  armco-iron and steels, 299
  experimental study, 298–301, *301*
  mechanism, 304–305
  melt, role of, *291*, 291–292
  solidification cracking, 298–301, *299*

Corundum particles, 207, 260
Covered-electrode arc welding
  acidic coating, 152
  drop formation, 137
  fluxes influence, 20
Cracks and crack-like defects, *see also*
      Defects; Solidification cracking
  apex, metal atoms interaction, *285*,
      285–288, *287*
  resistance, copper–steel pipes, *332*,
      332–333
  solidification cracking, 306–309,
      307–309
  strength deterioration, 160–161,
      161–162
  technologies, development,
      339–340
Craters
  formation, multilayer welding,
      201–202, *203*
  microgravity, 352
  strength deterioration, 162
Critical rate, deposition, 191–193,
      193–194
Crystal formation, 24
Crystallization phenomena, 28–33
Current, welding, 178, *178–179*
Current density, 125
Current polarity, 173–177, *174–175*
Cutting tools, deposited cutting edge,
      337–340

D

D.C. welding, *see* Direct current (d.c.)
      welding
Deep overcooling, 24
De Fries studies, 272–273
Defects, *see also* Cracks and crack-like
      defects; Porosity
  strength deterioration, 160–161,
      161–162
  weld and deposited material formation,
      198–204
Deformation, 231–233, *232*
Deformation rate–load graph, 53
Deposited cutting edge, cutting tools,
      337–340

Deposited metal
  surfacing, experimental findings, *326*,
      326–327
  weld and deposited material formation,
      189–197
Depth of penetration
  current polarity, 175–176
  electric arc welding, 10
  oxygen, 166–168
  wetting and spreading, 333–334
Deryabin and Popel' studies, 124
Deryabin and Saburov studies, 86
Deryagin and Sherbakov studies, 92
Deryagin studies, 92
Deterioration, 157–164, *158–160*,
      see also Defects
Devices, modern unit, 59–61
Deyev studies, 342
Different gases in pores, 242–245,
      244, *246*
Diffusion, metal vapors, 55
Diffusion phenomena, 265–268, *267*
Direct arcs
  electric arc welding, 9–10
  gas shielding, 11
  influence, wettability and spreading, 107
Direct current (d.c.) welding
  arc discharge action duration, 178
  comparison to a.c. welding, 4–7, *5*
  electric arc welding, 10
  electrochemical interactions, 125
  external electric field, 134
  slag, role, 22
  welding polarity, 153
Direct interaction, metal and gas, 14–17
Disjoining pressure, *197*, 223–225
Displacement speed, 194–197, *195–197*
Disruption, 231–233
Dmitriy, Baptizmansky, Bakhman and
      studies, 221
Dmukhovskaya, Popovich and studies, 287
Drop weighting
  polarity, 153
  surface properties and phenomena,
      44–45, 57
Drops contact method, 58
Ductility
  armco-iron losses, 296
  cracks, 286–287, 308, 312

ferrous-based alloys, 305
hydrogen, adverse influence, 13
melt, role of, 289–290
non-metallic melts, solid metals,
    296–298, *297*
technologies, development, 340, *341*
welded joint strength, 209
Dupre studies and equation, 37–38, 273
Duration, arc discharge action,
    178–179, *179*
Dyatlov, Leskov and, studies, 117
Dyatlov studies, 137, 147
Dzhemilev, Popel' and Tsarevsky
    studies, 66

E

Eager, Lin and, studies, 202
Earth conditions, 347
Efimov studies, 23
Electric arc welding, 2, *9*, 9–11
Electric field, porosity, 250–254
Electrocapillary curve, 42–43
Electrocapillary phenomena
    electrochemical interactions, 125
    surface properties and phenomena,
        59–64, *62–63*, 123–124
Electrochemical phenomena
    non-metallic inclusions, 214–216, *215*
    surface properties and phenomena,
        42–43, 124
Electrode manual arc welding, *359*
Electrode welding, consumable,
    172–173
Electrode-metal transfer
    aerodynamic forces, 138
    basics, 135–136
    current, 145–146
    detachment of drop, *140*, 140–141
    drop detachment, *140*, 140–141
    drop formation, 136–137, *137*
    drop transfer stages, 138–139
    electrode wire composition, 148–149,
        149
    electromagnetic force, 137
    filler metal, 149–150
    flux-cored wires and filler metal,
        149–150

gaseous medium, spatter and transfer,
    138–150
influencing forces, 135–138
interphase tension, 150–153, *151*
metal spatter formation, 141–142
molten electrode metal, slag contact,
    150–155
oxidizing gases, *71–72*, 146–148,
    146–148
part preheating, role, *83*, *142–144*,
    142–145
polarity, *152–153*, 153
reactive forces, 138
slag contact, molten electrode metal,
    150–155
spatter and transfer, electrode metal,
    138–150
underlying factors, 150
welding current, 145–146
welding polarity, *152–153*, 153
wettability, 145–146
Electromagnetic stirring, 25–26, 29–30,
    see also Stirring
Electron beam deflection, *336*,
    336–337
Electron beam oscillations, 194
Electron beam surfacing
    technologies, development,
        323–330
    weld and deposited material formation,
        191–194, *192*
Electron beam welding
    fusion welding process, characterization,
        7–8, 12–13, *13*
    gas influence, 3
    non-metallic inclusions, 205, 236
Electroslag remelting, 22
Electroslag welding
    contacting phase interaction, 2
    cracks, 309
    drop formation, 137
    electric field influence, 251
    electrochemical phenomena, 43
    fusion welding process, characterization,
        11–12, *12*
    non-metallic inclusions, 205
    polarity effect with flux, 155
El-Hebeary, Turi and, studies, 282
Elliot, Turpin and, studies, 211

Embrittlement, *see also* Brittleness
  corrosion, 292, 306, 308
  metal atoms, crack apex, 287
  welded joint strength, 210
Emelyanov studies, 183
Energy of formation, gas nucleus,
      252–254, *253*
Eremenko, Natanzon and Ryabov
      studies, 97
Ermakov, Bukarov and, studies, 147
Erokhin studies, 153, 186
Escape, gas bubbles, 269–278
Esche and Peter studies, 66, 70
Esin, Popel' and Nikitin studies, 83
Esin and Gel'd, Popel', studies, 86
Evaporation, 55–56
Experimental findings and observations,
      see also Study results, surface
      properties and phenomena
  chlorides/fluorides of fluxes, 123, *123*
  corrosion effect, 298–301, *301*
  Δ_m, wettability and spreading, *117*,
      117–118
  electron beam surfacing, 191–194, *192*
  FeO and FeS influence, 276
  film destruction, 274–275, *275*
  interphase tension, 86–88
  mechanical properties, solid metals,
      292–306, *293*
  surfacing, technologies development,
      324, 324–330
  technologies, development, 320–323,
      320–323
  weld and deposited material formation,
      177, 177–182
Extending time, weld-pool existence, 199,
      199
External electric field influence, 131–134,
      133
External fields influence, 92–95, *93*
External-diffusion stage, 14–15

F

Face-centered cubic (fcc) lattice, 35–36
Faraday number, 133
Fatigue life, undercuts, 161
Favier, Camel, Tison and, studies, 196

Fcc, *see* Face-centered cubic (fcc) lattice
Fedko and Sapozhnikov studies, 141
Fedko studies, 142
Ferrous-based alloys, *see also* Iron (Fe)
      and iron alloys
  alloy composition influence, 66
  carbon dioxide/air behavior, 73–74
  copper and copper alloys, 77
  ductility, 305
  electrochemical considerations,
      214–215
  film destruction, *275–277*, 275–278
  fluxes influence, 21
  gases influence, 67–68
  nickel, 81
  physical interactions, 261
  slag components, 86
  sulfur, 122
  ternary slags, 87
Ferrous metal–slag system, 58
Fick's laws, 16, 266
Figurovsky studies, 269
Film destruction, 273–278
Flame temperatures, 9
Fleming, Pfank, Chalmers and, studies, 31
Fluoc effect, *90*, 102–103
Fluorides
  slag, role, 22
  surface properties and phenomena, *122*,
      122–123
Flux- or flux paste-assisted welding, 236
Fluxes
  composition, *359*
  fusion welding process, 20–21, *21–22*
  surface properties and phenomena, *122*,
      122–123
Fominsly and Gorbunov, Skrinsky, studies,
      324
Formation, weld and deposited material
  base-metal surface properties, 166–168,
      166–168
  basics, 157
  butt weld formation issues, 182
  current polarity, 173–177, *174–175*
  defects, 198–204
  deposited metal formation, 189–197
  displacement speed, 194–197, *195–197*
  electron beam surfacing experiments,
      191–194, *192*

experimental findings, *177*, 177–182
graphical method, 182–184
gravitational force, 187–189
groove filling/unfilling, 198–200
inclination angle, 184–186, *185*
multilayer welding, 200–204
penetration zone, *165*
polarity, current, 173–177, *174–175*
serviceability, 157–164
shape defects, 198–204
strength deterioration, 157–164,
    158–160
surface phenomena defects, 198–204
surface tension, 187–189
surfactants, action, 168–169, *169*
temperature fields, 184–186, *185*
temperature gradient, 194–197,
    195–197
thermocapillary phenomena, 169–173
weld bead formation, 182–189
weld shape defects, 198–204
wetting angle, 189–190, *191*
Formations
    burns-through, *201*, 201–204
    craters, 201–202, *203*
    rolls, 202–204, *203*, 340
    solidification cracking, 280–284
Fourier equation, 143–144
Free energy
    coalescence and coagulation, 220–221
    gas bubbles, 229
    interphase tension, 36
    oxides, 221
    surface energy, 36
Frenkel studies and formula, 185, 211, 233,
    319
Frolov studies, 170
Fusion welding process, characterization
    basics, 1
    crystallization phenomena, 28–33
    d.c. *vs.* a.c. welding, 4–7, *5*
    direct interaction, metal and gas, 14–17
    electric arc welding, *9*, 9–11
    electron beam welding, 7–8, 12–13, *13*
    electroslag welding, 11–12, *12*
    features, 1–4, *2*
    fluxes influence, 20–21, *21–22*
    gas influence, 2–3
    gas-adsorption ability, 18–19

laser welding, 8–9, 12–13, *13*
physiochemical processes, 13–33, *14*
processes, 9–13
slag influence, 3–4
slag role, 21–23
solid-molten metal boundary, 23–26
σ_{sol-1} values, 26–28, *28*
surfactants influence, 17–18
vacuum influence, 19, *20*
weld formation, 1–2
welded joint formation, 13–33, *14*
welding heat sources, 4–9
Fuzi, Nogi, Aoki and, studies, 351

G

Gas bubbles
    free energy, 229
    gas influence, 3
    gravitational forces, 272
    growth, 262–269, 353–354
    kinetics, 277–278
    maximum pressure, 46, *47*, 48
    multiple bubbles allowance, 272
    non-metallic inclusions, 229–231,
        229–232
    nuclei formation, 239–248, *240*,
        248–262
    porosity in welds, 269–278
    technologies, development, 353–354
    transfer, 277–278
Gas flame, nature, *103*, *118*, 118–119
Gas nucleus, energy of formation,
    252–254, *253*
Gas-adsorption ability, 18–19
Gas shielding, 10–11, *10–11*
Gas welding
    contacting phase interaction, 2
    electroslag welding, 13
    fluxes influence, 20
Gas-shielded welding, 102
Gaseous environment, oxides effect, 222
Gases
    influence, 2–3
    surface properties and phenomena,
        67–82
    surfacing, experimental findings,
        325–326, *326*

Gas-slag-metal system, 20–21
Gasyuk and Sauchuk studies, 29
Geguozin studies, 36
Gel'd, Ayushina, Levin and, studies, 67
Gel'd and Akshentsev, Baum, studies,
    67, 70
Gel'd, Popel', Esin and, studies, 86
Georgiev, Yavojsky and Bobkova
    studies, 86
Gibb's adsorption equation, 32, 41
Gibb's energy, 217, 224
Girimadzhi, Olshansky, Gutkin and,
    studies, 170
Gorbunov, Skrinsky, Fominsly and,
    studies, 324
Gorshkov studies, 266
Grain boundaries
    corrosion, 304–305
    cracks, 279–280, 283
    thermal etching method, 54–56, *55*
Grain growth, 29–30
Graphical method, 182–184
Gravitational forces
    electrode-metal transfer, 136
    external fields influence, 94
    gas bubbles, 231
    multiple bubbles allowance, 272
    roughness, 91
    space conditions, 347
    weld and deposited material formation,
        187–189
    wetting angle, 190
Griffits equation, 305
Grigoryan, Shvindlerman and Aleev
    studies, 232
Grigoryants studies, 191
Groove filling/unfilling, 198–200
Groove smoothing method, 53–54
Growth, gas bubbles, 262–269
Gulyaev studies, 214, 339
Gurevich studies, 234
Gutkin and Girimadzhi, Olshansky,
    studies, 170

H

Halden and Kindgery studies, 66–67
Hardness, 337–339, *338–339*

Havalda, Sebo and, studies, 99
HAZ, *see* Heat-affected zone (HAZ)
Hcp, *see* Hexagonal close-packed (hcp)
    lattice
Heat sources
    electroslag welding, 11
    fusion welding process, characterization,
        4–9
    influence, 105–119
    laser welding, 8–9
    slag influence, 4
    welding, 4–9
Heat-affected zone (HAZ)
    electron beam welding, 8
    penetration depth, 165
    slag influence, 4
    transition angle, 188
Heiple and Burgard studies, 172
Helium (He), 66, 68
Heterogeneous environment, 248–262
Heterogeneous surfaces, 315–319, *317*
Hexagonal close-packed (hcp) lattice,
    35–36
Higbi studies, 15
High-carbon materials, 260–261
Hydrogen (H)
    adverse influence, 13–14
    alloy composition influence, 66
    cracks, 161
    electric field influence, 251
    fluxes influence, 21
    slag, role, 21
    surface properties and phenomena,
        67–69, *71–72*
    surface tension, molten and solid metals,
        67–69, *71–72*
Hydroxyl hydrogen, 21

I

Iida, Sato and Nagai studies, 162
Imperfections, *see* Defects
Inclination angle, 184–186, *185*
Indirect arcs
    electric arc welding, 9–10
    gas shielding, 11
    influence, wettability and spreading,
        107–109, *108–109*

Inert-gas welding, 236
Inhomogeneity, 4
Intercrystalline corrosion, 304–305
Interface shape, 25, *25*
Intermolecular adhesion forces, 248–250,
    249–250
Interphase surface energies, 56–59, 82–88
Interphase tension
    electrochemical phenomena, 42
    experimental determination, 86–88
    material composition influence, 91
    surface properties and phenomena,
        37–40, *39*
Iron (Fe) and iron alloys, *see also*
        Ferrous-based alloys
    alloy composition influence, 66
    carbon, 84
    copper and copper alloys, 78
    cracks, 306
    interphase surface energies, 85
    metallography observations, 299–300
    refractory impurities, 24–25
    substrate composition, 109
    sulfur, 84
    ternary slags, 88
Isothermal conditions, 95–99, 314
Ivanov studies, 319
Ivashchenko and Brodsky studies,
    139–140

J

John and Richards studies, 282
Joining, *see* Disjoining pressure
Joints, *see* Welded joints
Jones studies, 82
Joule's law, 342

K

Kalman, Kotze and, studies, 27
Kapustina studies, 269
Khlynov, Sorokin and Stratonovich studies,
    227
Khokonov, Shebzukhova and Kokov
    studies, 53
Khokonov, Taova and, studies, 28
Khollomon and Turnbull studies, 217, 241

Kindgery, Halden and, studies, 66–67
Kinetics
    ascending bubbles, 277–278
    nucleation, 211–212
Kingery, Kurkijian and, studies, 80
Kokov, Khokonov, Shebzukhova and,
    studies, 53
Kopan studies, 53
Kopersak and Solokha, Sivinsky, studies,
    87
Kortyuchenko, Ovsienko and, studies, 24
Kotze and Kalman studies, 27
Koz'min and Popel', Nemchenko studies,
    270
Kraus, Muu and, studies, 46
Kreshchanovski, Prosvirin and Zaletaeva
    studies, 68
Kreshchanovski and Kunin, Zaletaeva,
    studies, 29
Kudoyarov and Suzdalev, Russo, studies,
    165
Kunin, Zaletaeva, Kreshchanovski and,
    studies, 29
Kunin studies, 29
Kurkijian and Kingery studies, 80
Kurochkin and Umrikhin, Borodulin,
    studies, 67
Kushkov and Chernyak, Ryabtsev, studies,
    186
Kuzmenko studies, 176

L

Lacks-of penetration, 160–161, 200
Laptev studies, 215
Large drop method, *48*, 48–49
Laser welding
    fusion welding process, characterization,
        8–9, 12–13, *13*
    thermocapillary phenomena, 171–172,
        173
Latash and Medovar studies, 155
Lattices, surface energy, 35–36
Leontieva studies, 83
Leskov and Dyatlov studies, 117
Levin and Gel'd, Ayushina, studies, 67
Lin and Eager studies, 202
Lippman equation, 42, 131

Liquid movement, capillary forces, 317–318
Lithium (Li), 123
Local corrosion, 304
Lorentz forces, 352
Losses, *see also* Spatter
armco-iron, 296
electrode-metal transfer, 135–136
ferrous-based alloys, 276–277
Low-carbon steels, *see also* Steels
cracks, 159
disjoining pressure, 224
electric field influence, 251
gas flame, 118
non-metallic inclusions, 207
welded joint strength, 280
Low-melting melts effect, 306–312
Low-melting non-metallic melts effect, 292–306
Lubausky studies, 22

M

Macrowelding, 13
Magnesia, 207
Magnesium (Mg) and magnesium alloys
chlorides/fluorides, 123
gas-shielded welding, 102
interphase tension, 87
physical interactions, 261
ternary slags, 87–88
Malenkov's formula, 270
Manganese (Mn) and manganese alloys
alloy composition influence, 66
carbon dioxide/air behavior, 73
carbon monoxide behavior, 69
copper and copper alloys, 77–78
cracks, 281
current polarity, 173
gas-adsorption ability, 18
gas-shielded welding, 102
gases influence, 67
interphase surface energies, 85
interphase tension, 87
metal structure, 282
nickel, 80
physical interactions, 261
sulfur, 121

surface *vs.* volume concentration, 268
ternary slags, 87
Manganous oxide, 86
Manual arc welding, *359*
Manual stick-electrode arc welding, 10
Marangoni effect
convection, 352
electron beam surfacing, 197
gas bubbles growth, 353
thermocapillary phenomena, 170
Marangoni studies, 170
Maslov studies, 318
Mass transfer, 14–15
Material composition influence, 90–92
Mathematical treatment, porosity, 269–271
Maximum pressure, bubble, 46, *47*, 48
Mazel studies, 150, 221
Mazurin studies, 261
Mc Entee and Mysels studies, 273
Mechanical properties
experimental findings, 292–298, *293*
melt, role, 288–291, *289–290*
solidification cracking, 292–306, *293*
Mechanistic characteristics, 113, *115–116*, 115–119
Medovar, Latash and, studies, 155
Medzhibozhsky studies, 73
Melt–crystal interphase, 26
Melts
determination, 44–51
overcooling, 24
role of, 288–292
Mendeleev–Clapeyron equation, 265
Metal atoms, interaction, *285*, 285–288, *287*
Metal and gas, direct interaction, 14–17
Metal-arc welding, 350–351
Metal melting, pendant state, 49
Metal melts, solid surfaces, 88, 90–105
Metal stirring effect, 25–26
Metal structure role, 281–282
Metal temperature, *71–72*, *81–82*, *83*
Metallography observations
solidification cracking, 301–304, *303*
thermodynamic considerations, 217
Metals, composition, *356–358*
Metals effect, solid metal fractures, 284–292

Metal–slag systems, 82–88, 123–124
Microchemical observations, 217
Microgravity, 349–352, *350*
Micro-local corrosion, 304
Microwelding, 13
Miniwelding, 13
Missol studies, 305
Models, theoretical, 14–17, *15*
Modern unit device, 59–61, *60–61*
Molten metals
    gas bubbles escape, 269–278
    surface tension, 65–82
Molybdenum (Mo)
    alloy composition influence, 67
    arc discharge, 105
    interphase surface energies, 85
    material composition influence, 91
    substrate composition, 109
    sulfur, 121
    surface properties and phenomena, 85
Morozov studies, 22
Muller studies, 223
Multi-component alloys, 3, *see also specific type*
Multilayer welding, 200–204
Multiphase equilibrium method, 51–52, *52*
Multiple bubbles allowance, 271–273, *272*
Multiple gases in pores, 242–245, *244, 246*
Muu and Kraus studies, 46
Mysels, Mc Entee and, studies, 273

N

Nagai, Iida, Sato and, studies, 162
Naidich studies, 43
Najdich, Zabuga and Perevertajlo studies, 195
Naletov, Belchuk and, studies, 188
Nasarov, Shiganov and Raimond studies, 8
Natanzon and Ryabov, Eremenko, studies, 97
Nelson studies, 229
Nemchenko, Koz'min and Popel', 270
Nernst's diffusion theory, 29
Nickel (Ni) and nickel alloys
    alloy composition influence, 66
    binary gas mixtures, behavior, 80–81, *81–82*

sulfur, 121
surface *vs.* volume concentration, 268
Nikitin, Esin, Popel' and, studies, 83
Niobium (Nb)
    arc discharge, 105
    cracks, 281
    substrate composition, 109
Nitrides, 3
Nitrogen (N)
    current polarity, 175
    electric field influence, 251
    electrode welding, 172
    fluxes influence, 21
    gas influence, 3
    oxygen, 85
    slag, role, 22
    surface properties and phenomena,
        67–69, *71–72*
    surface tension, molten and solid metals,
        67–69, *71–72*
Nogi, Aoki, and Fuzi, 351
Non-isothermal processes, 100–105
Non-metallic inclusions
    adhesion forces, *221*, 225–226, *226*
    basics, 205
    calculations, 233–236, *235*
    coalescence and coagulation,
        219–220
    coarsening, weld pool, 218–222
    deformation and disruption, 231–233,
        232
    disjoining pressure, *197*, 223–225
    electrochemical considerations,
        214–216, *215*
    gas bubbles, 229–231, *229–232*
    influence, 208–210
    nucleation, weld pool, 211–218
    nuclei growth, 218–219
    oversaturation, 212–214
    oxides, *216*, 220–222, *221*
    porosity in welds, *254*, 254–259,
        257–258
    removal, weld pool, 227–236
    stirring effect, 222–223
    technologies, development, 353–354
    thermodynamic considerations, *216*,
        216–218, *218*
    types, 205–208, *206–207*
    welds, 205–210

Nonmetal melts, 119–123
Nucleation
    characterization, 214–215
    interface shape, 25
    kinetics, 211–212
    process, 23–24
    refractory impurities, 24–25
Nuclei growth, 218–219

O

Olshansky, Gutkin and Girimadzhi studies,
        170
Optical quantum generators, 12
Oscillations, *see also* Ultrasonic
        oscillations
    crystallization phenomena, 29–30
    electron beam type, 194
    porosity in welds, 264–265
    thermocapillary phenomena, 194
Overcooling, 24
Oversaturation
    electrochemical considerations, 215
    non-metallic inclusions, 212–214
    nucleation of pores, oxygen, 247–248
Ovsienko and Kortyuchenko studies, 24
Oxides and oxide substrates
    adhesion, 226
    film, 145, 148
    gas influence, 3
    material composition influence, 91
    non-metallic inclusions, *216*, 220–222,
        221
    oversaturation, 212
    slag components, 86
    slag influence, 3
    vacuum influence, 19
Oxygen (O) and oxygen alloys
    composite materials, 112
    copper and copper alloys, 77–78
    current polarity, 173–175
    electric field influence, 251
    ferrous-based alloys, 276
    fluxes influence, 21
    gas influence, 3
    gas-adsorption ability, 18–19
    interphase surface energies, 84–85

laser welding, 9
non-metallic inclusions, 207
nucleation of pores, 246–248
penetration depth, 166–168
physical interactions, 261
pore nucleation, 246–248
solidification cracking, 311–312
surface properties and phenomena,
        67–69, *71–72*, 84–85
surface tension, molten and solid metals,
        67–69, *71–72*
surface *vs.* volume concentration, 268
surfactants, 168, 245
vacuum influence, 19

P

Palladium (Pd), 102
Parabola form, 42–43
Parameters
    solidification cracking, *296–297*,
        305–306
    wetting and spreading, 334, *336*,
        336–337
Paton studies, 349
Patrov studies, 124
Peanes studies, 26
Peinter studies, 348
Pendant drop method, 44, *44*
Penetration depth
    current polarity, 175–176
    electric arc welding, 10
    oxygen, 166–168
    wetting and spreading, 333–334
Penetration zone
    calculations, *181*, 181–182
    weld and deposited material formation,
        165
Perevertajlo, Najdich, Zabuga and, studies,
        195
Permeation of surface, 15
Peter, Esche and, studies, 66
Peter, Esche and, studies, 70
Petrov studies, 146
Petukhov, Bobkova and, studies, 84
Pfank, Chalmers and Fleming studies,
        31

Phase transformation, 14–15
Phosphorus (P)
    interphase surface energies, 84
    surface properties and phenomena, 84
    surface *vs.* volume concentration, 268
Physical interaction forces, 261–262,
    262–263
Physiochemical processes, 13–33, *14*
Pipes, copper–steel, *324–325*, 331–333
Pirogov studies, 67
Plank's constant, 23
Plasma-arc welding, 7, 11
Pokhodnya studies, 18, 242
Polarity
    electric field influence, 251
    flux, *89–90, 132*, 155
    weld and deposited material formation,
        173–177, *174–175*
Polygonization cracks, 279, *see also*
        Cracks and crack-like defects
Popel', Deryabin and, studies, 124
Popel', Esin and Gel'd studies, 86
Popel', Nemchenko, Koz'min and, studies,
        270
Popel' and Nikitin, Esin and, studies, 83
Popel' and Tsarevsky, Dzhemilev, studies,
        66
Popel' studies, 18, 73, 84, 95, 102, 268
Popovich and Dmukhovskaya studies,
        287
Pores and performance, welded joints,
        237–239
Porosity, *see also* Defects
    advanced mathematical treatment,
        269–271
    allowance for multiple bubbles,
        271–273, *272*
    basics, 237
    different gases in pores, 242–245, *244,
        246*
    diffusion phenomena, 265–268, *267*
    electric field, 250–254
    escape, gas bubbles, 269–278
    film destruction, 273–278
    gas bubble nuclei formation, 239–248,
        240, 248–262
    growth, gas bubbles, 262–269
    heterogeneous environment, 248–262

intermolecular adhesion forces,
        248–250, *249–250*
molten metal, gas bubbles escape,
        269–278
    multiple bubbles allowance, 271–273,
        272
    multiple gases in pores, 242–245, *244,
        246*
    non-metallic inclusions, *254*, 254–259,
        257–258
    oscillations effect, 264–265
    oxygen effects, pore nucleation,
        246–248
    physical interaction forces, 261–262,
        262–263
    pores and performance, welded joints,
        237–239
    radius of curvature, 241–242
    stirring effect, 264–265
    surface concentration, components,
        268–269
    surface tension, 241–242
    thermodynamic considerations,
        263–264
    volume concentration, components,
        268–269
Potapov studies, 345
Potassium (K) and potassium alloys, 88
Potekhin studies, 202
Precipitation phenomena, *213*, 213–214
Preheating
    composite fiber-reinforced materials,
        321
    non-isothermal process behavior,
        103–104, 103–105
Privalov, Protsenko and, studies, 148
Problems, composite materials, 314–315,
        316
Processes
    fusion welding, 9–13
    interphase tension, 87–88, *89–90*
    surfactants, 17–18
Productivity, increasing, 341–343,
        342–343
Prokhorov studies, 279
Prosvirin and Zaletaeva, Kreshchanovski,
        studies, 68
Protsenko and Privalov studies, 148

Q

Quantities, expressions, 16
Quartz, 207

R

Rabkin studies, 251
Radius of curvature, 241–242
Raimond, Nasarov, Shiganov and,
    studies, 8
Rebinder effect, 284, 288, 338
Rebinder studies, 285
Refractory impurities, 24–25
Refractory metals, 91
Removal, weld pool, 227–236
Restoration of surface, 15
Reynold's number, 269
Richards, John and, studies, 282
Rolls
    ductility, 339–340
    multilayer welding, 202–204, *203*
    strength deterioration, 163–164
Roughness
    composite materials, 315
    groove filling and unfilling, *199*,
        199–200
    preheating, 142
Runge-Kutta method, 101
Russo, Kudoyarov and Suzdalev studies,
    165
Ryabov, Eremenko, Natanzon and, studies,
    97
Ryabtsev, Kushkov and Chernyak studies,
    186
Rybchinsky–Hadamard formula, 227

S

Saburov, Deryabin and, studies, 86
Safonov, Zubkova and, studies, 338
Sapiro studies, 245
Sapozhnikov, Fedko and, studies, 141
Sato and Nagai, Iida, studies, 162
Sauchuk, Gasyuk and, studies, 29
Sauerwald studies, 65
Sebo and Havalda studies, 99

Segregation processes, 30–33, *31*
Seidel method, 127
Selenium (Se), 67, 81
Self-diffusion, 55
Self-loading examples, 282–283
Semenchenko studies, 43
Semi-automatic welding processes, 10–11
Serviceability, 157–164
Sessile drop method
    electrocapillary phenomena, 62, 64
    electrochemical interactions, 124
    gases influence, 68
    surface properties and phenomena,
        45–46, *45–47*, 56–57
    x-ray photography, 58–59
Shape, interface, 25, *25*
Shape defects, 198–204
Shcherbakov studies, 26
Shchukin studies, 287
Shebzukhova and Kokov, Khokonov,
    studies, 53
Shenland studies, 8
Sherbakov, Deryagin and, studies, 92
Shiganov and Raimond, Nasarov, studies, 8
Shimmyo and Takami studies, 16
Shishkovsky studies, 41
Shorshorov studies, 281
Shvindlerman and Aleev, Grigoryan,
    studies, 232
Silicate inclusions, 214
Silicon (SI) and silicon alloys
    alloy composition influence, 66
    arc discharge, 106
    carbon dioxide/air behavior, 73
    carbon monoxide behavior, 69
    copper and copper alloys, 77–78
    current polarity, 173–175
    gas-adsorption ability, 18
    gas-shielded welding, 102
    gases influence, 67
    interphase surface energies, 85
    interphase tension, 87
    penetration depth, 166
    physical interactions, 261
    substrate composition, 110
    sulfur, 121
    surface properties and phenomena, 85
    surface *vs.* volume concentration, 268
    ternary slags, 87–88

Single-component coatings, 153–154, *154*
Sirgo studies, 238
Sivinsky, Kopersak and Solokha studies, 87
Sivkov studies, 96, 334
Skapski studies, 27
Skrinsky, Fominsly and Gorbunov studies, 324
Skvortsov studies, 29
Slags
    adhesion, 344–346, *344–346*
    composition, *359*
    heat source, 7, 11
    influence, 3–4
    interphase surface energies, 85–86
    role, 21–23
    surface properties and phenomena, 85–86
Slot capillary, 318–319
Smolukhovsky studies, 271
Sodium (Na) and sodium alloys
    chlorides/fluorides, 123
    current polarity, 173
    slag components, 86
    ternary slags, 87–88
Solid metal fractures, metal melt effect, 284–292
Solid metals
    mechanical properties, 292–306, *293*
    melt systems, 82–88
    nonmetal melts, 119–123
    surface properties and phenomena, 51–56
    surface tension, 65–82
Solid-molten metal boundary, 23–26
Solidification
    interphase tension, 26, 88, *90*
    slag influence, 4
Solidification cracking, *see also* Cracks and crack-like defects
    basics, 279–280
    carbon content, 280–281
    corrosion, experimental study, 298–301, *299*
    corrosion effect, 304–305
    crack apex, metal atoms interaction, *285*, 285–288, *287*
    crack formation, 306–309, *307–309*
    formation, 280–284
    low-melting melts effect, 306–312
    low-melting non-metallic melts effect, 292–306
    mechanical properties, 292–306, *293*
    melt, role of, 288–292
    metal atoms, interaction, *285*, 285–288, *287*
    metal metals effect, solid metal fractures, 284–292
    metal structure role, 281–282
    metallography observations, 301–304, *303*
    oxygen/sulfur combined effects, 311–312
    parameters, additional, *296–297*, 305–306
    solid metal fractures, 284–292
    solid metals, mechanical properties, 292–306, *293*
    surface phenomena role, *293–294*, 309–311, *310–311*
    susceptibility, 282–284, *283–284*
    test methods, 280–284
Solokha, Sivinsky, Kopersak and, studies, 87
Sorokin and Stratonovich, Khlynov, studies, 227
Space conditions, technologies, 347–354
Spatter, *see also* Losses
    ferrous-based alloys, 276
    gaseous medium, 138–150
    oxidizing gases, 147–148
Spreading
    arc discharge action duration, 179
    coefficient, 39
    composite materials, 315
    copper/aluminum, 99
    external fields influence, 93–94
    gas flame, 119
    heat source influence, 105–119
    interphase tension, 39
    mechanistic considerations, 115
    metal melts over solid surfaces, 88, 90–105
    preheating, 105
    surface properties and phenomena, 59–64

Spreading (*Continued*)
  technologies, development, 333–337,
    335
  titanium, 99
$\sigma_{\mathrm{sol}-1}$ values, 26–28, *28*
Steel aluminum, 96–97, *98*
Steel/copper and copper alloys, 95–96, *97*
Steel plates, dimensions and spacing,
    179–181, *180*
Steels, *see also* Armco-iron and steels;
    Low-carbon steels
  arc discharge, 105–106
  armco-iron, 76
  carbon-rich, 300–301, *300–301*
  composite materials, 113
  substrate composition, 109
Steg studies, 347
Stick-electrode arc welding, 10
Stikhin, Berezovskii and, studies,
    183–184, 186, 189
Stikhin and Bakshi, Berezovskii, studies,
    204
Stirring, 25–26
  crystallization phenomena, 29–30
  non-metallic inclusions, 222–223
  physiochemical processes, 25–26
  porosity in welds, 264–265
  removal basics, 227
Stock, Braun and, studies, 24
Stokes formula, 227, 269–271, 353
Storage batteries, 62
Stratonovich, Khlynov, Sorokin and,
    studies, 227
Strength deterioration, 157–164, *158–160,*
    see also Defects
Strontium (Sr), 123
Study approach, 124–125
Study results, surface properties and
    phenomena
  alloy composition influence, 65–67
  arc discharge influence, 105–113,
    106–107
  basics, 65
  cell dimension influence, 128–131,
    129–132
  chlorides influence, *122,* 122–123
  composition of system, 83–86
  electrocapillary phenomena, 123–124
  electrochemical interactions, 124

experimental determination, interphase
    tension, 86–88
experimental methods, 125–128, *126,*
    144
external electric field influence,
    131–134, *133*
external fields influence, 92–95, *93*
fluorides, *122,* 122–123
fluxes influence, *122,* 122–123
gases influence, 67–82
heat sources influence, 105–119
interphase surface energies, 82–88
interphase tension, 86–88
isothermal conditions, 95–99
material composition influence, 90–92
mechanistic characteristics, 113,
    115–116, 115–119
metal melts, solid surfaces, 88, 90–105
metal–slag systems, 82–88, 123–124
molten metals, surface tension, 65–82
non-isothermal processes, 100–105
nonmetal melts, 119–123
solid metals, 65–82, 119–123
solid-metal melt systems, 82–88
spreading, 88, 90–119
study approach, 124–125
sulfur and sulfides influence, 119–122,
    121
surface roughness influence, 92
surface tension, 65–82
system composition, 83–86
temperature of system, 86
welding heat sources influence, 105–119
wettability, 88, 90–123
Submerged-arc electroslag welding, 20
Submerged-arc welding
  current polarity, 175–177
  drop formation, 137
  electric field influence, 251
  electrochemical phenomena, 43
  non-metallic inclusions, 207
  polarity effect with flux, *132,* 155
  technologies, development, 341–346
  welded joint strength, 280
Subsolidus cracks, 279, *see also* Cracks and
    crack-like defects
Substrate composition, role, 109–110, *110*
Sulfur (S), sulfides, and sulfur alloys
  adhesion, 226

adsorption, 310
copper and copper alloys, 78
cracks, 280, 306
ferrous-based alloys, 276
interphase surface energies, 84
laser welding, 172
metallography observations, 299–300
metal structure, 282
nickel, 81
non-metallic inclusions, 207
penetration depth, 166–168
precipitation phenomena, 213
solidification, 214
solidification cracking, 311–312
surface properties and phenomena, 84,
    119–122, *121*
surface *vs.* volume concentration, 268
surfactants, 168
Surface concentration, components,
    268–269
Surface energy, 35–37
Surface properties and phenomena
adsorption process, 40–41
arc active spot, 49–51, *51*
autoelectronic emission method, 54
basics, 35
boundaries of contacting phases, 35–43
comparison of methods, 64
defects, 198–204
drop weight method, 44–45, 57
drops contact method, 58
electrocapillary phenomena, 59–64,
    62–63
electrochemical phenomena, 42–43
experimental findings, *324*
groove smoothing method, 53–54
interphase, surface energy
    determination, 56–59
interphase tension, 37–40, *39*
large drop method, *48*, 48–49
maximum pressure, bubble, 46,
    47, 48
melts, determination, 44–51
metal melting, pendant state, 49
methods comparison, 64
modern units, 59–61, *60–61*
multiphase equilibrium method,
    51–52, *52*
pendant drop method, 44, *44*

sessile drop method, 45–46, *45–47*,
    56–57
solidification cracking, *293–294*,
    309–311, *310–311*
solids metals, 51–56
spreading, 59–64
surface energy, 35–37
surface tension, 37, 49–51, *50*
thermal etching method, grain
    boundaries, 54–56, *55*
turned capillary method, 57–58
wettability, 59–64
x-ray photography, sessile drop, 58–59
zero creep method, 52–53
zone of arc active spot, 49–51, *50*
Surface properties and phenomena, study
    results
alloy composition influence, 65–67
arc discharge influence, 105–113,
    106–107
basics, 65
cell dimension influence, 128–131,
    129–132
chlorides influence, *122*, 122–123
composition of system, 83–86
electrocapillary phenomena, 123–124
electrochemical interactions, 124
experimental determination, interphase
    tension, 86–88
experimental methods, 125–128, *126,
    144*
external electric field influence,
    131–134, *133*
external fields influence, 92–95, *93*
fluorides, *122*, 122–123
fluxes influence, *122*, 122–123
gases influence, 67–82
heat sources influence, 105–119
interphase surface energies, 82–88
interphase tension, 86–88
isothermal conditions, 95–99
material composition influence, 90–92
mechanistic characteristics, 113,
    115–116, 115–119
metal melts, solid surfaces, 88, 90–105
metal–slag systems, 82–88, 123–124
molten metals, surface tension, 65–82
non-isothermal processes, 100–105
nonmetal melts, 119–123

Surface properties... (*Continued*)
  solid metals, 65–82, 119–123
  solid-metal melt systems, 82–88
  spreading, 88, 90–119
  study approach, 124–125
  sulfur and sulfides influence, 119–122, *121*
  surface roughness influence, 92
  surface tension, 65–82
  system composition, 83–86
  temperature of system, 86
  welding heat sources influence, 105–119
  wettability, 88, 90–123
Surface roughness
  groove filling and unfilling, *199*, 199–200
  influence, 92
Surface self-diffusion, 55
Surface tension
  porosity in welds, 241–242
  surface energy, 36
  surface properties and phenomena, 37, 49–51, *50*
  weld and deposited material formation, 187–189
Surfacing, experimental findings, *324*, 324–330
Surfacing powders, composition, *359*
Surfactants
  action, 168–169, *169*
  crystallization phenomena, 28–29
  influence, 17–18
  melt overcooling, 24
  oxygen, 245
Susceptibility, solidification cracking, 282–284, *283–284*
Suzdalev, Russo, Kudoyarov and, studies, 165
Suzdalev studies, 203
System composition, 83–86

T

Takami, Shimmyo and, studies, 16
Tamman studies, 52
Taova and Khokonov studies, 28
Taran, Chudinov and, studies, 187
Tarlinskii and Yatsenko studies, 87

Taylor studies, 26
Taylor's formula, 266
Technologies, development
  acceleration force, 347–349, *348*
  basics, 313
  composite fiber-reinforced materials, 313–323
  convection, 352–353
  copper–steel parts, 330–337
  cracks, 339–340
  cutting tools, 337–340
  ductility, 340, *341*
  earth conditions, 347
  electron beam surfacing, 323–330
  experimental findings and observations, *320–323*, 320–330, *324*
  gas bubbles, growth, 353–354
  hardness, 337–339, *338–339*
  heterogeneous surfaces, 315–319, *317*
  microgravity, 349–352, *350*
  non-metallic inclusions, 353–354
  pipes, copper–steel, *324–325*, 331–333
  problems, composite materials, 314–315, *316*
  productivity, increasing, 341–343, *342–343*
  slag adhesion, 344–346, *344–346*
  space conditions, 347–354
  spreading, 333–337, *335*
  submerged-arc welding and surfacing processes, 341–346
  surfacing, experimental findings, *324*, 324–330
  wetting, 333–337, *335*
Tellurium (Te)
  alloy composition influence, 67
  material composition influence, 91
  nickel, 81
Temperatures
  contacting phases interaction, 2
  d.c. *vs.* a.c. welding, 6–7
  fields, weld and deposited material formation, 184–186, *185*
  gas influence, 3
  gradient, weld and deposited material formation, 194–197, *195–197*
  laser welding, 9

|

metal, binary gas mixtures, behavior, *71–72*

splatter formation, 141

surfactants, 17–18

system, surface properties and phenomena, 86

Tensile strength, 280

Tensile stresses, 238, 282

Ternary slags, 87–88

Test methods, 280–284

Theoretical models, 14–17, *15*

Thermal etching method, grain boundaries, 54–56, *55*

Thermal spreading, 93

Thermal-structural factors, 208

Thermocapillary phenomena, 169–173

Thermodynamic considerations

non-metallic inclusions, *216*, 216–218, *218*

porosity in welds, 263–264

Thickness, electroslag welding, 12

Three-phase arcs, 10

Time, weld-pool existence

basics, 1–2

groove filling and unfilling, 198–199, *199*

Tison and Favier, Camel, studies, 196

Titanium (Ti) and titanium alloys

alloy composition influence, 66–67

arc discharge, 106

binary gas mixtures, behavior, *71–72*, 79–80, *79–80*

chlorides/fluorides, 123

composite materials, 111

cracks, 281

isothermal conditions, 99

slag components, 86

substrate composition, 109

surface properties and phenomena, 99

ternary slags, 87–88

Transition angle, 188–189

Tsarevsky, Dzhemilev, Popel' and, studies, 66

Tungsten (W)

alloy composition influence, 67

arc discharge, 105

composite materials, 111

interphase surface energies, 85

material composition influence, 91

substrate composition, 109

surface properties and phenomena, 85

Tungsten electrode welding, 351

Turbin, Tyulkov and, studies, 187

Turi and El-Hebeary studies, 282

Turnbull, Khollomon and, studies, 217, 241

Turnbull and Cech studies, 88

Turned capillary method

electrocapillary phenomena, 63–64

surface properties and phenomena, 57–58

Turpin and Elliot studies, 211

Tyulkov and Turbin studies, 187

U

Udin studies, 52

Ultrasonic oscillations, *see also* Oscillations

composite fiber-reinforced materials, 320

crystallization phenomena, 29–30

metal stirring effect, 25

nucleation of pores, oxygen, 247–248

surface *vs.* volume concentration, 268–269

Umrikhin, Borodulin, Kurochkin and, studies, 67

Undercuts, 161–162, *163*

V

Vacuum influence, 19, *20*

Vanadium (V)

alloy composition influence, 66–67

sulfur, 121

surface *vs.* volume concentration, 268

Van der Waals bonds, 42, 223

Ventsel–Deryagin equation, 315

Viscosity, 224–225

Volmer's equation, 261

Voloshkevich studies, 182

Volume concentration, components, 268–269

Volynov and Zholobov studies, 353

W

Wei studies, 247
Weld and deposited material formation
    base-metal surface properties, 166–168,
        166–168
    basics, 157
    butt weld formation issues, 182
    current polarity, 173–177, *174–175*
    defects, 198–204
    deposited metal formation, 189–197
    displacement speed, 194–197, *195–197*
    electron beam surfacing experiments,
        191–194, *192*
    experimental findings, *177*, 177–182
    graphical method, 182–184
    gravitational force, 187–189
    groove filling/unfilling, 198–200
    inclination angle, 184–186, *185*
    multilayer welding, 200–204
    penetration zone, *165*
    polarity, current, 173–177, *174–175*
    serviceability, 157–164
    shape defects, 198–204
    strength deterioration, 157–164,
        158–160
    surface phenomena defects, 198–204
    surface tension, 187–189
    surfactants, action, 168–169, *169*
    temperature fields, 184–186, *185*
    temperature gradient, 194–197,
        195–197
    thermocapillary phenomena, 169–173
    weld bead formation, 182–189
    weld shape defects, 198–204
    wetting angle, 189–190, *191*
Weld bead formation, 182–189
Weld shape defects, 198–204
Weld-pool existence, 199, *199*
Welded joints
    fusion welding process, characterization,
        13–33, *14*
    strength, 208–210, *209–210*
Weld formation, 1–2
Welding current, 178, *178–179*
Welding heat sources
    electroslag welding, 11
    fusion welding process, characterization,
        4–9
    influence, 105–119
    laser welding, 8–9
    slag influence, 4
    welding, 4–9
Welding processes
    basics, 9–13
    interphase tension, 87–88, *89–90*
    surfactants, 17–18
Welds, non-metallic inclusions,
        205–210
Wettability
    composite materials, 110–111, 314
    cracks, 339–340
    external fields influence, 93–94
    fluoc effect, 102–103
    gas atmosphere, 91
    heat source influence, 105–119
    indirect arc, 107
    material composition influence, 91
    mechanistic considerations, 115
    metal melts over solid surfaces, 88,
        90–105
    oxide films, 145
    preheating, 105
    productivity, increasing, 341
    roughness, 91
    solid metal by nonmetal melts,
        119–123
    surface properties and phenomena,
        59–64
    titanium, 99
Wetting
    angle, 189–190, *191*
    austenite, 95–96
    composite materials, 315
    copper/aluminum, 98–99
    gas flame, 119
    interphase tension, 39
    technologies, development, 333–337,
        335
Whiteway studies, 18
Work of adhesion, 37, 39
Work of cohesion, 37, 39
Wuelff's rule, 28

X

X-ray photography, sessile drop, 58–59

Y

Yakobashvili studies, 67, 87
Yatsenko, Tarlinskii and, studies, 87
Yavojsky and Bobkova, Georgiev, studies, 86
Yavojsky studies, 147, 261
Young's equation, 314

Z

Zabuga and Perevertajlo, Najdich, studies, 195
Zadumkin studies, 27

Zaletaeva, Kreshchanovski, Prosvirin and, studies, 68
Zaletaeva, Kreshchanovski and Kunin studies, 29
Zero creep method, 52–53
Zholobov, Volynov and, studies, 353
Zhuravlev studies, 259
Zinc (Zn), 77–78, 102
Zirconium (Zr)
    arc discharge, 105
    composite materials, 111
    cracks, 281
    substrate composition, 109
Zone of arc active spot, 49–51, *50*
Zubkova and Safonov studies, 338